Buhmann / Ervin / Pietsch (Eds.)

Peer Reviewed Proceedings of Digital Landscape Architecture 2013 at Anhalt University of Applied Sciences

Buhmann / Ervin / Pietsch (Eds.)

Peer Reviewed Proceedings of

Digital Landscape Architecture 2013

at Anhalt University of Applied Sciences

Wichmann

Cover: Image by Carl Steinitz (A Framework for Geodesign © 2012 Esri Press, Redlands, CA)

All explanations, data, results etc. contained in this book have been made by the authors to the best of their knowledge and have been approved with care. However, some errors could not be excluded. For this reason the explanations etc. are given without any obligations or guarantee by the authors, editors, and publisher. They cannot take over any responsibility for eventual erroneous contents.

Bibliographic information published by the Deutsche Nationalbibliothek
The Deutsche Nationalbibliothek lists this publication in the Deutsche Nationalbibliografie.
Detailed bibliographic data are available in the Internet at http://dnb.dnb.de.

ISBN 978-3-87907-527-0

Bismarckstr. 33, 10625 Berlin
www.vde-verlag.de
www.wichmann-verlag.de

Printed by DDZ – Digital-Druck-Zentrum GmbH, Berlin, Germany

Preface by the President of Anhalt University of Applied Sciences

On behalf of the editors I am happy to present you this current publication of Anhalt University. This year's peer-reviewed proceedings are the contributions to the 14th conference DLA 2013. During this year's international Conference on Computer Applications in Landscape Architecture we are also celebrating 20 years of the German language Bachelor Program in Landscape Architecture and 10 years of our English language Master of Landscape Architecture Program at our campus in Bernburg. The participants of the DLA conference will see landscape architecture interaction in the old part of Bernburg, called "Bernburg-Talstadt" along the Saale River next to the conference facilities at the Bernburg Kurhaus and INDIGO Innovationspark.

The series of conferences on Information Technology in Landscape Architecture is organized by Prof. Erich Buhmann. As well as being the Scientific Director of the Digital Landscape Architecture Conferences, Prof. Buhmann is also the Chair of the European Council of Landscape Architecture Schools (ECLAS) Committee on Digital Technology. The conference series is also supported by the Vice President for Information Technology of Anhalt University of Applied Sciences, Prof. Einar Kretzler, who now heads the International Master of Landscape Architecture MLA at our University of Applied Sciences.

I would like to thank the editors of these proceedings, Prof. Erich Buhmann, Dr. Stephan Ervin, and Matthias Pietsch, for processing all the submitted papers and compiling the top twenty six into a peer-reviewed scientific publication totaling 350 pages. The editorial work shouldered by these colleagues and the international review board for this peer review process is equal to a total of four months of editorial work over the period of half a year. We thank the thirty external reviewers who make this international annual scientific conference happen.

Thanks go to the Deutsche Forschungsgemeinschaft DFG for the financial support within the program for „Internationale Wissenschaftliche Tagungen“ and thanks to the funds provided by the Ministery for Science and Economy of the State of Saxony Anhalt. Special thanks to the sponsors from the software industry who have generously supported these conferences over the years.

We have the honor of welcoming a number of very notable keynote speakers in 2013. We rely on the continuing support of many outstanding scientists in the areas of geospatial planning and design. The long term and continuous support given by Professor Carl Steinitz, Harvard University was honored when we awarded him the Honorary Senator Award of Anhalt University in 2010.

We are also very thankful to Professor Dr. James Palmer, FASLA, Vermont, U.S.A for finding the time to present a keynote on his very extensive work on visual landscape evaluation of energy landscapes in New England, a topic which is specially crucial to our Anhalt region. It is a pleasure to have Professor Steinitz and Professor Palmer back in Bernburg, when we are able to celebrate the 10 years anniversary of the international

English Language Master in Landscape Architecture, as both of them helped Professor Buhmann design the original MLA curriculum based on their very long experience of curricula development in landscape architecture. Their original concept helped to develop one of the first international Master programs in Germany which is now an essential part of the European wide education in landscape architecture.

Anhalt University currently has over 8,000 students. With more than 25 percent international students, it continues to be one of the universities in Germany with the highest level of international enrollment. Holding this conference in the State of Saxony-Anhalt in Germany is a gesture to all the international collaborators working with Anhalt on the concept of worldwide training. The great number of international speakers indicates the level of international awareness the conference has developed. The conference series also serves as a conference of the ECLAS Committee Digital Technology and the IT working group of the European Network of Landscape Architecture Schools. This LE:NOTRE thematic network in landscape architecture already includes over one hundred universities in Europe and is cooperating with a number of selected schools outside of Europe.

International conferences open the window of the world for our students and researchers. Having positioned the international conference in Digital Landscape Architecture at our university fills us with pride. We have always given new technologies education special emphasis. Therefore we are very happy about the numerous research collaborations with the software industry. This international conference will strengthen the existing contacts and will be the platform for new partnerships world wide.

Finally, I do not want to miss this opportunity to thank Professor Buhmann for his strong personal commitment and for the work that he has done over the last 18 years in helping establish Landscape Architecture in Saxony-Anhalt, for which he has received national and international acknowledgment.

I wish the international group of landscape architects and environmental researchers an exciting and successful conference in Bernburg. You are always welcome to return to our Bernburg and Dessau campuses.

At the same time, it is a great honor for us to announce that next year's DLA will be hosted by the ETH Zurich, the Swiss Federal Institute of Technology from 21 May – 23 May, 2014 at their labs in the city of Zürich, Switzerland. Please find enclosed the early call for paper for the DLA 2014.

And please reserve the date DLA 2015, 28 May – 30 May, Dessau and Bernburg, Germany, when we invite you back to our university.

Bernburg, April 2013

Prof. Dr. Dr. habil. Dieter Orzessek
President Anhalt University of Applied Sciences

Foreword

These proceedings document the peer reviewed papers for the fourteenth Digital Landscape Architecture (DLA) conference, 2013, held in Bernburg, Germany, 6 – 8 June 2013, on the theme **Connectivity and Collaboration in Planning and Design**.

The DLA conference series has focused on teaching, practicing and research issues in the world of "Digital Landscape Architecture", by way of keynote addresses, lectures, workshops, and meetings, engaging an international audience.

As the conference theme addresses the entire field of landscape architecture under the special perspective of use of digital media, these proceedings document the most relevant current issues of this entire fast driving field. We are very happy about the growing number of worldwide contributions to these series of peer reviewed proceedings. Publishing a contribution, reviewed anonymously by very knowledgeable reviewers provides young scientists the opportunity to participate in our annual international conference.

When Anhalt University organized the first international conference on information technologies in landscape architecture fourteen years ago, computers had just started to replace drawing boards in landscape architecture practice. Today, landscape architects work digitally in all project phases, often in international and interdisciplinary settings. Responding to today's international planning and design practice, as well as to the international co-operation in software development for landscape architecture, the tenth annual Digital Landscape Architecture DLA Conference in 2009 was held for the first time outside the Anhalt campus. Local responsibility for that conference was assumed by the Faculty of the Built Environment at the University of Malta. The success of the first DLA held outside our home university may inspire other faculties to apply for hosting the conference for one year at their campus. We are pleased that the DLA 2014 will be hosted by the team of the ETH in Zurich in Switzerland.

Relevant research and development involving information and communications technology for landscape architecture is only possible when research and teaching units at universities of different disciplines cooperate with each other and with the software industry as well. The DLA conferences provide an international platform for this exchange of research and teaching. We are honored that so great a spectrum of our profession has spent so many hours of effort to document today's state of the art in digital work-flow practices, to share their own experiences, and to point out further needs for research and improvement.

A growing number of professionals are close to achieving a truly paperless digital workflow; able to analyze, design, review, construct and submit documents in digital form. Individual landscape architects and firms are routinely working around the world as web-connected design teams; CAD, GIS, and imaging software on a laptop computer or portable tablet with ample memory and a fast internet connection are the only requirements. Beyond simply replacing analogue utilities, digital tools enable us to confront the challenges of exploding complexity in 'geo-design' projects by taking advantage of powerful new tools for both creativity and collaboration. Those of us still accustomed to sketching ideas with pencil and paper or making corrections on printed plans today watch in astonishment as even these operations are accomplished by digital natives on hand-held tablet-computing devices using gestural-interfaces, open-source software mash-ups, and increasingly crowd-sourced data, etc.

One of the objectives of this conference series is to present these opportunities and challenges and to anticipate future developments. As these proceedings document, the digital landscape architecture work we can do today is outstanding. Exploring virtual worlds with geospatial tools and techniques reveals forms of creativity that are at once both new and familiar. Our challenge is to engage the best of each.

As introduction to these proceedings we have asked the keynote speakers for their contributions. Professor Carl Steinitz offered a selection of his "Wise Sayings" derived from many of his outstanding lectures. A different introduction is given by Professor Jörg Rekittke who uses his experience in the fast growing Asian countries to reflect on „Brawn under the Urban Canopy". And Professor James Palmer addresses how we change our visual landscape in order to convert to the concept of "energy sufficient" landscapes.

The peer reviewed selection of papers are categorized into the following themes:

- Environmental Information in Planning and Design
- Visualization of 3D Landscapes
- Modeling of 3D Landscapes
- Geodesign Applications
- Digital Communication in Landscape Architecture
- Collaboration in Landscape Architecture

As we are organizing a German language section within the DLA called "**DLA Forum 2013: Neue Technologien in der Landschaftsarchitektur und der Umweltplanung**", these contributions are in German.

In order to document an even wider area of relevant issues, we asked authors and research groups, both in universities and in the industry, for Invited Contributions. We also ask the coordinator of the workshops to document the main theme of their workshops to these proceedings.

And we are happy that we can also present the first accepted sets of poster contributions which we have received for our DLA 2013 Poster Competition: MY 3D LANDSCAPE.

Please see our call for papers for the DLA 2014 on the last pages.
Everyone of you can help define the focus of the next year's conferences by contributing your own papers and posters or by offering workshops. While giving a platform for the evolving concept of Geodesign in Europe, we are looking forward to contributions in all areas of digital landscape architecture research and applications.

Acknowledgments

Contributors to the DLA Conference
This conference is the fourteenth in a series of conferences headed by Scientific Director Prof. Erich Buhmann at Anhalt University of Applied Sciences. These conferences were initially organized in cooperation with the keynote speakers and the LE:NOTRE Information Technology working group. As documented in the closing chapters, a formal DLA

Program Committee has been established for the further scientific development of this international conference series. Members of the DLA Program Committee will also serve as local chairs for further DLA Conferences. More than thirty experts are working at the moment on the DLA Review Board. At the General Assembly Meeting of the 2009 ECLAS Conference, the LE:NOTRE Information Technology working group was additionally established as the ECLAS Committee on Digital Technology. Future DLA conferences will continue to focus on topics identified by ECLAS needs for improvements in teaching digital methods.

We would like to thank all of the authors, reviewers, moderators, and keynote speakers for their great efforts and contributions. We are also greateful for the support from the Deutsche Forschungsgemeinschaft DFG in the years 2009, 2010, 2012 and 2013 and the Ministry for Science and Economy (Ministerium für Wissenschaft und Wirtschaft) of the State of Saxony Anhalt in 2013. Additionally we are very thankful for the support in 2013 by the sponsors Esri and ComputerWorks which allows us to tender financial awards for the student poster competition.

We especially thank all of our contributors. Their assistance is greatly appreciated. Renate Geue made the printing of the program booklets possible just in time. Then there are those who are not named in this book but without whose help it would not have been possible. First we need to thank Prof. Joachim Kieferle of the Rhein-Main University in Wiesbaden, Germany for running the conference review system and answering email requests by authors and committee members 24 hours a day for many months during the last years. Second, we need to thank Sebastian Kluth for the internet processing of the conference contents and the design of the posters and handouts 2013, and our English language consultant Jeanne Colgan, and the layout team headed by the MLA students Toomaj Goudarznia and Michael O. Ajayi. The quality of all their work will be appreciated by many readers.

To all of them thank you for work, once again, above and beyond the call of duty!
And a big thank you to all our families for their patience and support.

Bernburg, April 2013 *Erich Buhmann, Stephen Ervin, and Matthias Pietsch*

Table of Contents

Introduction

Wise Sayings from my Collection

Carl STEINITZ

Introduction

I collect wise sayings and cartoons related to different aspects of design. During my years at Harvard University I taught a course called "Theories and Methods of Landscape Planning". It was my custom to have a slide of wise sayings related to the theme of the day's lecture or experiment visible as the students entered the room. It gave them a foretaste of what that day's activity would be about. I also liberally used cartoons to support points that I wanted students to think about. Some of the wise sayings and cartoons were purposely chosen to be serious, but many, especially the cartoons, were selected to be humorous and even cynical about that very seriousness. It has always been my view that we should take our work very seriously and engage in design for the improvement of significant societal and environmental conditions. However I also think that there is a role for the self-criticism that humor engenders.

I therefore decided to offer the editors of this volume of Digital Landscape Architecture papers a selection of wise sayings from my collection. I cannot publish the cartoons because of copyright issues, but I intend to use them in a different setting. The wise sayings are purposely organized in relationship to the major sections of the framework within which I organized my Harvard course and also my recent book "A Framework for Geodesign", Esri Press, 2012.

ON COLLABORATION

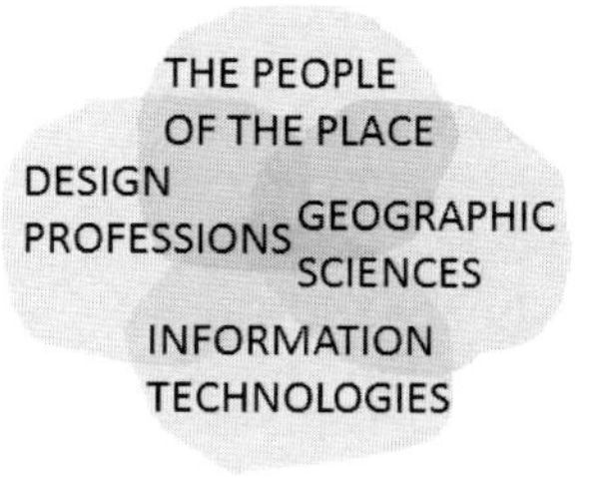

"Many ideas grow better when transplanted into another mind than the one where they sprang up."
Oliver Wendell Holmes

"Politeness is the poison of collaboration."
Edwin Land

"Guard against talking more clearly than you think."
anonymous

"The secret is to gang up on the problem, rather than each other."
Thomas Stallkamp

"The strength of the team is each individual member. The strength of each member is the team."
Phil Jackson

"No one can whistle a symphony. It takes a whole orchestra to play it."
H.E. Luccock

"The trouble with the rat-race is that even when you win, you're still a rat."
Lily Tomlin

A FRAMEWORK FOR GEODESIGN

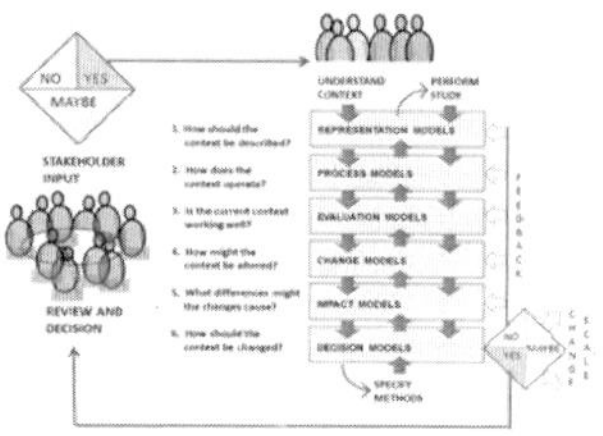

"Put in simplest terms, a theory explains, a model predicts and a framework organizes. A framework can be judged on its reasonableness and its utility, but claims no exclusivity vis-a-vis other frameworks".
Amos Rapaport

"The problem is not that we don't know; rather it is that we can't figure out and organize our questions, search out the answers (most of which exist), remember them and use them. If we could do this, we could then identify the truly missing overviews and parts, and work on these."
Carl Steinitz, 1972

"Grau, theurer freund, ist all theorie
und grun des lebens."
("All theory is gray and life is green.")
Goethe, *Faust*

A FRAMEWORK FOR GEODESIGN

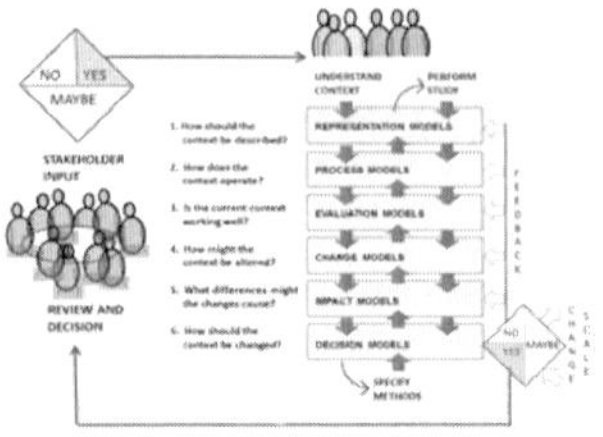

"In theory, there is no difference between between theory and practice.
In practice, there is no relationship between theory and practice."
Grant Gainey

"Furthermore, the gap between theory and practice in practice is larger that the gap between theory and practice in theory."
Jeff Case

Theory is when you know everything
but nothing works.
Practice is when everything works
but no one knows why.
In our lab, theory and practice are combined:
nothing works and no one knows why.
Sarah Steinitz

UNDERSTANDING THE CONTEXT

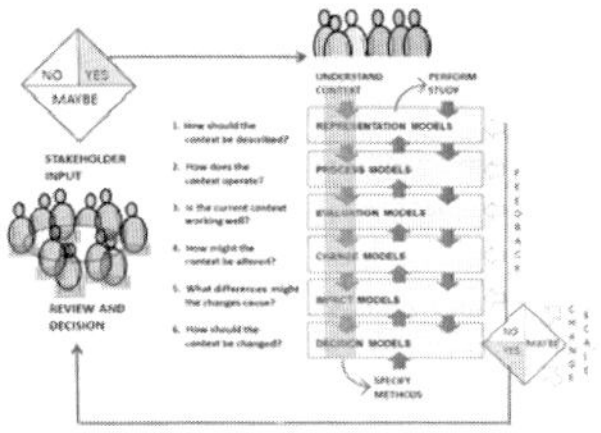

"The environment is full of 'To whom it may concern....' messages."
Norbert Weiner

"The significant problems we face cannot be solved at the level of thinking we were at when we created them."
Albert Einstein

"The judge held the orange in his hand while the two petitioners each insisted that he needed the whole thing. Unable to decide between then , the judge cut the orange in half, giving half to each petitioner.
Both went away very unhappy. Later, the judge found out that one had wanted the peel while the other had wanted the inside.
MORAL: Find out what the question is before you propose an answer."
as told by Sarah Steinitz

SPECIFYING THE METHODS

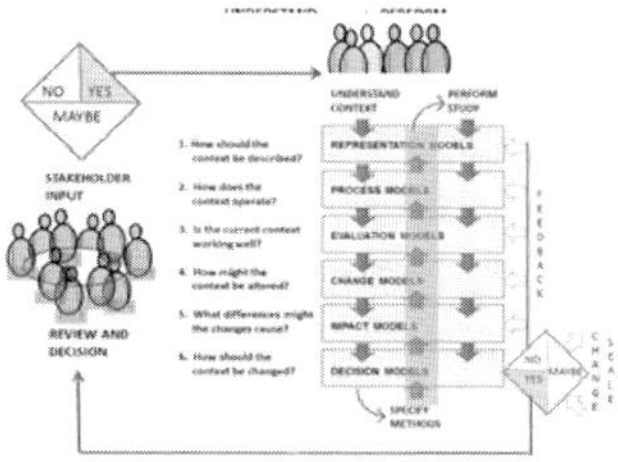

"He who seeks for methods without having a definite problem in mind seeks in the most part in vain."
David Hilbert

"By three methods we may learn wisdom:
First by reflection, which is noblest;
Second by imitation, which is easiest;
And third by experience, which is bitterest.
Confucius

"It would be naïve to think that the problems plaguing mankind today can be solved with the means and methods which were applied or seemed to work in the past."
Mikhail Gorbachev

DECISION MODELS

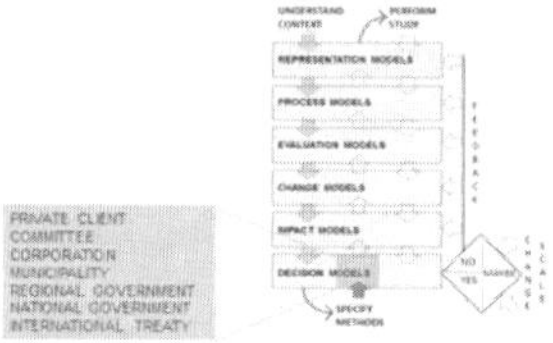

"Good decisions come from experience, and experience comes from bad decisions."
Author Unknown

"From experience of the past, the present acts prudently, lest it spoil future action."
Pierre Bersuire, medieval scholar

"The risk of a wrong decision is preferable to the terror of indecision."
Maimonedes

"As soon as questions of will or decision or reason or choice of action arise, human science is at a loss."
Noam Chomsky

"It's not hard to make decisions when you know what your values are."
Roy Disney

DECISION MODELS

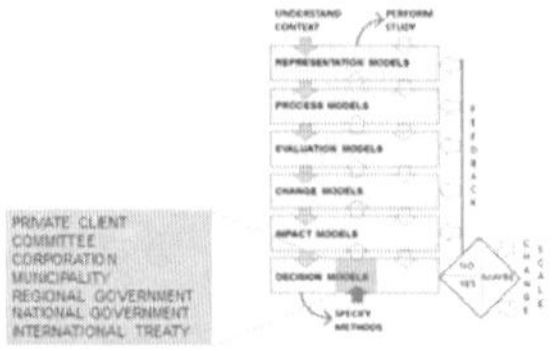

"We Athenians……take our decisions on policy and submit them to proper discussion: for we do not think there is an incompatibility between words and deeds: the worst thing is to rush into action before the consequences have been properly debated. And this is another point where we differ from other people. We are capable at the same time of taking risks and estimating them beforehand. Others are brave out of ignorance; and when they stop to think, they begin to fear. But the man who can be most truly be accounted brave is he who best knows the meaning of what is sweet in life and what is terrible, and then goes out to meet what is to come."

Pericles

IMPACT MODELS

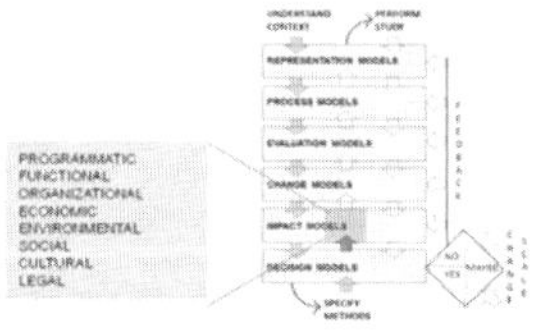

"Not everything which can be counted counts;
And many things which cannot be counted, count."

William Bruce Cameron

"I have found that all ugly things are made by those who strive to make something beautiful and that all beautiful things are made by those who strive to make something useful."

Oscar Wilde

"What's the use of a fine house if you haven't got a tolerable planet to put it on?"

Henry David Thoreau

"If you wish to make an impact for a year, plant corn; if you wish to make an impact for a generation, plant a tree; if you want to make an impact for an eternity, educate a child."

Author unknown

CHANGE MODELS

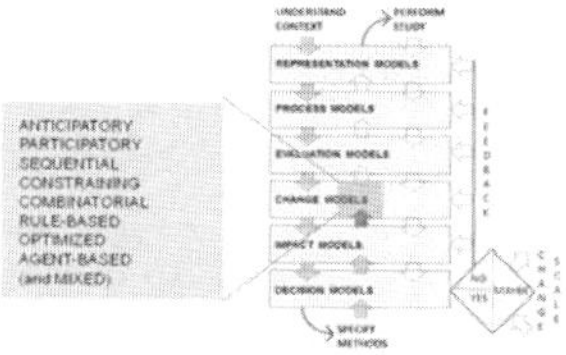

"Develop an infallible technique,
and then place yourself at the mercy of inspiration"
Zen saying, via Walter Gropius

"Everyone designs who devises courses of action aimed at changing existing situations into preferred ones."
Herbert Simon

"For every complex problem there is an answer that is clear, simple, and wrong."
H. L. Menken

"The true artist possesses the power to inspire others. A light that cannot kindle other lights is but a feeble flame. The core of art is its teaching and ennobling influence not only on other artists, but on humanity."
commentary on *Exodus XXV*

CHANGE MODELS

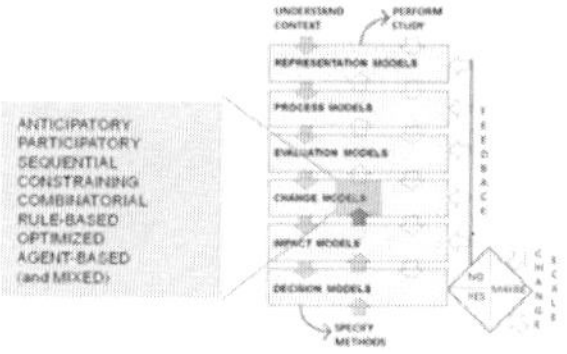

"Imagination is a good horse to carry you over the ground. Not a flying carpet to set you free from probability."
Robertson Davies

"The gift of imagination is not just the artist's—
it's a gift we all share."
Leonard Bernstein

'When creative genius neglects to ally itself to some public interest it hardly gives birth to wide or perennial influence. Imagination needs a soil in history, tradition, or human institutions, else its random growths are not significant enough and, like trivial melodies, go immediately out of fashion."
George Santayana

"If you want to make God laugh,
show him your plan."
heard on BBC4

EVALUATION MODELS

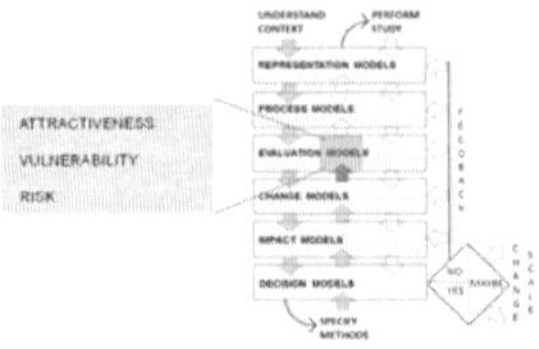

"True genius resides in the capacity for evaluation of uncertain, hazardous, and conflicting information."
Winston Churchill

"The most serious mistakes are not being made as a result of wrong answers. The truly dangerous thing is asking the wrong question"
Peter Drucker

We cannot discover what ought to be the case by examining what is the case. We must decide what ought to be the case. "
Paul Taylor

"Evaluation criteria should be expressed as the answers to three questions:
-What is important?
-How is it important?
-How important is it?"
Carl Steinitz

PROCESS MODELS

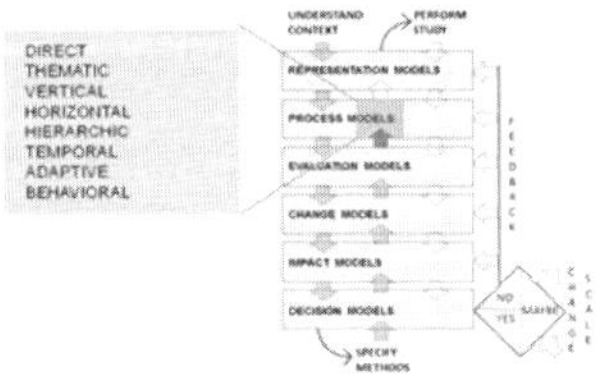

"Prediction is difficult, especially about the future."
Niels Bohr

"All models are wrong. Some are useful."
anonymous

"1--Guess
2—Compute consequences
3—Compare with nature and experience"
Richard Feynman

"What we observe is not nature itself, but nature exposed to our method of questioning."
Werner Heisenberg

"And they'll build systems dark and deep,
And systems broad and high;
But two of three will never agree
About the reasons why."
Peter Pindar, *The Wise Men of Gotham*

DATA/REPRESENTATION MODELS

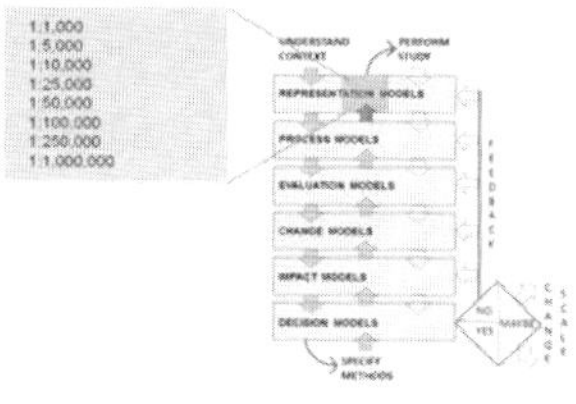

"I understand, therefore I see."
John Constable
(*not* "I see, therefore I understand." CS)

"To be conscious that you are ignorant of the facts is a great step to knowledge"
Benjamin Disraeli

"It is better to be approximately right than to be precisely wrong."
John Maynard Keynes

"It is the mark of an instructed mind to rest satisfied with that degree of precision which the nature of the subject admits, and not to seek exactness where only an approximation of the truth is possible."
Aristotle

"Be as simple as possible, but not more so."
Albert Einstein

PERFORMING THE STUDY

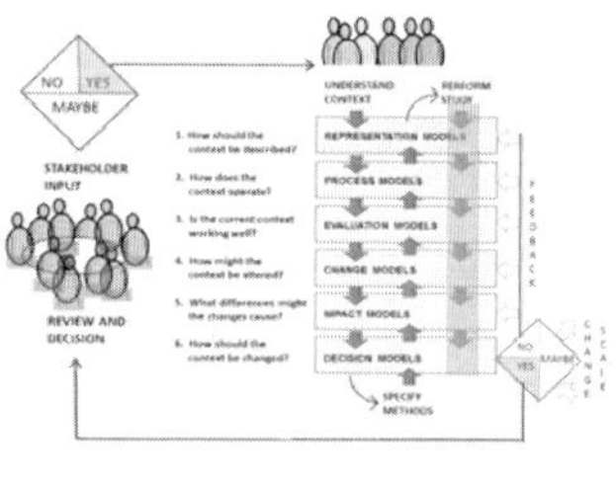

"There is no design without discipline. There is no discipline without intelligence."
Massimo Vignelli

"Design is the application of intent - the opposite of happenstance, and an antidote to accident."
Robert L. Peters

"Designers have a dual duty; contractually to their clients and morally to the later users and recipients of their work."
Hans Hager

" A good designer is like a good barber. There are three criteria. First, protect what you have. Second, edit carefully. Third, invent when necessary. And they are important in that order."
Laurie Olin

THE FRAMEWORK WILL NEVER BE LINEAR IN APPLICATION

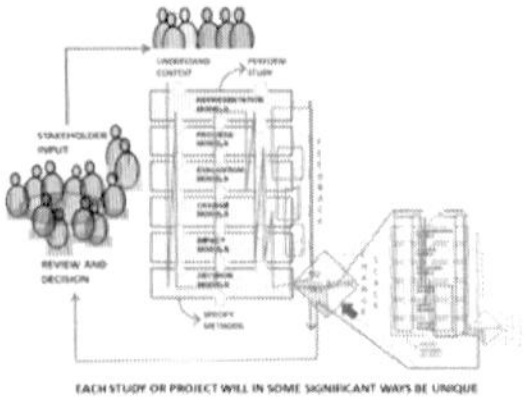

"In formal logic, a contradiction is the signal of defeat, but in the evolution of real knowledge it marks the first step in progress toward a victory."
Alfred North Whitehead

When adversity strikes, that's when you have to be the most calm. Take a step back, stay strong, stay grounded and press on."
LL Cool J

People take the longest possible paths, digress to numerous dead ends, and make all kinds of mistakes. Then historians come along and write summaries of this messy, nonlinear process and make it appear like a simple, straight line. "
Dean Kamen

ON EDUCATION FOR COLLABORATION IN GEODESIGN

"Learning is the organization of experience."
Kristin Hammond

"Fortune only favors the prepared mind."
Louis Pasteur

"I hear and I forget.
I see and I remember.
I do and I understand."
ancient Chinese proverb
"I do and I still don't understand."
Eleanor Duckworth

"It's what you learn after you know it all that counts."
Harry Truman

'A tragedy of our educational process--including graduate education--is the over-reliance on 'self discovery' by students of basic methods."
Carl Steinitz, 1967

References

Ervin, S. M. (2012), *Geodesign Futures – Nearly 50 Predictions*. In: Buhmann, E., Ervin, S. M. & Pietsch, M. (Eds.), Peer Reviewed Proceedings of Digital Landscape Architecture 2012 at Anhalt University of Applied Sciences. Berlin/ Offenbach, Wichmann Verlag, 22-30.

Miller, W. R. (2012), *Introducing Geodesign – The Concept*. In: Buhmann, E., Ervin, S. M. & Pietsch, M. (Eds.), Peer Reviewed Proceedings of Digital Landscape Architecture 2012 at Anhalt University of Applied Sciences. Berlin/Offenbach, Wichmann Verlag, 542-567.

Steinitz, C. (2012), *A Framework for Geodesign*. Esri Press.

Steinitz, C. (2012), *"Getting Started" – Teaching a Collaborative Multidisciplinary Framework*. In: Buhmann, E., Ervin, S. M. & Pietsch, M. (Eds.), Peer Reviewed Proceedings of Digital Landscape Architecture 2012 at Anhalt University of Applied Sciences. Berlin/Offenbach, Wichmann Verlag, 2-21.

Brawn and Technology under the Urban Canopy

Joerg REKITTKE, Philip PAAR and Yazid NINSALAM

1 Base of Operations

Considering the growing variety of electronic tools being deployed by city and geo designers and researchers in fieldwork operations, we distinguish two main groups of users. The first group we refer to as 'aviators'. Namely: research teams who profit directly from a large range of high-end technology – satellites and their data, all sorts of flying super-technology and respective precise aerial imagery. The second group we refer to as 'foot soldiers' (REKITTKE, PAAR & NINSALAM, 2012). At this point we feel up to put forward the alternative term 'earthlings', due to political correctness and peer scrupulosity. Developing our research and fieldwork practice as data gathering and 3D modelling earthlings in the third year of a related research project, we continue in improving and optimizing our light, portable and inconspicuous equipment. We stand by our claim – despite all the theoretically available remote sensing technology – that in order to build complete and highly detailed digital 3D models of complex terrain and urban territory, direct contact to ground and detail remains indispensable. Compared to the sophisticated machines and efficient methods in remote sensing, the craft of the earthling appears to be more intricate and laborious – but it can deliver unique results, which we subsume under the term of 'Grassroots GIS' (REKITTKE & PAAR, 2010). By now, we can be certain that the scientific and methodological niche of the GeoDesign earthling is located under the *urban canopy* and in *urban canyons*, where remote sensing technology is proven to be blind and ineffective. In this paper, we meticulously describe our most recent advances and let downs in the development of a method of on-site data and image gathering. This evidently creates a new kind of high precision three-dimensional puzzle, which makes a widely inaccessible and undocumented piece of *under the urban canopy* terrain visible, understandable and designable. The work we present focuses on the space defined by the Ciliwung River in the Kampung Melayu-Bukit Duri district, an informal urban segment in Jakarta, Indonesia. We bring ourselves, together with students, to this area in the course of the research module Landscape Ecology in the 'Future Cities Laboratory' (REKITTKE & GIROT, 2012), hosted and managed by the Singapore ETH Centre for Global Environmental Sustainability (CAIRNS et al., 2012). With our feet in the sludge, waste and faeces of the Ciliwung River, the help of a lot of brawn and some new tools, which are typically used by free fallers, bungee jumpers and other contemporaries tired of life, we also experimented with toy drones, featuring inbuilt cameras and representing the approaching armada of flying devices for the purpose of image data gathering in science, research and design.

2 Earthlings at Work

The described experiments have been conducted during a fieldwork campaign over a period of six days and invariably focussed the Ciliwung River and its edges. Our mission was to document, model and design the course of the river between 6°13'28. 07"S, 106°51'49. 35"E and 6°13'2. 23"S, 106°51'28. 75"E – a distance of 2.86 km. Initially we conducted our data gathering process by walking, picture taking and filming along the banks. This proved to be incredibly tedious when our feet sank too deep into the sludge, and finally came to a halt when the banks became too steep to continue. But to this point we could successfully deploy some of our newest equipment (Fig. 1) and give some new methodological ideas a try.

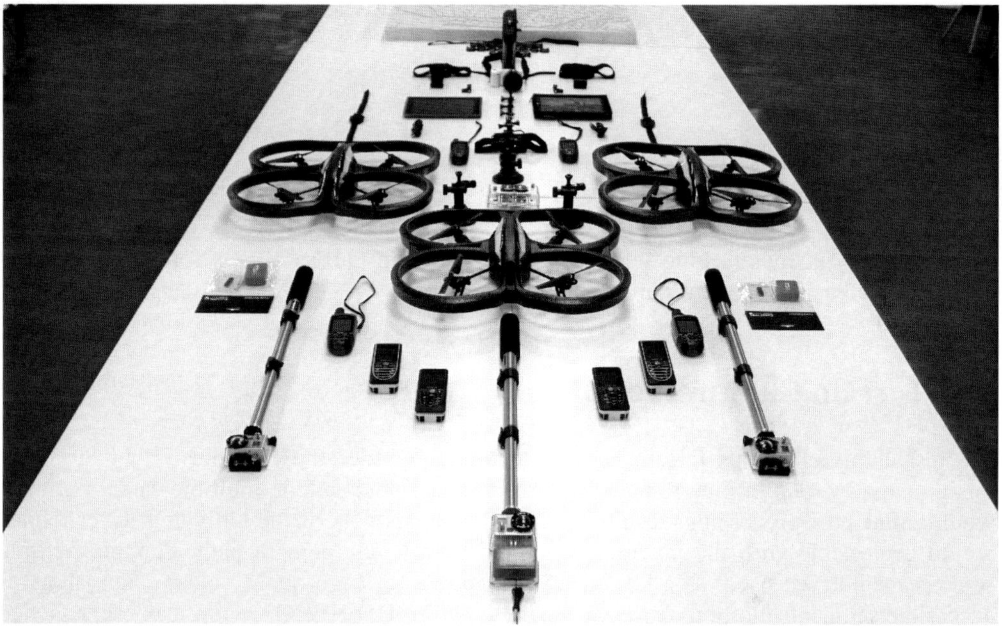

Fig. 1: Technological arsenal (2012): Multiple sets of the GoPro HD HERO2 action camera Outdoor Edition; Parrot AR Drones 2.0 inclusive accessories and spare parts; Leica Disto D8 laser meters; Garmin GPS-MAP 60CSx handheld tracking devices; Apple iPhones and iPads for drone flight control; additional DSLR cameras; camera poles etc. (Photo: Rekittke)

Earthlings roam as backpackers, they have to be able to carry all their equipment in the field. The technological arsenal of the fieldwork at hand felt considerable. We carried along three sets of the GoPro HD HERO2 action camera Outdoor Edition – video cameras fitted in waterproof housings and attached to 3-way pivots on a telescopic mast; three iPad-controlled Parrot AR Drones 2.0 – remote controlled by the gratis AR.Free Flight App – inclusive a good deal of accessories and spare parts. These inexpensive plaything quadrocopters, equipped with a built-in on-board video camera for wireless video recording and transmission via a picture resolution of 1280 x 720 dpi, enabled us to also visually illuminate completely inaccessible or hidden spots of the site. In addition we carried several Leica Disto D8 laser meters as well as Garmin GPS-MAP 60CSx handheld tracking devices as well as Apple iPhones and iPads for further navigation and measuring applications. For our photographic work, we would always carry high-capacity DSLR cameras. Two purchases turned out to be unexpectedly useful for our mission. Firstly, a six meter long telescopic camera mast, model Kamkop GoPro, made of GFK and weighing only 850 grams, allowed us to generate unique image material from bird's eye view (Fig. 2). The long pole almost beats flying devices concerning precision and ease of handling. Secondly, an offhanded purchased rubber boat, a Challenger 4 Inflatable – the mother of all dinghies, finally brought us onto the river. The originally supplied plastic oars broke during the first of six trips down the river but the craft with the capacity of four people, sustaining a load of up to 400 kg, transformed bogged earthlings into successful river rafters. All our items are consumer grade, easily accessible and comparatively low cost.

Fig. 2: The six meters long telescopic mast Kamkop GoPro allows for work under as well as above the urban canopy (Photo: Paar)

3 Terrain Capture Techniques

We tested miscellaneous terrain capture techniques under the hypothesis that the combination and overlap of picture material generated by independent, multiple camera sources would, after post-processing, result in better quality models of the Ciliwung River than it would be feasible with the material generated by a single camera path. For the multiple sources method we used: a) a Canon DSLR camera for high-quality close-range shots of accessible segments along the river; b) a single GoPro HD HERO2 action camera, attached to the end of the Kamkop GoPro telescopic mast, in order to capture images through a *swaying* motion across river and terrain (Fig. 3); c) the built-in front camera of the Parrot AR Drone 2.0 (Fig. 4), delivering a sequence of pictures from the captured video; d) a set of three GoPro HD HERO2 action cameras, mounted on a pole for three-directional simultaneous dolly shots, carried down the river in a rubber boat mission (Fig. 5).

Fig. 3: *Swaying* a GoPro HD HERO2 action camera, attached to the Kamkop GoPro telescopic mast, in order to capture river and terrain

Fig. 4: Flying the Parrot AR Drone 2.0 over river and terrain – under the canopy

Fig. 5: Three GoPro HD HERO2 action cameras, mounted on a pole for three-directional simultaneous dolly shots – on the river

For the single camera path method we merely used the above mentioned set of three pole and boat mounted GoPro HD HERO2 action cameras, adjusted for three-directional simultaneous shots.

4 Postprocessing of Image Material

To come to the point, our hypothesis – the more the better – crystallised being not axiomatically true. We will try to explain this conclusion in conjunction with the explanation of our postprocessing methods. The hardware we use for postprocessing represents common standards in computer gamer circles – Grassroot GIS asks for affordable off-the-shelf articles: a laptop with a 2.9Ghz 3rd Generation Intel Core i7-3920XM Processor, 32GB RAM running 64-Bit Windows 7. All software we apply for the described work is free. For the postprocessing of former fieldwork material we gained experience with the web-based software Autodesk 123D Catch that only allows for a maximum upload of less than one hundred photos at once. For the described mission we experimented with processing of up to three thousand photos in one session, which can be done with a combination of the free programmes VisualSfM version 0.5.20 and CMPMVS version 4.0. VisualSfM is a graphics user interface (GUI) application of Structure from Motion (SfM) (WU, 2011). The system is based upon the measured correspondence of image features, such as points, edges and lines – inferring camera poses from a selection of respective photos (WANG, 2011). Subsequently the measured camera poses are processed by CMPMVS, a multi-view reconstruction software, which generates a textured mesh and re-constructs the surface of the final 3D model (HAVLENA et al., 2010). VisualSfM attends to the needs of both basic and advanced users, the application runs by executing the supplied scripts and parameters. The resulting product of VisualSfM – a point cloud – then can be processed with the help of CMPMVS. The standard workflow (Fig. 6) for VisualSfM and CMPMVS can be abstracted as follows:

1) Import of all the images into the SfM workspace as a readable image file in jpg-format.
2) Feature detection and full pairwise image matching, this generates a set of measured image features and correspondence (WU, 2007).
3) Incremental reconstruction, this will start the multicore bundle adjustment, which calculates the 3D point positions and camera parameters from the set of measured image features and correspondence (WU et al., 2011)
4) Dense reconstruction by using a multi-view stereo (MVS) application – CMPMVS – this performs the surface reconstruction from the point clouds (HAVLENA et al., 2010).

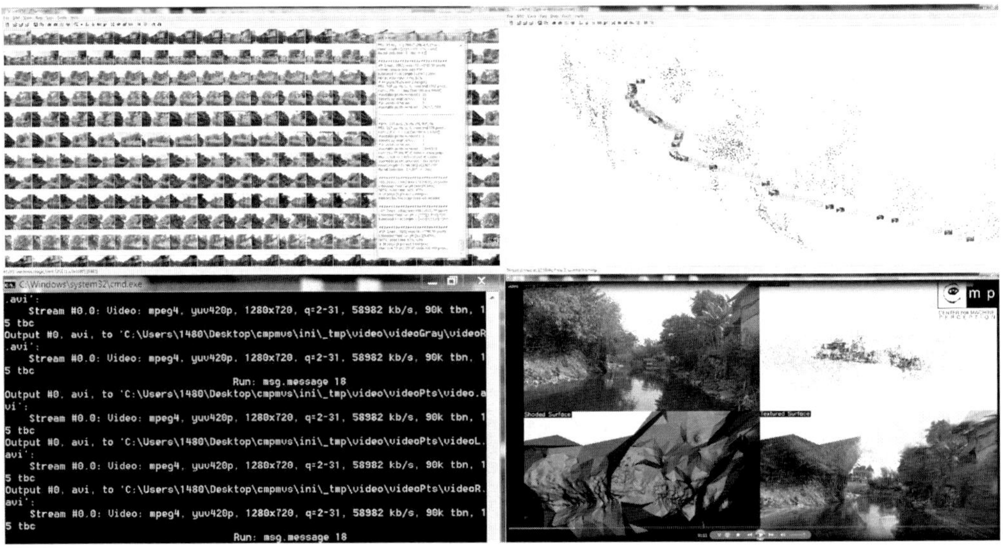

Fig. 6: VisualSfM workflow: Feature Detection and Bundle Adjustment (top l.+r.); Dense and Surface Reconstruction in CMPMVS and video output (bottom l.+r.)

The Autodesk software 123D Catch offers limited functionality concerning a detailed determination of the resulting model, but it is the best choice for basic users. Features that hitherto overtrump VisualSfM include the implementation of a reference scale allowing scaling of the entire 3D model, taking of reference measurements, animation of the 3D model in the software directly, and a complete 3D surface reconstruction pipeline. The academic project VisualSfM thrives on constructive feedback by an active user community that constantly seeks advice and makes suggestions on a Google Groups discussion thread. VisualSfM stores and processes the selected image material on a local folder. This bypasses any time consuming and often impractical web-based upload action and enables the user to check feature matching and sparse reconstruction processing already in the field. A critical issue during time-constrained field campaigns, because in semester-long academic design studios a return visit is often not an option. Main goal of our research is to achieve the best model quality possible. Therefore, VisualSfM currently turns out to be the optimal cost-free option (Fig. 7). It allows a flexible addition of selected photos to integrate and improve existing models, and a manually performed initialization of the automatic feature detecting process by the determination of a proper pair of photographs. A 'Pause' and 'Run' button facilitates the integration of further photos during the feature-matching phase. Such sophisticated fine-tuning is not possible in 123D Catch, it prompts the user to manually define at least three corresponding points on the sequential images and does not allow pausing a running reconstruction process.

Fig. 7: Screenshot of a 3D model generated via Autodesk 123D Catch (left); screenshot of the same model generated via VisualSfM (right)

5 Less Input, more Model

The comparison between the postprocessing results of single and multiple source images for the reconstruction of terrain (modelling) along the Ciliwung River will illustrate the advisable balance between image input and model output. Before looking into the results we need to distinguish the different methods of image processing. We work with two methods, both based on manually selectable settings within the VisualSfM platform. The first method makes use of an *exhaustive pairwise matching* technique, the second method makes use of *sequential pairwise matching*. We fed the material of our multiple sources capture technique (see point 3) into VisualSfM by selecting the exhaustive pairwise matching option. This option attempts to match every single image and their features to all corresponding features of all available images within the supplied dataset. From video sequences one frame of every 30th frame was extracted, in order to have regular image coverage of the environment. Images from the Canon DSLR shot in RAW were downsized to match the resolution of the extracted video images. In our tests of the exhaustive pairwise matching option we processed a total of 543 images along a segment of the river (Fig. 8). For the processing of the material gained via our single camera path capture technique (see point 3) we selected the sequential pairwise matching option in VisualSfM. This option is optimised for the processing of images that had been taken in a continuous sequence (path). To compare the results of the sequential pairwise matching option with those of the exhaustive pairwise matching option, we selected – for the same river segment – a total of 100 sequential images from our HD resolution video footage, by extracting one frame of every 30th frame.

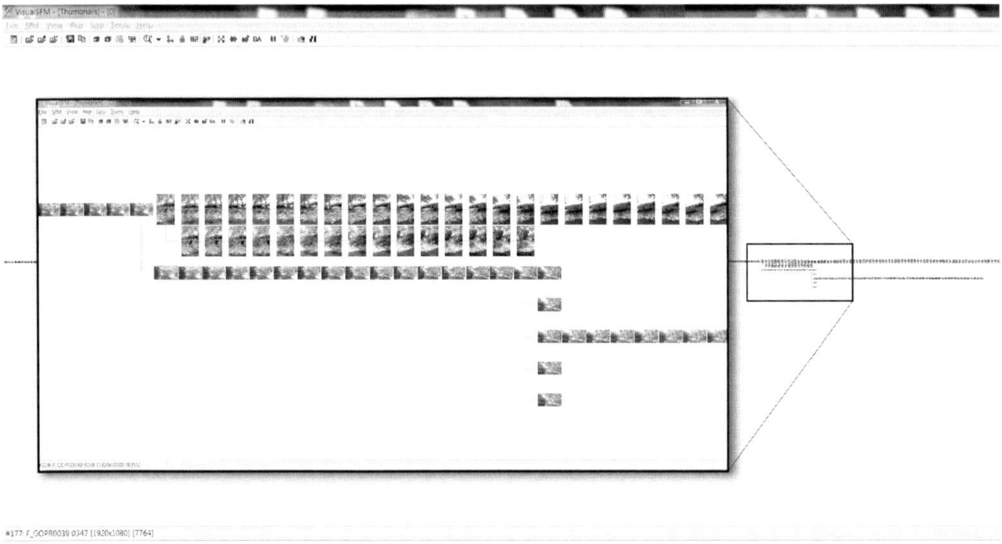

Fig. 8: VisualSfM *spanning tree*, indicating the corresponding matches between images from multiple sources. The highlighted window shows an instance, where 72 (of a total of 543) images matched, originating from the AR Drone 2.0 and Canon DSLR camera.

As a result we observed that in the reconstruction of the river corridor sequence, the sequential pairwise matching option outplays the exhaustive pairwise matching option of VisualSfM in both the quality of the resulting model and time taken to post process the images. Although some model sets successfully integrated images from various sources – the model quality is similar to that of the sequential pairwise matching operation as it includes the same main image source (GoPro HERO2 HD cameras), the most coherent models result from single source sequential images (Fig. 9). However the exhaustive pairwise matching option can improve segments of the geometry where the model lacks data due to insufficient image coverage. The quality of the model is determined by the quality and integrity of the images from the source. Images from multiple sources can be added successfully to a single sequential dataset as a model improver in place of using it as the primary model generating method. This is especially relevant for our fieldwork environments where physical constraints like low-hanging branches or other urban canopy may hinder the continuous filming of the entire river sequence and contribute to intermittent recording. An issue worth mentioning concerning the multiple sources input in the VisualSfM exhaustive pairwise matching workflow is the fact that the subsequent surface reconstruction procedure of the CMPMVS software is unable to generate geometry from image sources that do not share the same pixel dimensions.

Fig. 9: Three-dimensional puzzle of partial terrain models, positioned in Google Earth (top). Video capture (middle) of the river terrain *under the urban canopy*, and textured model of the same site (bottom).

On these grounds it is most advisable to use cameras with same specifications (and settings), when adding images to supplement the model generating process via exhaustive pairwise matching. Maybe the processing time proves crucial for the method of choice. For the same river segment sample, the completion of the feature matching process in VisualSfM, using the exhaustive pairwise matching option and a total of 543 images, took six days of uninterrupted computation. The sequential pairwise matching method, using 100 omnidirectional images, took three hours for the same task.

6 Homework

While the described fieldwork mission took quite a bit of brawn, it remained rewarding and evolved our understanding of the technology related challenges facing Grassroots GIS. The quality of image and data material is key and time is on our side. The craft of an earthling's fieldwork will continue to require rolling up their shirtsleeves, however precision, accuracy and dependability of available equipment will rapidly advance. Coherence as well as complexity of our technique and technology will have to be increased. The next generation of outdoor action cameras and camera carrying flying devices is in the stores yet. Our low-budget approach sets us certain financial limits but we begin to leer at portable 3D scanner systems with real-time 3D reconstruction capability and will have to decide if such technology will be part of our future baggage. These scanners are quickly becoming affordable and are paired with software development kits, which enable developers to create various new applications (ANDERSEN et al., 2012). Another future standard for the work in data poor environments will be made up by extremely precise portable satellite navigation systems with global coverage (GNSS) and enhancements by Differential Global Positioning Systems (DPGS). Eventually we will have to experiment with still expensive Light Detection And Ranging (LIDAR) technology, which is being integrated into standard GIS software (JORDAN, 2012). The trend points towards a conflation of GIS data sets and their corresponding image-based geometries. Conflation – in GIS – is defined as the process of combining geographic information from overlapping sources in order to retain accurate data, minimize redundancy, and reconcile data conflicts (LONGLEY et al., 2001). We undertake not to reduce our commitment in brawn but we might have to increase our investment in technology.

7 Acknowledgements

Special thanks go to Professor Christophe Girot, in his capacity as head of the research module "Landscape Ecology" of the Future Cities Laboratory, Singapore ETH Centre for Global Environmental Sustainability. Our work is embedded in the Future Cities Laboratory, in form of Design Research Studios as well as in form of PhD research. In the same way we would like to thank Professor Wong Yunn Chii, head of the NUS Department of Architecture, for his unorthodox support – in face of our out of the ordinary travel preferences.

References

Andersen, M. R., Jensen, T., Lisouski, P., Mortensen, A. K., Hansen, M. K., Gregersen, T. & Ahrendt, P. (2012), *Kinect Depth Sensor Evaluation for Computer Vision Applications* (Technical Report ECE-TR-6). Department of Engineering, Aarhus University, 1-39.

Cairns, S., Gurusamy, S., Prescott, M. & Theodoratos, N. (Eds.), *Rivers in Future Cities: The Case of Ciliwung, Jakarta, Indonesia.* Module VII: Landscape Ecology, FCL – Future Cities Laboratory Gazette, 12, Singapore ETH Centre for Global Environmental Sustainability.

Havlena, M., Torii, A. & Pajdla, T. (2010), *Efficient Structure from Motion by Graph Optimization.* In: Computer Vision – ECCV 2010. Berlin/Heidelberg, Springer-Verlag.

Jordan, L. (2012), *Integrating LiDAR and 3D into GIS.* Available from: http://www.youtube.com/watch?v=0USiCcsiFqE (accessed 25 November 2012).

Longley, P. A., Goodchild, M. F., Maguire, D. J. & Rhind, D. W. (2001), *Geographic Information Systems and Science.* John Wiley & Sons, 134.

Rekittke, J. & Girot, C. (2012), *Cali Ciliwung Experiment.* In: S.L.U.M. Lab. Sustainable Living Urban Model, Edition 7, 68-69.

Rekittke, J., Paar, P. & Ninsalam, Y. (2012), *Foot Soldiers of GeoDesign.* In: Buhmann, E., Ervin, S. M. & Pietsch, M. (Eds.), Peer Reviewed Proceedings of Digital Landscape Architecture 2012 at Anhalt University of Applied Sciences, Berlin/Offenbach, Wichmann Verlag, 199-210.

Rekittke, J. & Paar, P. (2010), *Grassroots GIS – Digital Outdoor Designing Where the Streets Have No Name.* In: Buhmann, E., Pietsch, M. & Kretzler, E. (Eds.), Reviewed Proc. of Digital Landscape Architecture 2010 at Anhalt University of Applied Sciences. Berlin/Offenbach, Wichmann Verlag, 69-78.

Wang, Y.-F. (2011), A *Comparison Study of Five 3D Modeling Systems Based on the SfM-Principles* (Technical Report No. Visualsize Inc. TR 2011-01), Goleta, USA, 1-30.

Wu, C. (2007), *SiftGPU: A GPU implementation of Scale Invariant Feature Transform (SIFT).* Available from: http://cs.unc.edu/~ccwu/siftgpu (accessed 1 November 2012).

Wu, C. (2011), *VisualSFM: A Visual Structure from Motion System.* Available from: http://homes.cs.washington.edu/~ccwu/vsfm/ (accessed 3 November 2012).

Wu, C., Agarwal, S., Curless, B. & Seitz, S. M. (2011), *Multicore Bundle Adjustment.* Presented at the IEEE Computer Vision and Pattern Recognition (CVPR), 3057-3064.

Maine's Experience Evaluating Wind Energy Development

James F. PALMER

1 Introduction

On April 18, 2008, Governor John Baldacci signed *An Act to Implement Recommendations of the Governor's Task Force on Wind Power Development* (the Act). It establishes a favorable State policy encouraging grid-scale wind energy development in appropriate locations. In particular, it designates a large portion of the state for expedited grid-scale wind energy development. While most environmental impacts within the expedited area are evaluated in the same manner as previously, special provisions are made for scenic impacts.

Since passage of the Act, four visual impact studies for grid-scale wind energy projects have conducted surveys of users at a viewpoint within a scenic resource of state or national significance (SRSNS) where the proposed project's wind turbines are expected to have clearly visibility. At a minimum, respondents were asked to rate the existing view, a photograph of the existing view and a photosimulation of the view with the project turbines using a rating scale of lowest to highest scenic value. Additional questions asked about how the proposed project would affect their recreation enjoyment at the viewpoint, and whether they would return to the area to recreate if the project is built. One survey included a photograph and photosimulation from a second viewpoint. An additional web-based survey for one project also included two impacted viewpoints as well as two additional photographs, but of course could not include ratings of the view in the field.

This paper presents the results of an independent analysis of the data from these user surveys conducted for grid-scale wind energy projects being permitted under the Act's Evaluation Criteria.

2 Methods and Materials

A total of seven viewpoints from four wind projects have been evaluated in the user surveys conducted by Market Decisions and Portland Research Group.

2.1 Perception of Scenic Value and Scenic Impact

Most of the data analyzed here were collected by intercepting people engaged in recreation pursuits at or near the viewpoint being evaluated. The major exception is the web-based survey for Highland, which used a random sample of people from a marketing firm's panel of outdoor recreation participants in northern New England. Survey respondents were shown photographic representations of the existing condition and a photosimulation of the visual condition if the proposed project were built. They were instructed to view these representations from a specific distance, so that they would appear in proper perspective. Scenic evaluations were made using a standard 7-point or 10-point rating scales. In all cases, a rating of 1 is given to the lowest scenic value. A measure of scenic impact is obtained by subtracting the scenic value of the view without the project from the scenic value of the view with the project. However, the meaning of the ratings and the raw impact values are difficult to compare since two different rating scales are used. In addition, the impact value provides no intuitive sense of when the impact is Unreasonably Adverse. While *t*-tests show that these perceived scenic impacts are statistically significant, this is not an indication that the impacts reach the threshold of Unreasonably Adverse. [1]

These shortcomings are common to most scenic impact assessments. However, they can be overcome using the procedures described in the remainder of this section. First the reliability of the data that were collected is considered. Then two possible ways of describing change in scenic value are described: percent change has more intuitive appeal, while effect size has become the preferred way to report findings about change in the scientific literature.

2.2 Visual Simulations

All the images used in these surveys are based on photographs captured with high resolution digital cameras. Each photo was taken using a "normal" lens focal length, though some of the images used in the survey were panoramas created by stitching together two or more normal photos.

2.3 Effect size

The current best practice in scientific analysis and reporting is to use effect size as a way to report the strength of the relationship between the means of two variables measured on the same scale (e.g., APA 2010, p. 33). The statistic used in this paper is HEDGES' (1985) *g*, which estimates the effect size based on the difference between means. Effect size is also comparable across all viewpoints because it is not affected by the rating scale.

STAMPS (1997, 2000) describes how to conduct a study investigating scenic impacts, and how a local Development Review Board might use the effect size results to determine whether or not these impacts are acceptable. After investigating thousands of paired landscape scenes, STAMPS (2000, page 162) has adopted the effect size thresholds suggested by COHEN (1988, pp. 24-27): when $d = 0.2$ it is too small to be noticed, $d = 0.5$ is a medium effect size that is "large enough to be visible to the naked eye," and $d = 0.8$ is large enough to be "grossly perceptible."[2] However, COHEN warns that "the terms 'small,' 'medium,' and 'large' are relative, not only to each other, but to the area of behavioral science or even more particularly to the specific content and research method being employed in any given investigation" (COHEN 1988, p. 25). Fortunately we have measures

1 Statistical significance means the results are unlikely to occur by chance, not that the magnitude of the difference is large enough to be important or even noticeable in the everyday world.

2 "Grossly" has a variety of meanings, but I believe the intent here follows the online Oxford English Dictionary's seventh definition: Indelicately, indecently.

for two of the Act's criteria that can be used to appraise these thresholds for how the scenic impact might affect the enjoyment of using a SRSNS or the continued use of SRSNSs.

3 Results

3.1 Reliability

The first thing to consider is whether the respondents' scenic ratings are reliable – that is if another survey is conducted in a similar manner, how confident should we be that the results would be the same. Reliability can range between 0 and 1. NUNNALLY (1978) states that reliability coefficients of 0.70 or 0.80 are normally acceptable for research purposes, but that reliability should be 0.90 or higher in situations where the measurements are the basis of important decisions.

The intraclass correlation coefficient is the standard measure of reliability for individual respondents, as well as for the group of respondents (PALMER & HOFFMAN, 2001). The individual reliabilities are quite low – they average 0.30 for the intercept surveys and 0.01 for the web survey. However, reliability can be improved by averaging the responses for a group of respondents; normally the more respondents, the higher the reliability. The group reliability for the intercept surveys is quite high (0.91), though the average for the web surveys is still rather low (0.44). In general, the ratings with and without the proposed project from each viewpoint are very reliable for the intercept surveys, though it may be necessary to survey more than 35 respondents to reach reliabilities of 0.90 or higher.

The reliability of the web survey is very low, even though they included 104 respondents who indicated that they recreated in western Maine. In other contexts, web surveys have been shown by others to be an effective tool for public use to evaluate scenic impacts (ROTH 2006, WHERRETT 1999). However, web surveys do not provide an opportunity to experience the context within which the scenic change is viewed (i.e., realism validity). Nor can a web survey provide data useful to estimate the nature, extent and duration of use at a scenic resource, as does a survey that intercepts users at potential viewpoints.

3.2 Scenic Impact

The average effect size from Table 1 for the intercept surveys is -1.21; for the web survey it is -0.27. The result suggests that respondents at the viewpoint are much more sensitive to the potential scenic impact than are the respondents to the web survey.

3.3 Effect on Enjoyment

The Act requires that permitting agencies consider "the potential effect of the generating facilities on the public's… enjoyment of the scenic resource of state or national significance." On a 10-point scale, a mean response of 5.5 indicates that the presence of the turbines would have no effect on the enjoyment of the scenic resource. The mean value for all four studies is 5.055, indicating that on average the presence of the wind turbines would have a slight negative effect. This effect is more negative (and statistically more significant) for the respondents to the Bull Hill and Highland surveys than it is for those to the Saddleback Ridge and Spruce Mountain surveys.

Table 1: Effect Size of Scenic Impact

Case: Project	Pre– $\bar{x}$	Post– $\bar{x}$	Pooled *s*	Effect size
1. Bull Hill (Black Mtn)	8.861	6.003	2.048	-1.395
2. Bull Hill (Donnell Pd)	7.750	6.426	1.853	-0.715
3. Highland (Hikers@9D)	7.514	5.405	1.857	-1.136
4. Highland (Web@9D)	7.971	7.605	1.751	-0.209
5. Highland (Web@4)	7.490	7.308	1.888	-0.097
6. Saddleback Ridge	7.682	5.841	2.606	-0.706
7. Spruce Mountain	6.300	4.950	1.577	-0.856

Using these data, it is possible to demonstrate how one might establish a scenic impact threshold based on when the effect on enjoyment is unacceptable. While only four data points is admittedly quite limited for this task, it is nonetheless still instructive to demonstrate the procedure and consider the results.

The statistical procedure used is linear regression analysis. Fig. 1 shows a scatter plot where the x-axis is effect on enjoyment (or continued use) and the y-axis is effect size. Regression analysis determines the line that best fits the data points in these plots. There is an equation that describes this line in the form of:

$$Y = (b*X) + a \tag{1}$$

"Y" is the dependent variable and the equation will be used to determine unknown values for it. The dependent value effect size. "X" is the independent variable that is used to determine the unknown value of Y for a known value of X. The independent variable is effect on enjoyment (or continued use). "b" is the beta coefficient or slope, and it describes the amount of change in Y for one unit of change in X. "a" is the constant or y-intercept, and it is the value where the line intersects the y-axis when the value of x is zero.

The regression equation that describes the line plotted for effect on enjoyment in Figure 1 is:

$$\text{Effect Size} = (0.7 * \text{Effect on Enjoyment}) - 4.6 \tag{2}$$

There is a strong relationship between effect on enjoyment and effect size (adjusted R^2 = 0.875). Though there are only six data points, such a high relationship strongly suggests that there is a valid connection between scenic impact, as measured effect size, and the effect on enjoyment of a SRSNS.

This equation can be used to determine the values of the effect size for specific values of effect on enjoyment, as shown in Table 2. The Act does not provide guidance for when a negative effect on enjoyment is unacceptable. Must the mean rating be 1.5, or could it be 2.0 or 3.0? However, if 3.0 is established by the permitting agency as the appropriate threshold on a 10-point scale, then the threshold for an Unreasonably Adverse scenic impact would be a size effect of -2.5.

3.4 Continued Use

The Act requires that permitting agencies consider "the potential effect of the generating facilities on the public's continued use... of the scenic resource of state or national significance." On a 10-point scale, a mean response of 5.5 indicates that the presence of the turbines would have no effect on the likelihood that a respondent would continue to use the scenic resource. The mean value for the intercept surveys is 5.504, indicating that the presence of the wind turbines would have essentially no effect on respondents' continued use. However, this effect is substantially greater (and statistically significant) for the respondents to the Highland web survey, who indicate that they would be more likely to return if the project were built ($\bar{x} = 7.1$).

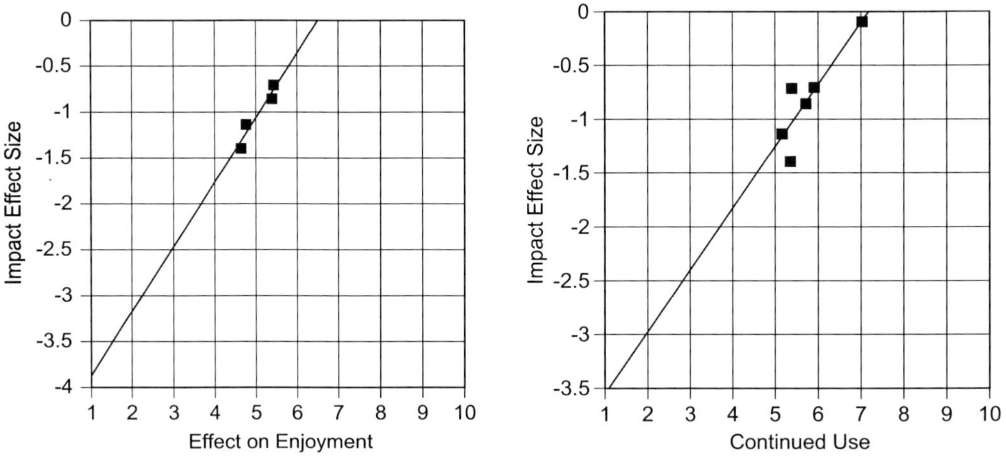

Fig. 1: The relationship between continued use and two measures of scenic impact

Regression analysis is used again to identify the thresholds where the effect size of the scenic impact become Unreasonably Adverse as indicated by users' ratings of whether they would continue to use the SRSNS. The scatter plot and regression line is shown in Fig. 1 There are now severn points and they are more dispersed along the line. However, this analysis must still be treated more as a demonstration of the approach rather than a definitive establishment of thresholds. The regression equation that describes the line plotted in Fig. 1 is.

$$\text{Effect Size} = (0.52 * \text{Continued Use}) - 3.84 \tag{3}$$

There is a strong relationship between continued use and effect size (adjusted $R^2 = 0.782$). Though there are only seven data points, such a high relationship strongly suggests that there is a valid connection between scenic impact, as measured by effect size, and the continued use of a SRSNS.

This equation can be used to determine the value of the effect size for specific values of continued use, as shown in Table 2. If 3.0 is established as the appropriate threshold on a 10-point scale where the Act considers the negative effect on the likelihood of returning to be Unreasonably Adverse, then the threshold for an Unreasonably Adverse scenic impact would be a -64.4 percent change in scenic value or a size effect of -2.3.

Table 2: Effect Size Values for Thresholds of Effect on Enjoyment and Continued Use

	Extremely Negative (1.5)	Very Negative (2.0)	Negative (3.0)	No Effect (5.5)	Positive (8.0)
Effect on Enjoyment	-3.5	-3.2	-2.5	-0.7	1.0
Continued Use	-3.1	-2.8	-2.3	-1.0	0.3

4 Discussion

These results present an opportunity to discuss when the scenic impacts from grid-scale wind energy projects create an Unreasonable Adverse scenic impact. The threshold where the scenic impact becomes Unreasonably appears to be approximately a -2.5 effect size.[3]

This value is admittedly higher than one might expect as the beginning point for such a discussion. For instance, STAMPS (2000) suggests that an effect size threshold of 1.1 be used to identify very large scenic impacts, and the effect size recommended by this study is much larger than that. However, they are based on the judgments of people actually using the affected SRSNSs, and the data appear to be both statistically reliable and valid.

On the other hand, it is acknowledged that there are relatively few data points and these thresholds will need to be recalculated as more surveys are conducted. In addition, it is necessary to include scenic evaluations where the impacts are clearly Unreasonably Adverse. The thresholds suggested here are based on data from viewpoints where users of SRSNS did not think the scenic impact was Unreasonably Adverse.

As more intercept studies targeted to the Act's scenic impact evaluation criteria are conducted a conscious attempt needs to be made to investigate a wider range of users of SRSNSs. The Act requires that the "nature" of the use be considered – hikers may be more or less sensitive to scenic impacts than snowmobilers or people fishing. It also requires that the "duration" of use, and therefore the length of time they are potentially exposed, be considered. To date, all the intercept surveys have evaluated the scenic change at specific viewpoints. It may be that the cumulative exposure to multiple views of wind energy development during a day's outing will result in Unreasonably Adverse threshold levels that are higher or lower than those identified here.

A word of caution about the use of intercept studies may be prudent at this point. While there can be little doubt about the validity of a well conducted intercept study, there is the potential to introduce bias into this method. It is important that the selection of respondents to intercept studies continue to represent the people typically found using SRSNS. To date

[3] This is based on the assumption that a mean rating of 3.0 for effect on enjoyment and continued use is the threshold for unacceptability. It may be that 1.5 or 2.0 is more appropriate.

the intercept studies have been conducted over a weekend, or perhaps a couple consecutive weekends. It is advisable to increase sampling throughout the season of use at specific SRSNSs. In addition, it is also important to be vigilant that interest groups not learn of the dates and places of the intercept surveys, since if they do it may result in a "call to action" that in effect results in stuffing the ballot box. If this situation occurs, the respondents would no longer represent the evaluations of "typical" SRSNS users and the results should be discounted.

5 Summary Conclusions

This paper has demonstrated how to identify when users of a SRSNS find a scenic impact to be Unreasonably Adverse, based on user ratings of (1) a photograph of the actual view, (2) a photosimulation of how the view will appear if the wind development is constructed, (3) how the wind development will effect enjoyment of their use of the SRSNS, and (4) whether the wind development will affect their continued use of the SRSNS. Specifically, these data can be used to:

1. Evaluate the reliability of the scenic value ratings.
2. Calculate the effect size of the potential scenic impact.
3. Determine the threshold of Unreasonable Adverse scenic impact based on an assumed level where the project's effect on enjoyment of the SRSNS is unacceptable.
4. Determine the threshold of Unreasonable Adverse scenic impact based on an assumed level where the project's effect on the continued use of the SRSNS is unacceptable.

The data from four intercept surveys were reanalyzed, and their results were found to be both valid and reliable. A web-based survey was also reviewed and the reliability of the responses was lower than for the intercept surveys. However, the web survey does provide an opportunity for a more diverse public to contribute to the scenic impact assessment process. It is recommended that the effectiveness of using web-based surveys be further evaluated.

This paper is intended to initiate a discussion on how to identify the threshold between Adverse and Unreasonable Adverse scenic impacts. While the results appear impressive, they are based on relatively few data points. Future applications for wind energy development should include intercept surveys from more viewpoints that provide a greater range in the scope and scale of visible generation facilities, and are frequented by people engaged in a greater diversity of activities. Methods also need to be developed and validated to evaluate the effect of multiple exposures to scenic impacts while using SRSNSs. Intercept surveys should also be conducted in a manner that provides for estimating the extent, nature and duration of use at a viewpoint. Web surveys might be used to supplement intercept studies at locations with very few users, where there is the expectation of great controversy, or other situations where the scenic impact evaluation would benefit from a greater number of responses from potential users.

Finally, it is recommended that post-construction studies be conducted to monitor the actual scenic impact. In particular, intercept surveys should be conducted at the same locations as reviewed in this paper and include an evaluation of the perceived veracity of the photosimulations as a tool for evaluating scenic impacts.

References

American Psychological Association (2010), *Publication Manual of the American Psychological Association* (6th ed.). Washington, DC, APA.

Cohen, J. (1988), *Statistical Power Analysis for the Behavioral Sciences* (2nd ed.). New York, Psychology Press.

Expedited Permitting of Grid-Scale Wind Energy Development (2008), MRSA Title 35-A, Chapter 34-A. http://www.mainelegislature.org/legis/statutes/35-A/title35-Ach34-A.pdf (accessed February 23, 2010).

Mildner, C. & MacBride, J. (2010), *Research Report: Mt. Blue-Saddleback Ridge Wind Power Project Intercepts*. Portland, ME, Market Decisions.

Mildner, C. & MacBride, J. (2010), *Spruce Mountain Wind Project Intercepts Research Report*. Portland, ME, Market Decisions.

Nunnally, J. C. (1978), *Psychometric Theory* (2nd ed.). New York, McGraw-Hill.

Palmer, J. F. & Hoffmann R. E. (2001), *Rating reliability and representation validity in scenic landscape assessments*. In: Landscape and Urban Planning, 54 (1-4), 149-161.

Portland Research Group (2011), *Highland Wind Project Survey of Hikers March 11: Report of Trail Intercept and Web Survey Research Conducted Summer and Fall 2010.* Portland, ME, Portland Research Group.

Robertson, B. & MacBride, J. (2010), *Research Report: Bull Hill Wind Power Project Intercepts*. Portland, ME, Market Decisions.

Roth, M. (2006), *Validating the use of Internet survey techniques in visual landscape assessment – An empirical study from Germany*. In: Landscape and Urban Planning 78, 179-192.

Stamps, A. E. (2000), *Psychology and the Aesthetics of the Built Environment*. Boston, Kluwer Academic Publishers.

Wherrett, J. R. (1999), *Issues in using the Internet as a medium for landscape preference research*. In: Landscape and Urban Planning, 45, 209-217.

Environmental Information in Planning and Design

Photo-based Terrain Data Acquisition and 3D Modeling

Howard HAHN

1 Introduction

When performing hydrologic and geomorphologic analyses for small area sites, it is desirable to capture terrain data for 3D modeling at higher resolution and lower cost than typical LiDAR imaging provides. This is especially desirable when assessing frequent or detailed terrain changes due to erosion or human activities such as agriculture, development grading, and mining. This paper explores recent advances in the application and effectiveness of relatively low-cost software that relies on modified automatic aerial triangulation techniques for modeling 3D terrain based on overlapping, multiple photographs (DUNN, 2009; REMONDINO et al., 2011).

The long term goal of this exploration is to determine if this new, relatively low-cost software might replace manual field techniques being used by landscape architects and other university researchers conducting hydrologic analyses. Gully erosion is being monitored and studied within agricultural fields and a restricted military base. In all instances, major erosion is interfering with land uses and contributing to stream sediment. Gully prediction, both in terms of formation and geomorphology, relies on accurate terrain modeling that is geo-referenced enabling periodic measurements and comparisons over time (CASTILLO et al., 2011; GESSESSE et al., 2010; SHRUTHI et al., 2011).

Preliminary investigations were conducted using three sites at three different scales to assess general applicability of photo-based 3D terrain modeling. The impetus for this initial exploration is threefold: 1) For sites 250 ha or less, determine if drainage patterns can be identified and mapped with greater precision than available from public sources offering 2 m LiDAR resolution, and lower cost than contracted LiDAR flights of higher precision or frequency; 2) Test the suitability of two different unmanned aerial platforms for low altitude data acquisition; and 3) Test whether derived terrain models are accurate enough to detect minute surface variations of 1 cm or less that might support erosion estimates.

2 Software Tool: Agisoft PhotoScan Pro

The software chosen for this investigation is PhotoScan Pro (AGISOFT, 2012). This Russian software is preferable over other 3D photo-modelers because it supports geographic or projected coordinate systems which make it suitable for photogrammetric and remote sensing applications. It is also relatively affordable: 420 Euro ($549 USD) for the academic version and 2,680 Euro ($3,500 USD) for the commercial version. FRANKENBERGER et al., (2008) and others have shown that photo-modeling software can approach the accuracy of terrestrial based LiDAR scanners.

PhotoScan is designed to work with digital cameras. If the software cannot automatically determine the optical and format parameters of the camera lens, Agisoft also offers free software to calculate these parameters through a lens calibration routine. A camera resolution of five megapixels or greater is recommended. The software accepts the following image formats: JPEG, TIFF, PNG, BMP, JPEG Multi-Picture Format (MPO) [and RAW saved as a TIFF]. Major processing steps include photo import, photo alignment and point cloud matching, surface geometry building, texture building, ground control adjustments/model updating, and exportation of 3D models, DEMs, and orthophotos.

3 Unmanned Aerial Systems

To increase image acquisition efficiency for larger area sites, two types of unmanned aerial systems were used for aerial photography: the RiteWingRC Zephyr (flying wing) (RiteWing, 2012) and the DJI S800 Hexacopter (UAV Products, 2012). Both are small radio controlled vehicles and can be flown interactively or through programmed flight paths (Fig. 1). Flight duration is brief (12-18 minutes) due to battery power limitations. Both vehicles carried aloft a lightweight Canon Powershot S100 (12.1 MP) digital camera. Images were captured every four seconds and saved in both .JPG and .CR2 formats.

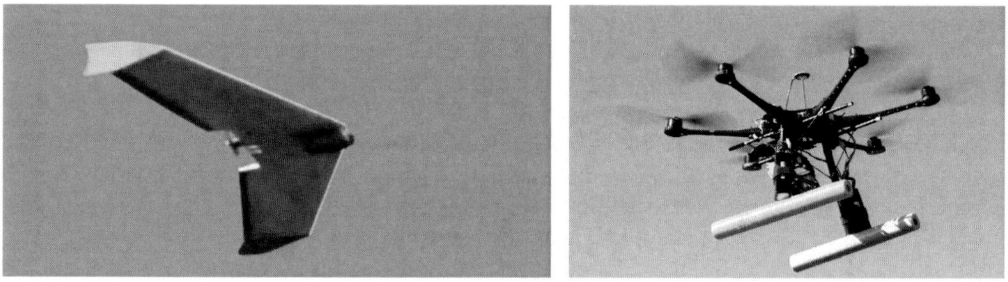

Fig. 1: RiteWingRC Zephyr flying wing and the DJI S800 hexacopter (Dr. Kevin Price)

In the United States, the Federal Aviation Administration (FAA) currently restricts the commercial use of unmanned aerial systems (UAS) until standards, certification, and operating procedures are in place by 2015 (Faa, 2011). After this date, widespread adoption of this technology for close range remote sensing will become more widespread. At this time, public agencies and universities can operate a UAS if a Certificate of Authorization (COA) is obtained per UAS per site, and proper flight protocols are followed.

4 Methods

To investigate the three research objectives defined in Section 1 Introduction, three test sites (TS) near Manhattan, Kansas were selected. Basic site parameters, lighting/weather conditions, and the number of ground control (GC) points used are summarized in Table 1. Table 2 provides additional data of the image collection phase.

Table 1: Test Site Summary

Test Site (TS)	Coverage Area	Land Cover	Imaging Date/Time	Images Used	GC Pts.
TS1–Field	129,112.0 m^2 (12.9 ha)	Short grass & tree patches	Nov 16/ 1:30p	9	6
TS2–Pasture Gully	792.0 m^2	Grass/dirt	Dec 4/1:00p	16	3
TS3–Eroded Slope	6.7 m^2	Bare dirt	Jan 7/1:45p	9	2

Weather for all imaging times was clear and sunny, with winds 0-5 km/hr. TS1 and TS2 were chosen because of prior COA flight clearances.

Table 2: Imaging Summary

Test Site (TS)	Acquisition Method	Camera	Altitude or Camera Dist.	Ground Resolution
TS1–Field	Zephyr	Canon S100*	195.41 m	68.85 mm/pix
TS2–Pasture Gully	Hexacopter	Canon S100*	16.06 m	4.87 mm/pix
TS3–Eroded Slope	Ground Photo	Nikon D50**	2.76 m	0.85 mm/pix

*Canon S100: 5.2 mm focal length; 4000 x 3000 resolution setting.
**Nikon D50: 18 mm focal length; 3008 x 2000 resolution setting.

TS1-Field: This site was selected to investigate if PhotoScan could detect and model subtle drainage patterns from imagery collected at low altitudes using the unmanned Zephyr flying wing. A programmed flight path provided uniform aerial coverage. Because of the ~ 60 km/hr flight speed, a limited number of images were acquired due to the four second image interval. Ground control was provided by using UTM projected Bing imagery available through ArcOnline (ESRI, 2012) for XY positioning, and using 2 m LiDAR data available through the Kansas GIS Data Access & Support Center (DASC, 2012) to estimate Z coordinates. After error correcting the 3D terrain model using ground control and applying image texturing, the model was exported in DEM and orthophoto formats. A comparison was then made in a GIS environment between the newly acquired terrain and LiDAR terrain. A sophisticated analysis was not deemed necessary for this preliminary investigation; rather, a simple visual comparison was made between hill-shaded surfaces.

TS2-Prairie Gully: A typical prairie gully section was chosen as a site to test the efficiency and effectiveness of the low flying hexacopter for image collection. The hexacopter was flown interactively down the gully section (~ 80 m long with 0.30-0.75 m vertical head cuts). Precise ground control using a surveyor's total station was not used. A sub-objective of this investigation was to determine if PhotoScan could rapidly model gullies in sufficient detail and accuracy without making future field work unduly complicated or require survey personnel. PhotoScan model accuracy was compared with the real-world gully using ten spot check measurements taken on a return field visit.

TS3-Eroded Slope: In the last investigation, a very small eroded slope was chosen to determine how well PhotoScan could resolve small surface variations over time. Aligning surfaces and calculating volume differences would enable estimating quantities of eroded soil. Procedures began by embedding two control markers, spaced 1.93 m apart, in the slope that would be included in the photographs to serve as a reference scale. Existing slope conditions were then photographed from a distance of approximately 2.76 m. The undisturbed slope was photographed a second time to serve as a control surface. Using a trowel, small quantities of soil were scraped from the slope and collected for volume measurement. Scraping depth was approximately 1-2 cm and the total soil collected was 8,000 cm^3. The scraped slope was then photographed to document "eroded" conditions.

Three surfaces were built in PhotoScan corresponding to the three sets of slope photographs. Surfaces were then aligned through the resident PhotoScan feature: "Align Chunks..." using the "Point based" option. By visual inspection, the surfaces appeared to be translated and rotated to achieve a best fit. Surfaces were then exported in both .3ds model format (triangulated mesh) and DEM raster format. All surfaces, in both formats, were imported into ESRI ArcMap software (ESRI, 2012). All surface rasters were clipped to the exact same rectangular extent to enable volumetric calculations. Any calculated differences between the two control surfaces would indicate alignment, feature resolution, or replication errors –"error noise".

5 Results

TS1-Field: Even with minimal ground control and aerial coverage photographs, the orthophoto mosaic produced through PhotoScan horizontally matched the GIS reference sources within ± 1.5 m (spot checks), although PhotoScan resolution was better than 7x7 cm. The Zephyr flying wing was also an extremely efficient method of acquiring aerial imagery (~1.4 km^2/min) with minimal preparation and setup time. With the methods employed, topographic modeling using PhotoScan was far inferior compared the lower resolution bare-earth LiDAR terrain (Fig. 2). PhotoScan failed to pick up drainage subtleties and dirt road tracings present in the LiDAR terrain. This is likely attributable to vegetation which cannot be processed out of photo-based data, too few photographs, and relatively inaccurate Z-control. In patch areas of leafless deciduous trees, accuracy was poor, or the terrain surface contained holes. The Z-difference mean between the PhotoScan and LiDAR surfaces was 1.08 m.

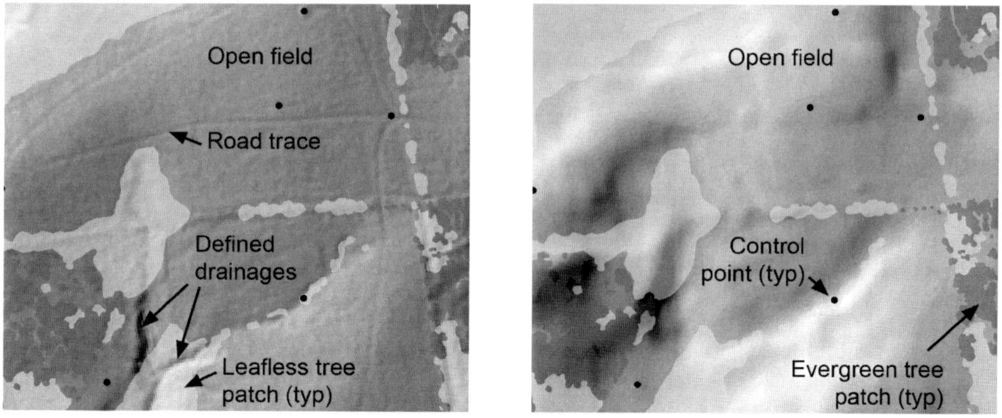

Fig. 2: Comparison of 2 m LiDAR surface (left) with 7x7 cm PhotoScan surface (right)

TS2-Prairie Gully: The lower altitude and slower speed of the hexacopter produced far better terrain modeling results than TS1 (Fig. 3). Ten random field check measurements in all XYZ orientations and distances resulted in a mean error of 3.53 cm. This is certainly within the probable error of establishing the field markers to match model reference points. At this relatively close imaging range, the 7-10 cm high grass cover did not appreciably affect relative terrain accuracy (little variation in grass height). Compared to the manual method of taking cross-sectional measurements at limited intervals along the gully flowline, the PhotoScan method offers several significant advantages:

- Fully continuous surfaces are depicted and resolved down to millimeters of detail.
- Terrain surfaces are fully textured mapped with high resolution imagery that serves as good documentation. Texture maps can be optionally turned off.
- Using the texture map as a visual guide, additional gully measurements can be made off-site at any future time.
- Digital terrain models can be imported in a GIS environment and combined with other data for a variety of analyses across different time scales.
- Texture mapped terrain models can be emailed as 3D .pdf files which can be interactively manipulated and viewed from any angle.

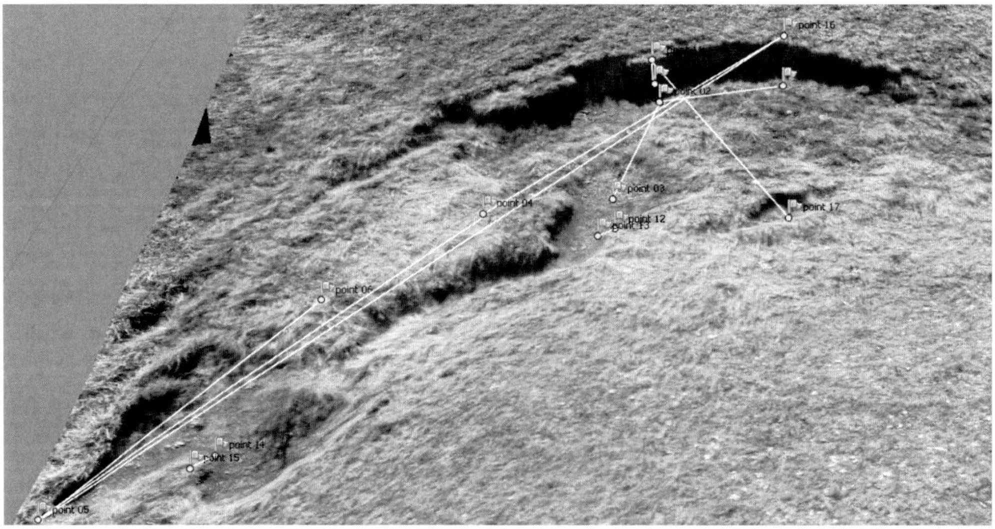

Fig. 3: PhotoScan terrain model of prairie gully with field measurement spot checks

TS3-Eroded Slope: At a close camera range of 2.76 m, PhotoScan is capable of producing a 3D model surface with a Z-resolution of 5 mm or less for bare earth conditions (Fig. 4). Eroded conditions are easily photo captured and modeled.

Subsequent volume calculations referenced the 8,000 cm^3 of scraped soil that was collected. Volumes were calculated for both the soil scraped and collected as "erosion", and for "error noise" between two control surfaces.

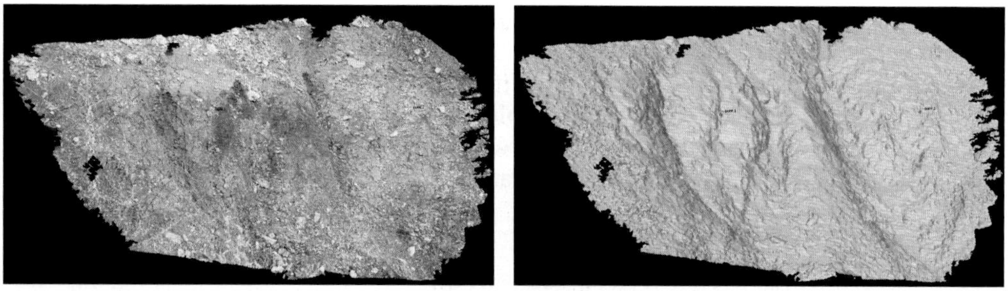

Fig. 4: PhotoScan 3D model displayed with textures (left) and as a bare surface (right)

Subtracting the scraped surface from the undisturbed surface resulted in a raw calculated volume of 10,740 cm^3 (74.5% accuracy) depicted in Fig. 5a. Direct error compensation was conducted by generating a histogram of Z differences per raster cell displayed with a natural breaks (Jenks) distribution. Filtering out apparent error noise ranges resulted in a corrected volume of 9,390 cm^3 (85.2% accuracy).

Alternatively, a second correction method was also applied. Subtracting volumetric control surfaces generated from supposedly identical photo/model sets of the same undisturbed condition, revealed a mean z-difference of 0.47 mm per raster cell. Aggregated over the entire clipped surface area, total error volume serving as the control amounted to 1,420 cm^3 (Fig. 5b). Accounting for errors revealed by the volumetric control, the net "eroded" volume is 9,320 cm^3 (85.8% accuracy).

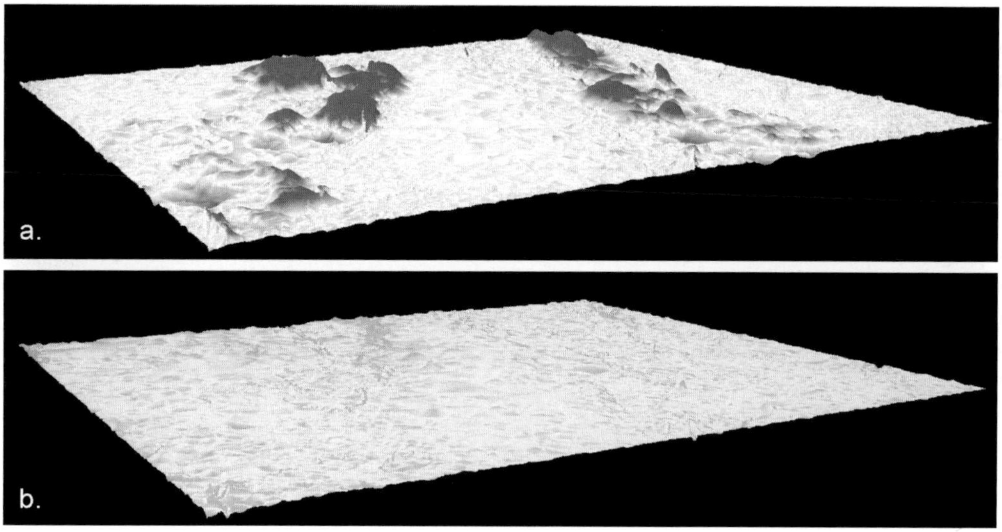

Fig. 5: Volume differences for detected "eroded" soil (5a) and error "noise" between control surfaces generated from the same undisturbed condition (5b)

Overall, the accuracy of the calculated "erosion" volume was a little lower than expected, but certainly superior to what could be achieved through manual slope measurements of 3D surfaces for such minute erosion quantities – if it could be done at all. Volume inaccuracies are likely attributable to a small sample volume relative to error accumulation, minute surface misalignment, or slight inaccuracies when measuring the reference distance in the field or setting marker targets within PhotoScan.

6 Conclusions and Future Research

Initial results indicate that the utility of photo-based orthomapping and 3D terrain modeling is well-suited for some types of hydrology work, but is not a replacement for terrestrial or aerial LiDAR for vegetated sites. One overall advantage of photo-based terrain modeling techniques is the inclusion of photo texturing.

TS1-Field: In this investigation, the Zephyr flying wing used in combination with PhotoScan software to model drainage networks for sites less than 250 hectares, did not yield better results than 2 m LiDAR data at a coarser XY resolution. Because of its potentially high utility, further investigations will be conducted to determine if PhotoScan accuracy can be improved, most notably by using more aerial photographs from multiple flight passes and better ground control. Unlike bare-earth processing available in LiDAR, vegetation may be an insurmountable issue for some sites if using PhotoScan.

TS2-Prairie Gulley: Using a low flying hexacopter proved to be an efficient method to collect aerial imagery for gully modeling. From low altitude (16 m) photographs, PhotoScan was capable of producing terrain models within a mean ±3.53 cm of spot field measurements. Future investigations will deploy this new tool to regularly scheduled gully research work. More rigorous testing and positional control consistently applied across erosion monitoring events will be compared to manual cross section techniques. Again, vegetated gullies are anticipated to be problematic, but attempts will be made to efficiently edit point clouds to remove vegetation from the terrain models.

TS3-Eroded Slope: Even with minimal control, PhotoScan is capable of resolving minute changes (±5 mm) in terrain surfaces useful for monitoring erosion. Initial investigations demonstrated that accuracy of at least 85.8% can be achieved when estimating erosion quantities. Future investigations will review the procedure protocol to isolate error sources and seek to improve accuracy to 95% or better. Larger and more complicated gully configurations will also be tested.

Acknowledgement

The author would like to thank Dr. Kevin Price and Dr. Deon van der Merwe who directed the unmanned aerial systems, shared aerial imagery, and provided consultation.

References

Agisoft, PhotoScan Pro, v. 0.9.0. Available at: http://www.Agisoft.ru/products/photoscan (accessed December 22, 2012).

Castillo, C., Pérez, R., James, M., Quinton, J., Taguas, E. & Gómez, J. (2011), *Comparing the accuracy of several field methods for measuring gully erosion.* In: Soil Science Society of America Journal, 76, 1319-1332.

DASC (2012), *2006 River Corridor LiDAR Project (2m data).* Available at: http://maps.XXXXXgis.org/dasc/ims/dasc.cfm?servername=maps.XXXXXgis.org&servicename=dasc_LidarTile (accessed December 22, 2012).

Dunn, M. (2009), *Recent Developments in Close Range Photogrammetry for Mining and Reclamation.* In: Barnhisel, R. (Ed.), Revitalizing the Environment: Proven Solutions & Innovative Approaches. Proc. of 2009 Meeting of the American Society of Mining and Reclamation, 30 May – 5 June, 2009 – Billings, MT, ASMR, Lexington, KY, 390-399.

ESRI, *ArcMap 10.1 and Microsoft Bing Aerial Imagery available through ArcOnline.* Available at: http://www.esri.com/software/arcgis (accessed December 22, 2012).

FAA (2011), *Unmanned Aircraft (UAS) Questions and Answers.* http://www.faa.gov/about/initiatives/uas/uas_faq (accessed October 31, 2012).

Frankenberger, J., Huang, C. & Nouwakpo, K. (2008), *Low-Altitude Digital Photogrammetry Technique to Assess Ephemeral Gully Erosion.* In: The Next Generation. Proc. of 2008 IEEE International Geoscience and Remote Sensing Symposium, 6-11 July, 2008 – Boston, MA, IV: 117-120.

Gessesse, G., Fuchs, H., Mansberger, R. & Rieke-Zapp, D. (2010), *Assessment of erosion, deposition and rill development on irregular soil surfaces using close range digital photogrammetry.* In: The Photogrammetric Record, 25 (131), 299-318.

Remondino, F., Barazzetti, F., Nex, F., Scaioni, M. & Sarazzi, D. (2011), *UAV Photogrammetry for Mapping and 3D Modeling: Current Status and Future Perspectives.* In: International Archives of the Photogrammetry, Remote Sensing and Spatial Information Sciences, Proc. from Conference on Unmanned Aerial Vehicles in Geomatics, 14 – 16 Sept., 2011 – Zurich, Switzerland, 38-1/C22 UAV-g2011, 1-7.

RiteWing, RC Zephyr. Available at: http://www.ritewingrc.com (accessed Dec. 22, 2012).

Shruthi, R., Kerle, N. & Jetten, V. (2011), *Object-based gully feature extraction using high spatial resolution imagery.* In: Geomorphology, 134, 260-268.

UAV Products, DJI S800 Hexacopter. Available at: http://www.uavproducts.com (accessed December 22, 2012).

Tactile Models Carved by Binary Numbers – Connecting Digital and Physical Design Methods

Dietmar STRAUB and Anna THURMAYR

1 Introduction

Landscape architecture, as with any other design discipline, has always required representation. Before an idea can be realized it must be communicated by words and visualization. Efficient methods of exhibiting, explaining and selling a design are sketch, collage, section, layout plan, digital fly-through, perspective and computer-generated or physical models. "Of all these forms of representation, the model is the only physical, three-dimensional realization of the architect's idea – which, after all, is ultimately intended to be a physical, three-dimensional thing" (MOON, 2005, 11). Even the so-called 3D modeling done on the computer screen is finally represented two-dimensionally.

From a historical perspective there has always been a fascination with scaled-down representations of objects and buildings. Models were used as grave goods in ancient Egypt while in medieval times masons were working with templates and full-scale mock-ups as learning tools. The architectural model, as it is used still today, was invented during the Renaissance (MORRIS, 2006). But do tactile models play a role in design education today? And if so, what are the benefits?

This paper explores parallel teaching strategies that address both digital and analogue design methods in undergraduate design studios. The relationship between the physical model and the computer takes centre stage.

2 Role and Benefits of Physical Models

Nowadays the teaching and learning of digital techniques and their application is considered an essential component of basic design education. While using digital tools and methods is beneficial to students it is also important to include physical, three-dimensional visualization techniques in the early stages of design education. The following advantages are reasons for integrating model-making in undergraduate design studios at the University of Manitoba:

- Model-making fosters creativity. Three-dimensional activities made by real hands help students to overcome mental block and actuate creative thinking.
- Physical models can be used as a tool in the design process. It is "a means of inventing, searching and investigating" (JANKE, 1978, 10). It helps students make design decisions.
- To read a three-dimensional model requires less training than to understand an architectural drawing. "The scale model … persists as a kind of universal represen-

tation, intelligible to non-architects. CAD drawings and renderings remain, like blueprints, professionally codified." (MORRIS, 2006, 166).

- A three-dimensional model can effectively communicate students' designs.
- Model-making fosters spatial thinking. It is challenging, especially in the case of beginners, to master the relationship between design and spatial impact. Doing drawings digitally and zooming in and out on a computer screen adds to this challenge, as it can actually hinder spatial comprehension. In contrast to this the to-scale model reveals the spatial consequences of a design idea.
- A model helps students to focus on the 'big idea' instead of falling in love with details. "The whole is apprehended in advance of the parts" (MORRIS, 2006, 166).
- The model suggests the design is almost real. It fosters students' pride and engagement.

3 Relationship of Digital and Analogue Design Methods

Nowadays, new technologies such as digitally fabricated models cut by laser or router or those built up by a three-dimensional printer require not only new content but different ways of teaching. "The fact that a three-dimensional model can now be electronically modified, refined, and digitally transformed back into a new physical existence brings a new and exciting dimension to the design previewing skill of the architect" (PORTER & NEALE, 2000, V). In an attempt to foster three-dimensional thinking the authors challenge their students to work with a variety of both digital and analogue design tools. The following teaching methods utilize synergies of two-dimensional, three-dimensional, digital and analogue visualization tools. Employing the mutual interaction of these methods constitutes a timely approach to landscape design teaching in undergraduate programs.

4 Interaction of Hand Drawing and Digital Drawing

Students at the University of Manitoba are introduced to hand and digital drawing through specific workshops as part of their design studios. The digital drawing workshop instructed by one of the authors (Fig. 1) provides an introduction to computer-aided design and computer graphics. The objective of this workshop is to teach students how to draw, illustrate and layout their ideas using digital as well as analogue media. During this first week students get the chance to experiment with the computer as a medium through which to express and present ideas.

Depending on the skills of the students this workshop demonstrates the basic concepts of VectorWorks such as 2D drafting, file organization, basic structure and operation of landscape projects, interaction with photo editing and layout programs, principles of data exchange and data saving, simple 3D modelling to verify the setting and composition, and digital and analogue possibilities of presentation.

Fig. 1: Students attending the workshop "Digital Drawing"

5 Interaction of Hand Models and Digital Drawing

The digital drawing workshop serves as a base to do further experiments connecting analogue to digital design methods. At the University of Manitoba the authors push their students to explore and refine hand models with the help of digital tools.

At the beginning of the design process students are asked to work with their hands in order to reveal design options. They create and constantly revise the hand model in order to test their design idea as it emerges. During this process certain discoveries or decisions may arise that challenge the students' ability to compose and sculpt spaces but with this dynamic tool both the students and the instructor can be actively involved in the design process together.

Fig. 2: Designing with salt

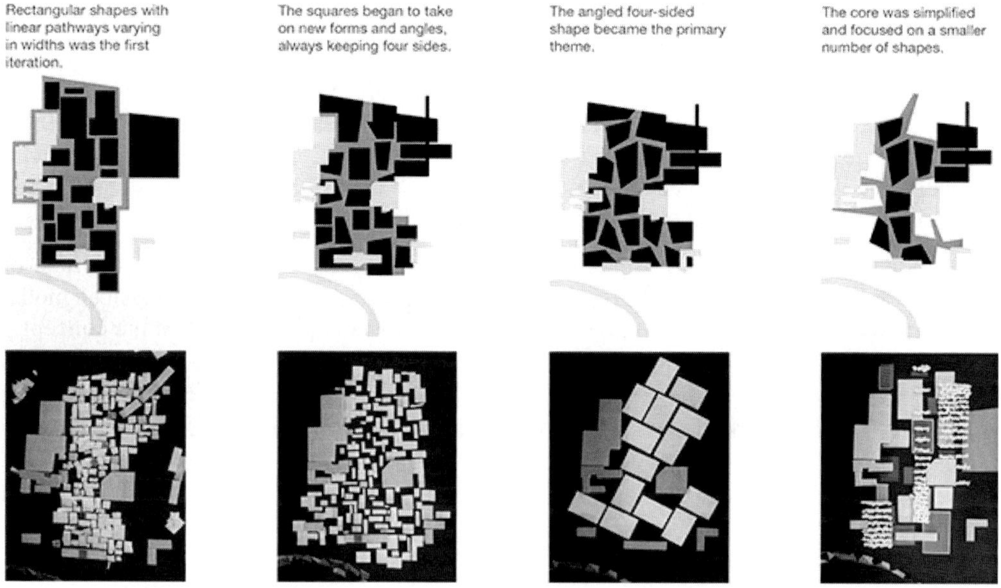

Fig. 3: Design iterations using paper & cardboard versus digital sketch by students Taylor Bishop & Roxane Gratton, 2011

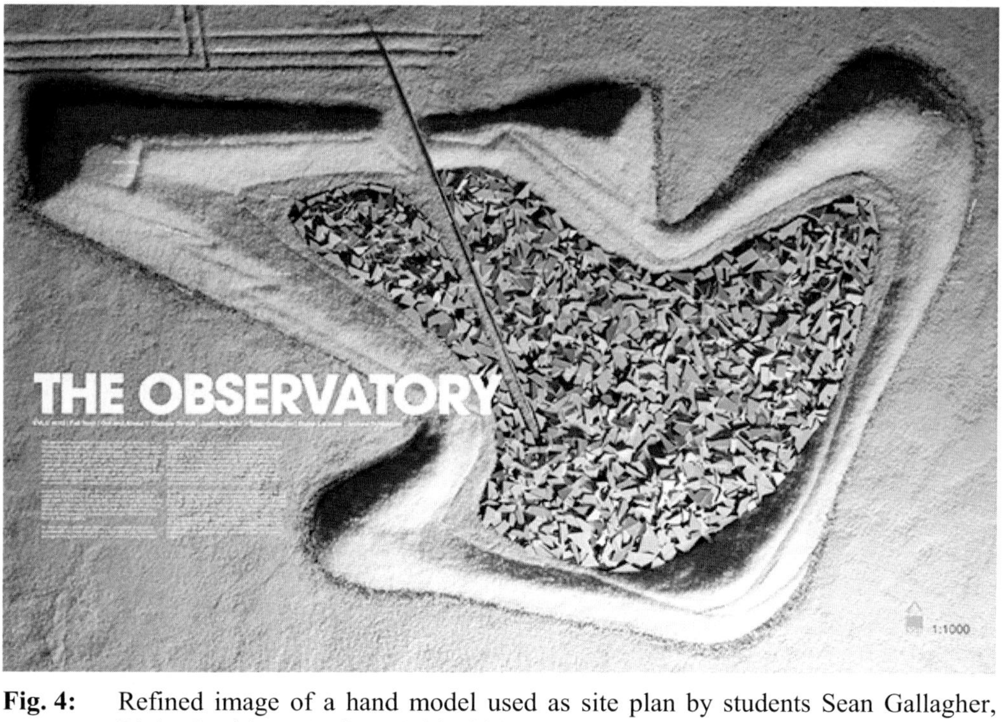

Fig. 4: Refined image of a hand model used as site plan by students Sean Gallagher, Blaise Lachiver, Andrew Schluckbier, 2011

Suitable materials for modelling landscape or urban designs in the early stages include paper, cardboard, fabric, snow, salt (Fig. 2), and clay. Materials should be inexpensive, quick to assemble and supportive to any design changes (MOORE, 1990). In contrast to hand drawing or computer modelling this technique requires a minimum of technical skill. Students benefit from going back and forth between physical model and digital sketch or physical model and hand drawing during the design process. Refining the hand model with the help of a different tool generates new perspectives (Fig. 3). The authors consider this switch between media to be hclpful in learning about design and spatial impact.

Hand models can also be used for final two-dimensional representation. With the help of photo editing and digital layout programs students refine pictures of the physical model (Fig. 4) and use them on their design boards instead of the site plan to explain the concept.

6 Interaction of Digital Drawing and Laser Cut Model

Finally, the digital drawing workshop creates the opportunity to explore interactions between the digital drawing and laser cut model. At the University of Manitoba the authors ask their students to provide a digitally fabricated model, cut by laser, for final representation. The main objective is to confirm design decisions and to communicate with an audience who was not part of the underlying design process.

Before laser work for the final model can be done a CAD drawing has to be prepared by students. Specific requirements are to be followed. Similar to acquiring language students have to learn how to communicate with the machine when producing laser-cut models. It is a salutary experience when they learn that neither the computer nor the mechanical cutter machine is automatically designing for them.

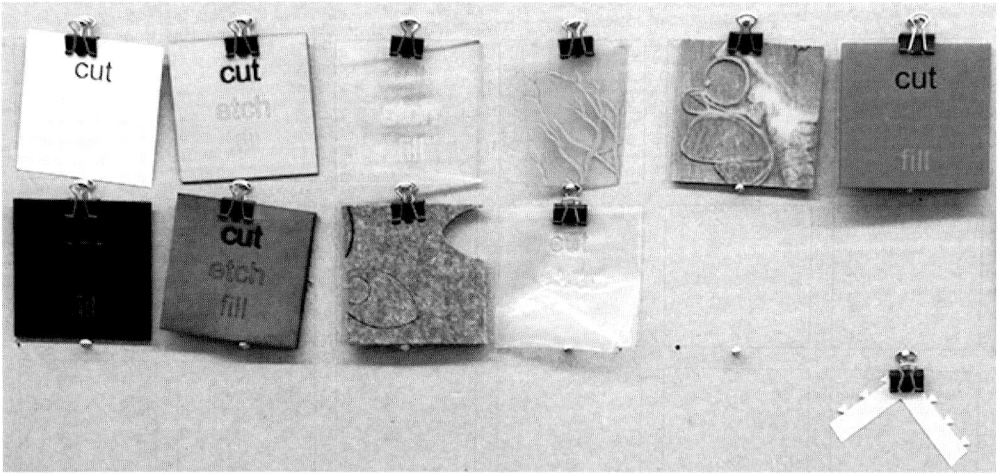

Fig. 5: Laser cut samples, Workshop at the Faculty of Architecture, University of Manitoba

Certain procedures have to be followed in order to successfully operate the machine and students need to learn the various skills of cutting, etching, snapping, filling and editing prior to exporting the product to a dwg.file (Fig. 5). They also need to appreciate that decisions must pass through the computer and how any gaps in a project's development will be revealed. Once mastered this knowledge and these skills become essential to the design process.

7 Laser Cut Model and Analogue Exploration

Use of the laser cutter to create models offers up the opportunity of further creative exploration. "Finished models are typically built as monochromatic constructions made from one material… The blank, abstract treatment allows the model to be read in many ways" (MILLS, 2005, 18). The authors ask their students to do further exploration of the unfinished casting through a variety of additional processes such as threading, dusting, burning, candying, and wrapping. These processes are done by hand after laser work has been done and might include additional material that pushes the model beyond a 'monochromatic construction'. In contrast to being blank the students' models become enriched and pictorial in order to represent and punctuate the envisioned character. The plain laser cut pieces serve as a base for additional analogue investigation expressing individual design ideas in a refined way.

Students build the final model with due care and great attention to craftsmanship (Fig. 6). Usually they become very proud of their final laser-cut model as it emerges in three-dimensions. The model suggests that the design is close to reality, a reality that is not apparent in the two-dimensional plan.

Since the laser-cut model is a small-scale copy of the envisioned project discussions about attention to material are inevitable and important. The model brings back memories of childhood and associations with train models or other detailed assembling toys. The challenge is to appreciate the potential of the design model to represent the whole in advance of details.

Fig. 6: Students working on their final model

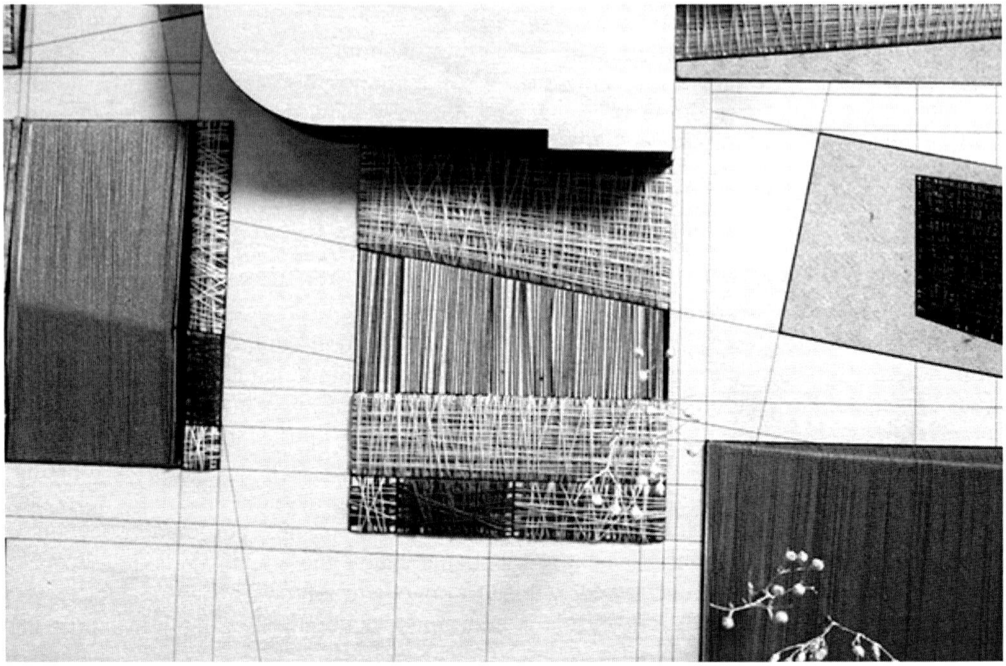

Fig. 7: Threading by students Taylor Bishop & Roxane Gratton, 2011

Fig. 8: Candying by students Andrea Linney & Lee Ann Bobrowski, 2011

In this spirit the authors ask their students not to do literal simulation in order to avoid distraction. Once students understand that the 'big idea' and its representation is the main focus there are no bounds to creativity. Thus students zero in on how to represent topography, buildings, water, and plants in a representational rather than literal way (Fig. 7 and 8).

8 Conclusion and Outlook

Students benefit from working with models both as part of their design process and as part of their final presentation. Model making takes time and money but more importantly, for someone who is starting a design career, it fosters spatial thinking. With increasing demands for students with expertise in both digital drawing and spatial thinking, it seems essential to promote a parallel education that fosters not only CAD skills and physical hands-on experiences of representation but an understanding of how to glean most use from both of them. The key to success in undergraduate design courses is for students to become capable of understanding spaces and their dimensions.

Some might think the binary revolution of images would have rendered the physical model obsolete. It seems to be the very reverse. New technologies expand the possibilities of designing with scale models. The "tactile model is enjoying a comeback" (MORRIS, 2006, 8) – and with the benefits of digital technology it is more useful than ever before.

References

Janke, R. (1987), *Architectural Models*. New York, Architectural Book Publishing Co.

Mills, C. B. (2005), *Designing with Models*. Second Edition. New Jersey, John Wiley & Sons, Inc.

Moon, K. (2005), *Modeling Messages*. New York, The Monacelli Press.

Morris, M. (2006), *Architecture and the Miniature*. Wiley-Academy, Chichester UK.

Porter, T. & Neale J. (2000), *Architectural Supermodels*. Oxford. Architectural Press.

University of Manitoba (2012): *Laser Cutting Tutorial*. http://umanitoba.ca/faculties/architecture/facilities/workshop.html.

Simple, but Therefore Clever Idea – An Effective Dendrological Inventory Process for Landscape Architecture and Design Projects

Heiki KALBERG, Nele NUTT, Sulev NURME and Mart HIOB

1 Introduction

Good solutions do not necessarily need to be highly complicated or costly. This article introduces a simple way to improve the economic efficiency of the digitalization of dendrological inventory data. The application developed in this study makes it possible to effectively link drawing and database programmes so that data can be utilized in both CAD (computer-aided design) and GIS (geographic information system) systems. Furthermore, no additional investments in training or technology are necessary, as this application uses software already featured in most plan design offices.

In Estonia, more than four hundred parks are protected areas (SINIJÄRV, 2008), but there are many more historical parks (HEIN, 2003). After Estonia regained independence (1991), people grew more passionate in restoring and rebuilding historical parks, which explains why landscape architects deal with compiling restoration projects on a daily basis. Compiling park design projects involves the inventory of park trees (EESTI STANDARD, 2010), and provides a much needed overview of park conditions along with the data needed for choosing the best restoration methods that promote sustainable design (NURME, 2008). Dendrological inventory is a time-consuming process due the extensive fieldwork involved and the detailed data digitalizing and systemizing processes that follow.

2 Application for inventory data integration

2.1 Reasons and causes of application development

There are two kinds of dendrological inventory data: primary data and collected data. *{Primary data involves geodetical base plans that have been compiled during surveys (RTL, 2007) on which locations of tree trunks are marked, together with the area layout that links to the L-EST97 coordinate system.}* For compiling digital base plans, the CAD system software is suitable; in Estonia the most commonly used are vector graphics programmes MicroStation (dgn) and AutoCad (dwg). The geodetical measurements must be taken with professional equipment, usually an automated theodolite operated by one geodesist. GPS measurements are not suitable as they are too inaccurate. Another component of inventory data consists of data collected during fieldwork. In Estonia, no specific methodology has been established for dendrological inventory (KUPPER, 2012). Trees are most commonly taken into inventory as individual items. The species of each tree is determined and certain parameters are measured (trunk diameter, crown projection area, tree height). Landscaping value (TALLINNA LINNAVALITSUS, 2006) and condition of the tree (NURME, 2008) are also analyzed, with damages described when necessary. Two people are needed for data collection, with one researcher taking measurements as the other records the information.

During fieldwork, this data is recorded by hand into inventory data tables, while geodetical base plans are used to mark tree locations. Data from the tables and trees on the plan are systematically linked to each other with numbers. Data collected during fieldwork is then entered into a spreadsheet table (Microsoft Excel is used primarily) and marked on a plan in a vector graphics programme, but this table and plan are not digitally linked to each other. This process is extremely time-consuming and creates excessive data entry after concluding field research.

Fig. 1: Determining the damage on a tree trunk (Researcher is shown holding a calliper, a device for measuring trunk diameter)

2.2 Methodical basis of the application and description of the work process

The objective was to develop an application that would:

1. be easy to use, save time and resources, work on basic programmes found in offices that deal with compiling plans (MS Excel, MicroStation);
2. combine MS Excel data tables with digital plans into a linked spatial database that would make it possible to automatically read data from the database and mark the information down on the plan, while also adding coordinates from the plan to the database;
3. create the necessary structure of layers on the plan and allow data to be added to these layers;

4. allow adding information derived from the data collected during fieldwork to the database;
5. allow for flexible data interpretation regardless of the inventory methodology used;
6. perform independently if necessary, but could also be combined with GIS programmes (MapInfo Professional or ArcGIS).

In 2012, the landscape architecture office Artes Terrae started looking for ways to optimize the inventory process by utilizing already existing tools. They focused on two objectives: Digitally recording data collected in the field through the use of existing tools such as phones and tablet computers, while also working with licensed programmes already in use to further refine the data analysis process.

First, an Android-based Smartphone (Samsung GALAXY S III) was used to record measurements. As a result, data entry took slightly more time during fieldwork than it had when recording the data onto tables by hand; however, the lengthy process of transferring the data after completing fieldwork into MS Excel tables was avoided. Researchers also utilized tablet computers (Samsung GALAXY tab 2 10.1) for entering plan data. The tablet computer with its larger screen was more practical to use during fieldwork, but had issues with processing the plan information entered due to the low productivity of the CAD software used in tablets, making it impractical for drawing plans. Researchers then started looking for other ways to increase the economic efficiency of the project.

Several existing applications such as Ezytreev Tree Management software, Treeworks, and Tree Tracker offer good options for arborists, dendrologists, and other specialists involved in tree inventory, evaluation, and planning maintenance (RUNNEL, 2012), but they are not suitable for the plan projection process. MicroStation, the CAD system vector graphics programme used by Artes Terrae, allows for entering various types of data in addition to plan information by using tags, like GIS systems. Still, using this during fieldwork can be impractical and time-consuming compared to MS Excel. Furthermore, GIS programmes such as MapInfo Professional are too inaccurate for practical application to the plan projection process. For this reason, to satisfy the needs of plan compilers, a macro working on MicroStation basis that could share data with MS Excel was developed.

During the first stage of the application development process, researchers defined the structure of data used for dendrological inventory (order of fields, data type of entries, abbreviations of entries, etc.) then used these for creating the table structure of the MS Excel programme along with the MicroStation macro. *{During the second stage, a technical solution was developed, but its specifics will not be discussed in this article. However, the usage purposes of the application will be described}*.

To use the application, the plan in the MicroStation programme must be opened; symbols of trees are linked to coordinates and the Excel table containing fieldwork data. When clicking on a symbol representing a tree, the programme automatically transfers its coordinates to the corresponding row in the Excel data table, and therefore links the plan to the table. It is important to make sure that when clicking on tree symbols on the plan, they are chosen according to their numbers.

The programme automatically chooses the parameters of the correct tree from the MS Excel table, and with the next click a circle with the necessary diameter is created on the plan. In drawing the diameter of the tree, the programme chooses data from the MS Excel table in a certain order. It first chooses measurements of the crown projection area or diameter, or when no data is given, uses the diameter of the trunk. When the necessary data is not supplied (for instance, diameters of shrubberies, individual shrubs, and dense groups of young trees are not measured), the manual drawing programme is activated and the image of the crown is mechanically marked on the plan using fieldwork data. If necessary, the digital drawing can be manually corrected using the fieldwork plan because the fieldwork plan is visible on the desktop as the bottom layer.

Moreover, the application simultaneously organizes database entries by adding related information, such as the full names of tree species in both Estonian and Latin. Abbreviations are converted into text with the help of special support files which have been created for this purpose.

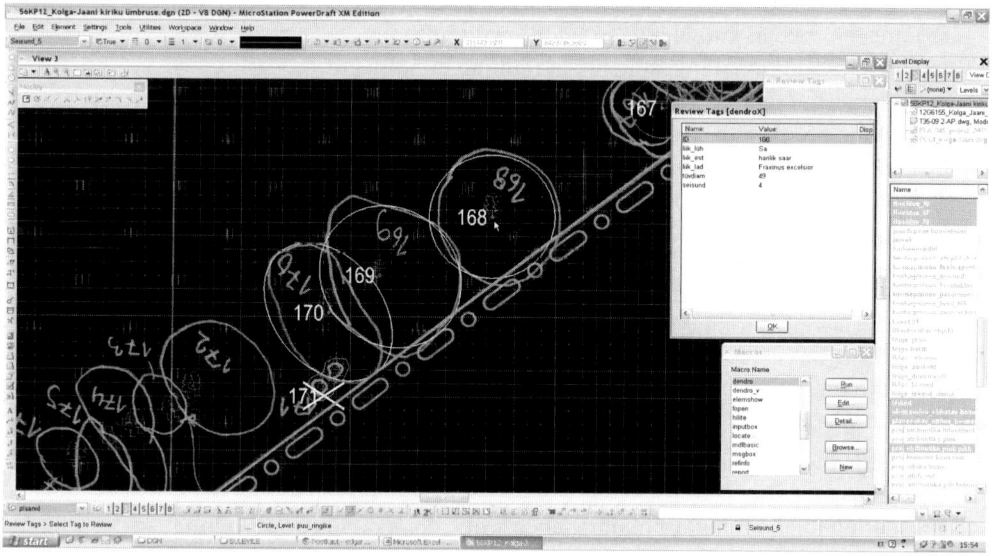

Fig. 2: When clicking on a symbol, the number of the tree and a circle marking the crown diameter automatically appear. The accuracy of the data can be easily verified with the help of the fieldwork plan stored in the bottom layer.

3 Results

The application was first tested in August 2012 with the inventory of Toila Oru grove park, which covers 105 hectares and contains over 6000 individual trees that were inventoried (NURME et al., 2012). The application worked well and no technical problems occurred. Half as much time was spent on data digitalizing, and data conversion into MapInfo went smoothly. Some problems arose when reading the data table, and therefore the format of the data needed alterations (abbreviations were needed, order of letters needed changing, etc.) Legends for abbreviations were created, which thanks to their universality can also be used in the future.

As expected, developing a technical solution (macro programming) for the application was cheaper in comparison to the cost of the previously mentioned programmes. This application fulfilled the developers' expectations, and with further development its economic efficiency can be maximized.

4 Conclusion

In conclusion, it can be said that the application developed in this study:

- is very easy to use;
- can be used on the basis of existing programmes;
- requires no special knowledge or additional tools
- requires no additional investments in training or software
- makes managing large amounts of data easier;
- saves time previously spent on digitalizing and analyzing data;
- makes for cleaner analysis of various types of data such as volume of tree-felling, species composition and landscape condition by systemizing different types of information (placing on different layers);
- allows for conversion of the spatial database into CAD and GIS system programmes;
- can be used to create a database that compiles information on all parks in Estonia for use in renewing national registers.

Goals for the future:

- making analysis and summarization processes more automated;
- automatically detecting possible errors made during the data entry process by setting size limits;
- calculating total amount of maintenance work, for example, tree-felling;
- automatically converting chosen codes and abbreviations into text and vice versa;
- converting data for different geodetic systems (WGS->L-EST);
- possibly adding a date to the fieldwork data to make it ready for database storage.

In the future, the dendrological data could be recorded during the geodetical survey of the area. This adds to the costs of the survey, for in addition to a geodesist, an arborist or landscape architect must be present to evaluate measures such as health, condition, species or damages, but would further improve efficiency in the compilation of park design plans.

The results – time expenditure in different work phases using different methods and additional characteristics of the methods – are summarised in the following figure.

Traditional method		New method		Possible future method	
Geodetical survey	1	Geodetical survey	1		
Dendrological fieldwork, two specialists using paper-table and paper-plan	2	Dendrological fieldwork, two specialists using tablet	2	Geodetical survey and dendrological fieldwork, two specialists using self-registering tacheometer	2
Spreadsheet data transfer to computer	1				
Map data transfer to computer	1	Map data transfer to computer	0,2	Map data transfer to computer	0,06
Total working days	5	Total working days	3,2	Total working days	2,06
Possibility to put data to different layers depending of condition-class of tree	No	Possibility to put data to different layers depending of condition-class of tree	Yes	Possibility to put data to different layers depending of condition-class of tree	Yes
Coordinates from digital map to MS Excel	Additional work	Coordinates from digital map to MS Excel	Automatic	Coordinates from digital map to MS Excel	Automatic
Names of species to the map	Additional work	Names of species to the map	Automatic	Names of species to the map	Automatic
Transfer of species' abbreviations to full names	Additional work	Transfer of species' abbreviations to full names	Automatic	Transfer of species' abbreviations to full names	Automatic

Fig. 3: An example of work process, a 3 hectares area with 400 trees. The numbers indicate the number of working days per person.

References

Eesti standard EVS 907:2010, *Rajatise ehitusprojekt. (Facility building project.)* Estonian Centre for Standardisation Tallinn, 9.

Ehitusgeodeetiliste uurimistööde tegemise kord. RTL 2007, 71, 1231. https://www.riigiteataja.ee/akt/12861144.

Ezytreev Tree Management software. http://www.ezytreev.com/.

Hein, A. (2003), *Eesti mõisaarhitektuur historitsismist juugendini (Estonian Manor Architecture from Historicism to Art Nouveau)* Hattorpe, 11).

Kupper, K. (2012), *Comparison of various research methods of allees according to the example of the avenue on Vanapargi street in Pärnu.* Tallinn University of Technology, Tartu College. Tartu, Estonia, 15-20.

Nurme, S. (2008), *Ajaloolise haljasala puistu inventeerimine. (Inventory of a Historical Green Grove).* Parkide restaureerimine (Restoration of Parks), Tartu College of Tallinn University of Technology, Tartu, Estonia, 285-306.

Nurme, S. & Kalberg, H. (2012), *Reconstruction project of Toila Oru park.* Artes Terrae Ltd, Tartu, Estonia.

Puittaimestiku ja haljastuse inventeerimise kord (Requirements for inventories of deciduous plants and urban greenery). Tallinna Linnavalitsus 03.05.2006 määrus number 34.

Runnel, H. (2012), *Comparative study of tree management software*. Project GreenMan (ELRI-177) Kino Ltd, Tartu, Estonia.

Sinijärv, U. (2008), *Pargid kaitsealuste objektidena (Parks as Objects of Protection)*. Parkide restaureerimine, (Restoration of Parks) Tartu College of Tallinn University of Technology, Tartu, Estonia, 179-182.

Treeworks. http://www.kenersongroup.com/treeworks.aspx.

Tree Tracker. http://www.treetrackersoftware.com/.

09:03 A Dozen Years to Make a State Go Virtual – LIDAR Data Use for 3D Visualisation of the Maltese Islands

S. FORMOSA

Abstract

Creating a new modus operandi for 3D data analysis that covers an entire state is a endearing task which required considerable funding and the integration of various thematic domains that lent an operational hand to spatial analysis in the Maltese state. An ERDF project taken up in Malta saw the integration of various environmental themes together with the creation of baseline surveys that serve the state as a launching pad for strategic analytical processes. The Malta study comprises the 3D component of the project which enabled the seamless integration of Terrestrial LiDAR, bathymetric LiDAR and bathymetric acoustic scans up to one nautical mile from the baseline coast, The aim of this process was to ensure that the integration of the datasets conformed to the requirements of the EEA (European Environment Agency) dataflow process (2012), the INSPIRE Directive (OJ, 2007), the Aarhus Convention (OJ, 2003a), the Freedom of Information Act (OJ, 2003b) and the Public Sector Information Directive (2003c), In addition, this project aimed to be the first to test the Shared Environment Information System (FORMOSA, SCIBERRAS, FORMOSA PACE, 2013; BORG, FARRUGIA, 2010).

1 Introduction

Nine years ago, embarking on a project to upgrade and enhance mapping in the Maltese Islands was deemed as a veritable cliff-hanger decision if not a potentially impossible-to-manage project. Perseverance and a drive to tackle the different data sources in a holistic approach has resulted in the capture of a nationwide series of datasets and information systems, which systems would deliver a new modus operandi for spatial analysis in the Maltese state. The aim was to base the whole process on the integration of the requirements of the EEA (European Environment Agency) dataflow process (2012), the INSPIRE Directive (OJ, 2007), the Aarhus Convention (OJ, 2003a), the Freedom of Information Act (OJ, 2003b) and the Public Sector Information Directive (2003c), In addition, this project aimed to be the first to test the Shared Environment Information System (FORMOSA, SCIBERRAS, FORMOSA PACE, 2013; BORG, FARRUGIA, 2010).

The project[1] was also aimed at creating an effective tool for the monitoring of and reporting of environmental information in the air, water, noise, radiation and soil domains, which

[1] Following an extensive analysis of the requirements for spatial data enhancement in the physical, social and natural environments, the Maltese Islands are in the final stages of acquiring a comprehensive set of data that will serve as a basis for cross-thematic research. This is being made possible through the creation of fundamental datasets that will bring Maltese terrestrial and

outputs would have ensured the creation of baseline studies for future trend analysis (Malta Environment and Planning Authority, 2009).

The results have yielded international feedback (Martirano, Bonazountas, Formosa, Nolle, Sciberras & Vinci, 2012) at the INSPIRE 2012 Conference and at the ELMF 2012 LIDAR[2] Mapping Forum (Hoffmann, Formosa, Sciberras, Alberts & van der Vegt, 2012), with outputs breaking some records on data returns. This paper concentrates on the 3D data deliverables emanating from this process in 3d employing high resolution data emanating from the ERDF project.

2 Processes and Methods

2.1 Input Data and Technologies

The project delivered terrestrial and bathymetric data at high resolutions suitable for detailed environmental modeling and for EU reporting purposes. Dealing with the dearth of information that is currently the status in both academia and in the industry, which situation hinders comprehensive land use planning, environmental monitoring and the resultant socio-security impacts, the project takes into account the various dataset requirements to ensure an overarching approach to analytical research and legislative implementation.

Due to the various national/EU environmental obligations, MEPA is committed to upgrade Malta's environmental regulatory capacity, including efforts to ensure full compliance with relevant Community Directives as well as national legislation. However, environmental monitoring and reporting is hampered by an incomplete monitoring strategy, lack of baseline environmental data on ambient conditions, lack of monitoring infrastructure and modern monitoring equipment and limited human resources. The project was aimed at ensuring the enhancement of national monitoring programmes in the five environmental themes through the identification of information gaps in monitoring processes and filling data gaps, the carrying out environmental baseline surveys, the procurement of monitoring equipment & information management systems and the training of experts. Each of the main five domains identified were investigated in depth for their data dearth's.

This is especially true due to a lack of high quality 3D spatial data, where the need was felt to commission such a project that covered the terrestrial area up to 1 nautical mile from the baseline coast (Figure 1a), the marine area registering a depth of 0-15m (Figure 1b) and the bathymetric area up to 200m depth (Figure 1c).

---

bathymetric baseline free information to the public domain. Such was made possible through an initiative as part of a €4.6 million project, entitled Developing National Environmental Monitoring Infrastructure and Capacity, which involves monitoring of air, water, soil, radiation and noise and 3D terrestrial and bathymetric surveys. This project co-financed by the European Regional Development Fund, which provides 85% of the project's funding and the Government of Malta, which finances the rest under Operational Programme 1 – Cohesion Policy 2007-2013 – Investing in Competitiveness for a Better Quality of Life.

2 Light Detection and Ranging (LiDAR).

Fig. 1a: Terrestrial Area and baseline coast up to 1 nautical mile

Fig. 1b: Coastal water areas of depths between 0m to 15m

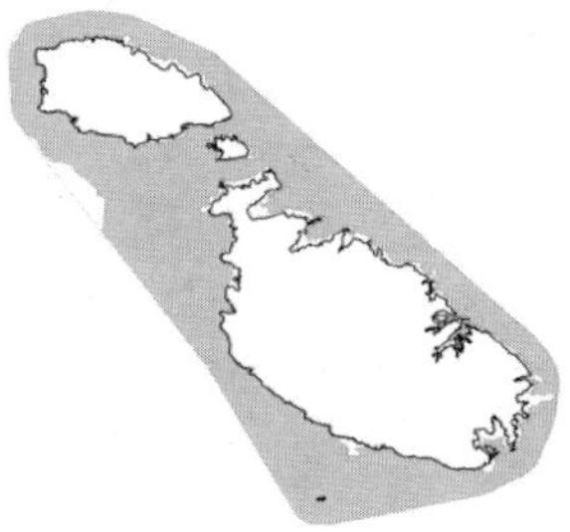

Fig. 1c: Coastal zone of depths between 15 m to 200m

The data incoming from the project included a high resolution 3D terrestrial data coverage for the Maltese Islands using a combination of oblique aerial imagery (Figure 2) and Light Detection and Ranging (LIDAR) data (Figure 3), as well as through a bathymetric survey of coastal waters within 1 nautical mile (nm) radius off the baseline coastline, using a combination of bathymetric LIDAR surveys (Figure 4), acoustic scans and a physical grab sampling survey (Figure 5). These outputs, together other fieldwork technologies as are handheld scanners and GIS handhelds, aid the researchers with a launching pad for the diverse physical, environmental and social studies that are undertaken in relation to the strengthening of social and environmental fabric. The basis for the future analysis of the multi-thematic analysis was laid through the acquisition of various deliveries that include the following 5-point outputs by technology, delivery and area:

- LIDAR Scan: Terrestrial (Topographic Light Detection and Ranging (LiDAR)) Digital Surface Model (DSM) and Digital Terrain Model (DTM) (316 km.sq).
- Bathymetric LIDAR aerial survey – depths of 0 m to 15m within 1 nautical mile from the Maltese coastline (38 km.sq).
- Bathymetric Scan: Acoustic (side scan sonar). Digital Surface Model and an acoustic information map of sea bed (361 km.sq).
- High resolution oblique aerial imagery and derived orthophoto mosaic and tiled imagery of the Maltese Islands (316 km.sq).
- Satellite imagery (GeoEye, RapidEye, Quickbird) (316 km.sq).

Fig. 2: Satellite and oblique imagery

Fig. 3: LiDAR data

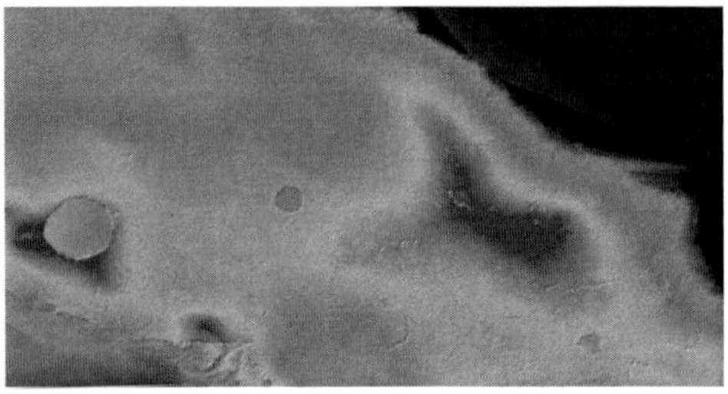

Fig. 4: Bathymetric LiDAR data

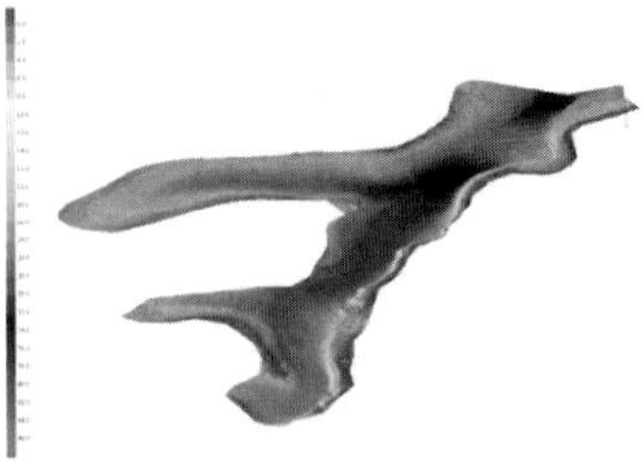

Fig. 5: Acoustic Scan data

2.2 Modelling approach

Malta's target to acquire LIDAR scans as well as a full bathymetric survey has been carried out for all the Maltese terrestrial areas and the coastal waters. Whilst various terrestrial exercises were carried out for development planning and environmental protection purposes, various bathymetric surveys were carried out for specific projects and research work, however, the latter were ad hoc and highly localized with rare attempts at a strategic methodology.

Mapping the terrestrial and seabed topography has various applications. It will provide a basemap from which various environmental studies can be carried out. These include studies as required by the Corine Landcover Initiative, the Water Framework Directive and the Marine Strategy Framework Directive which are aimed to protect more effectively the territorial landmass and the marine environment and in turn effect the health of the region's population. It will also serve as a tool for impact analysis of development planning, pollution and chemical prediction modeling, climate change monitoring as well as render recognition of features. It will also serve to aid enforcement officers to analyse the trends in security breaches and illegal development as well as plan for potential health hazards. In addition, it will allow thematic experts to study, monitor, analyse and protect those areas that are vulnerable to degradation and exposure. Spin-offs from the results include updated nautical charts, viewshed analysis and cross-thematic studies in the physical, social and environmental domains.

The 3-D aerial surveys using Lidar technology surveys were carried out by the main contractor TerraImaging, whilst the aerial Lidar bathymetric scan was carried out by Peldryn (UK) and the vessel-based bathymetric surveys are carried out by AquaBioTech Group. Following these aerial surveys, users will be able to generate digital terrain 3-D models which can be used for various applications such as urban and transport planning, environmental impact assessments, infringement analysis, security review, modeling of runoff water, monitoring of and enforcement of land use activities, amongst others.

The original terrestrial LiDAR scan was targeted at acquiring a minimum of 1 point per square meter but resulted in a 4.3 point per meter square outcome, which resulted from the use of an AGL Sensor: IGI Litemapper 6800 (Riegl 680i) at a Fh of 950 m. The acquisition was carried out together with the simultaneous image acquisition of a mosaic with GSD of 16cm employing and IGI DigiCam. The DSM and DTM outputs at 4.3 Pts/m^2 had a height accuracy of >5cm.

The bathymetric LiDAR employed a HawkEye IIb (AHAB), operating at frequencies of 1Khz for the bathymetric areas and at 8Khz for the topographic areas. What is unique in the Malta scan is that the bathymetric LiDAR achieved higher depths that the 15m designated for the project and achieved reliable returns at 50m depths.

The bathymetric acoustic scans were carried out using sidescan sonar with main issues encountered related to unstable weather, turbidity, high waves and wind. Ground truthing of bathymetric surveys was carried out to determine the type of sea bottom, whilst grab sampling samples are taken where the ground changes. Accuracies were achieved as per IHO requirements.

The oblique imagery and Orthoimage mosaic had a resolution of 15cm and an accuracy of ±1-2 pixels. The technology used was through a VisionMap A3 Camera, where more than 2000 photos were taken to cover the islands at a Fh of 3000m.

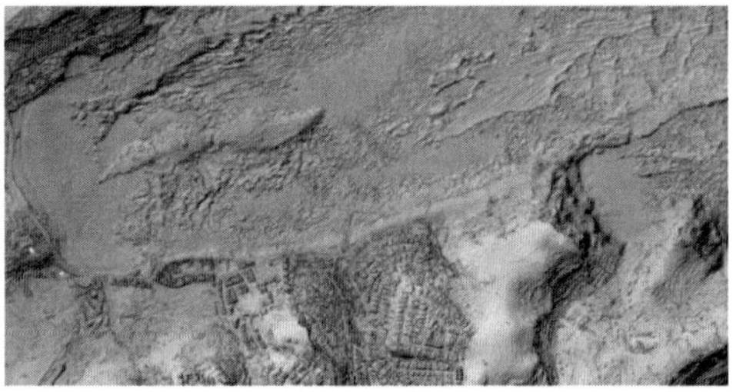

Fig. 6: Seamless LiDAR integration

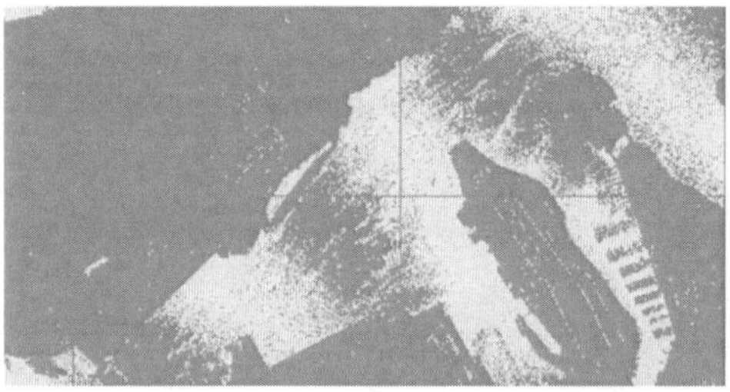

Fig. 7: DEM

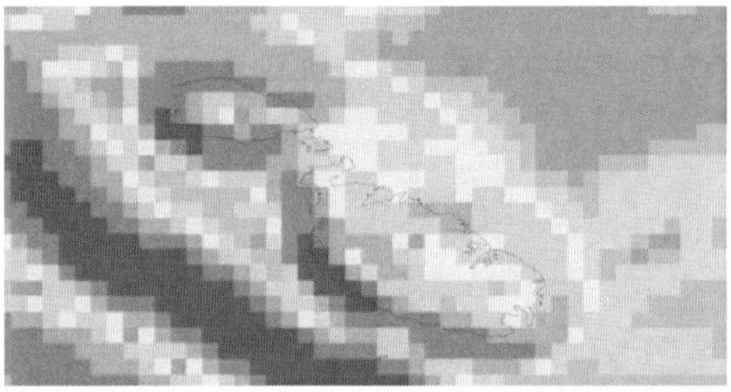

Fig. 8: Slope analysis

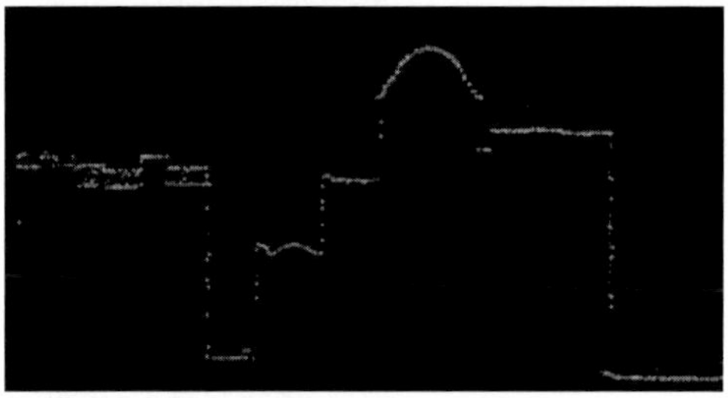

Fig. 9: Cross-Sectional analysis

Merging the three scans together proved challenging (Topographic LiDAR data + Bathymetric LiDAR data + Vessel bathymetric data) due to the varying coverage (data overlaps), data accuracies ranging from 5-50cm RMS varying by depth, as well as the different acquisition windows for the three technologies. TerraImaging carried out an advanced interpolation technique that allows one to weigh the different properties of the data sets (e.g. accuracy, resolution). This resulted in a seamless transition (Figure 6) where shallow areas the data fitted very well (within 10 cm) with clear data boundaries of each data source being defined and systematic effects being detected and corrected (ALBERTS, 2012).

The results show that as the data is integrated into one system, the outputs aid the researchers to carry out detailed analysis such as the identification of vulnerable areas affected by climate change. This includes analysis of marine and urban areas that will be effected due to their topographic location using integrated LiDAR datasets (figure 6), building Digital Elevation Models of the zones and the infrastructure in order to identify those areas that could for example experience sea-level rise (Figure 7), employing height/depth interpolation and slope analysis (Figure 8) as well as carrying out cross-sectional analysis of infrastructure (figure 9). Thesc initial outputs show that with the setting up of an integrated system which adheres to official environmental requirements, results are readily available, pending the further scenario testing as new knowledge is made available.

3 Results

Deliveries included a terrestrial LIDAR Scan (Figure 10) which resulted in a baseline map for the Maltese Islands infrastructure and landcover/landuse analysis which is required for the monitoring of structures that impact on noise levels, enforcement issues, resource monitoring and risk prediction, amongst others.

Fig. 10: Integrated Terrestrial and Bathymetric Data

The bathymetric LIDAR aerial survey for depths of 0 m to 15m (Figure 10) and the ship-based bathymetric scan will enable the creation of new nautical charts as well as bathymetric outputs that will help in marine spatial planning.

3.1 Initial Analytical Outputs and Dissemination

An analytical process has been initiated to develop the first strategy maps and tools for policy and decision makers. The contouring of the bathymetric data in the vicinity of the Azure Window in northwest Gozo has resulted in the identification of anomalous structures that are too linear to be natural and are being further investigated (Figure 11). In addition, the use of a 3D printer, resultant from the same project has enabled the decision makers to objectively visualise potential developments prior to their being approved as against potentially subjective photomontages (Figure 12). This data was also enhanced using landscape renderers such as Bryce resulting in a photo-realistic interpretation of the area under study, in this case the Dwejra Cliffs in Gozo (Figure 13).

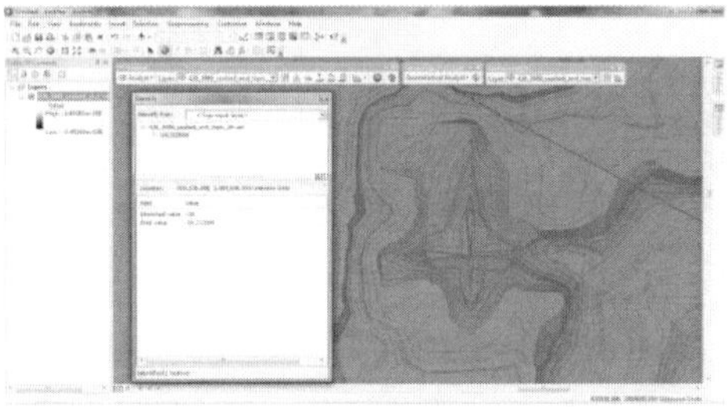

Fig. 11: Satellite and oblique imagery

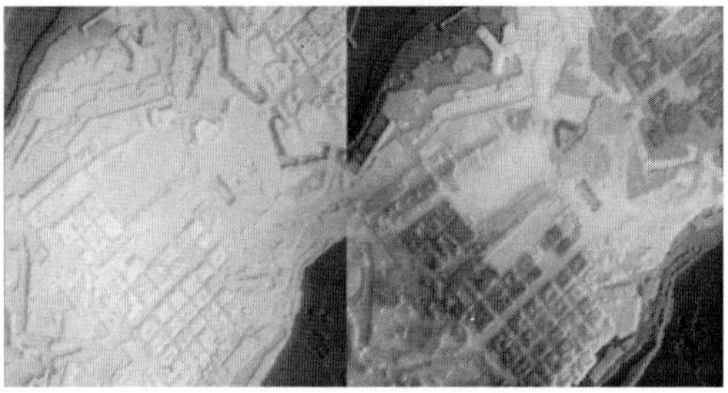

Fig. 12: 3D printing of models

Fig. 13: Landscape rendering

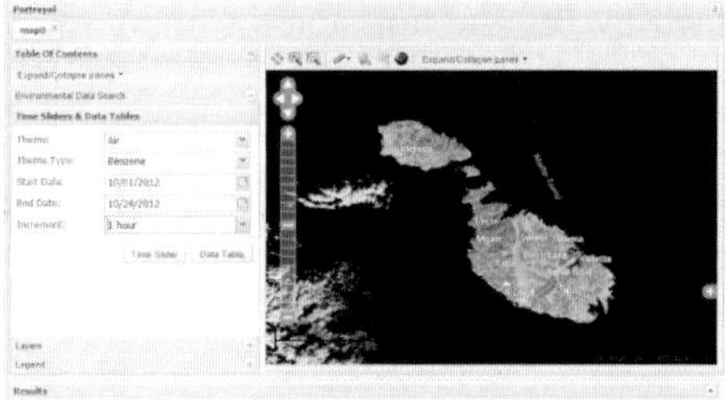

Fig. 14: SEIS interface

An interesting study that as also carried out on the initial data refers to the integration of the project data with data from international sources (Global Elevation Data extracted from dataset ETOPO1_Ice_c_Geotiff1.tif[3], Coastlines: GSHHS dataset[4], and country administrative regions: glntlbnds.shp (created by the Global Rural-Urban Mapping project,

[3] 1 arc-minute global relief model of the Earth's surface that integrates land topography and ocean bathymetry. Grid of the Earth's surface depicting the top of the Antartic and Greenland ice sheets. www.ngdc.noaa.gov/mgg/global/relief/ETOPO1/data/ice_surface/grid_registered/xyz/.

[4] Global Self-consistent Hierarchical High-resolution Shorelines version 2.0 July 15, 2009, WESSEL, P. & SMITH, W. H. F. (1996), *A global, self-consistent, hierarchical, high-resolution shoreline database*. In: J. Geophys. Res., 101 (B4), 8741-8743.

Version 1 (GRUMPv1) National Administrative Boundaries[5]). The study sought to understand the extent of the Maltese landmass in ancient periods, circa 12,000 years in the past when the waters from the glacial period had yet to melt and subsequently submerge the land between Malta and Sicily eventually rendering the Maltese Islands into its current form (Figure 15).

Fig. 15: Ancient Maltese Landscapes: Pre-Glacial Melting circa 12,000 Years Ago

In terms of dissemination of the results, the project, having been structured around a number of international directives, aimed to ensure the free delivery of all data to the general public. This was the result of an integrated exercise to adhere to the requirements as outlined by the Commission's Communication COM (2008) 46 Final "Towards a Shared Environmental Information System", the INSPIRE Directive (Directive 2007/2/EC) and the Aarhus Convention.

All data from this project is being made viewable and disseminated through a web portal, currently being designed, known as a Shared Environmental Information System (SEIS) (Figure 14).

4 Conclusions and Outlook

Scanning technologies may yet make inroads in the social and planning operational fields that have to date been reluctant to take up the outcomes of the spatial technologies. However, the Maltese scenario is ripe to take off through the employment of the various 3D technologies employed through the ERDF156 project. The terrestrial LIDAR scan combined with the bathymetric acoustic scan up to 1 nautical mile from the baseline coast and the inclusion of bathymetric LIDAR ensure seamless analysis of both domains. Combined with oblique imagery takes and in-situ data trawling, the end result has enabled

[5] Center for International Earth Science Information Network (CIESIN), Columbia University; International Food Policy Research Institute (IFPRI); the World Bank; and Centro Internacional de Agricultura Tropical (CIAT). 2011. Global Rural-Urban Mapping Project, Version 1 (GRUMPv1), National Administrative Boundaries. Palisades, NY: Socioeconomic Data and Applications Center (SEDAC), Columbia University.
http://sedac.ciesin.columbia.edu/data/dataset/grump-v1-national-admin-boundaries (June 2012).

the production of a unique model of the islands for use across the scientific platforms. The free dissemination of the data through Aarhus-compliance to the general public and the scientific community should enable the launching of integrated research across the spatial and social themes. Urban and environmental planners can undertake urban modeling scenario-building for strategic, regional and local areas, with the add-on integration of social, economic and criminological issues leading to healthier and safer localities. In addition, the data can be integrated with impact-analysis of baseline studies for air, water, noise, soil and radiation themes and their impacts on high population/urban density Malta.

The paper investigated the process taken to implement such a multi-tech process, the proposed data integration processes and the eventual studies that will be launched to ensure sound urban, social and environmental studies over the next decade. Together with such technologies as remote GPS cameras, 3D scanners and mobile GIS handheld loggers as well as other spatial data trawlers, the future for landscape assessment will be drastically enhanced in terms of provision of data and analytical information resources.

Nine years in the making, but the most crucial will be the next three years were the data will be integrated into a wide range of systems as disseminated to the public and private arenas as well as the non-governmental organisations. The technology required to analyse the massive volumes of data may not be available to all but with open-source applications and access to hardware from the University of Malta and the environmental agency, such analysis would be made possible.

References

Alberts, B. A. (2012), *Quality Report: Terrestrial 3D Surveys and Coastal Water Bathymetric Surveys of the Maltese Islands*. Utrecht, TerraImaging.

Formosa, S., Magri, V., Neuschmid, J. & Schrenk, M. (2011), *Sharing integrated spatial and thematic data: the CRISOLA case for Malta and the European project Plan4all process* In: Future Internet, 2011, 3(4), 344-361. doi:10.3390/fi3040344. ISSN 1999-5903.

Formosa, S., Sciberras, E. & Formosa, J. (2012), *Taking the Leap: From Disparate Data to a Fully Interactive SEIS for the Maltese Islands*. In: Murgante, B., Gervasi, O., Misra, S., Nedjah, N., Rocha, A. M., Taniar, D. & Apduhan, B. O. (Eds.), Computational Science and its Applications – ICCSA 2012, LNCS 7334, Heidelberg, Springer. ISBN 978-3-642-31074-4.

Hoffmann, A., Formosa, S., Sciberras, E., Alberts, A. & van der Vegt, J. W. (2012), *LIDAR mapping and coastal water bathymetric surveys of the Maltese Islands*. European Lidar Mapping Forum ELMF, December 2012, Salzburg.

Malta Environment & Planning Authority, (2009), *Developing National Environmental Monitoring Infrastructure and Capacity*. MEPA, Floriana, Malta.

Martirano, G., Bonazountas, M., Formosa, S., Nolle, M., Sciberras, E. & Vinci, F. (2012), *INSPIRE EF Data Specifications to develop the SEIS-Malta Geodatabase for the Air Quality data management*. INSPIRE Conference June 2012, Istanbul, Turkey.

Official Journal of the European Union, (28 January 2003a), *Directive 2003/4/EC of the European Parliament and of the Council of 28 January 2003 on public access to environmental information and repealing Council Directive 90/313/EEC*, L 041, 14/02/2003.

Official Journal of the European Union, (26 May 2003b), *Directive 2003/35/EC of the European Parliament and of the Council of 26 May 2003 providing for public participation in respect of the drawing up of certain plans and programmes relating to the environment and amending with regard to public participation and access to justice Council Directives 85/337/EEC and 96/61/EC - Statement by the Commission*, L 156, 25/06/2003.

Official Journal of the European Union, (17 November 2003c), *Directive 2003/98/EC of the European Parliament and of the Council of 17 November 2003 on the re-use of public sector information*, L 345 , 31/12/2003.

Official Journal of the European Union, (25 April 2007), *Directive 2007/2/EC of the European Parliament and of the Council of 14 March 2007 establishing an Infrastructure for Spatial Information in the European Community (INSPIRE)*, L108, Vol. 50.

The Place for Information Models in Landscape Architecture, or a Place for Landscape Architects in Information Models

Andrew NESSEL

1 Introduction

At Harvard University's Landscape Infrastructure symposium in March of 2012, Pierre Belanger's keynote began the proceedings with a dire warning. In a particularly chilling chart, Mr. Belanger illustrated how small the role of landscape architects in the design and construction industries truly is. In the US, the Department of Labor estimates that in 2010 there were 113,000 architects, 262,000 civil engineers, 523,000 construction managers, and only 21,000 landscape architects then employed. When you consider what Ahmad and Aliyu noted in 2012, that the UK Landscape Institute is noticing "major concern in the landscape industry regarding BIM use, that lack of expertise in the use of Landscape Information Modeling could effectively remove landscape architects from the supply chain", the outlook for landscape architecture looks grim indeed.

The Architecture, Engineering, and Construction (AEC) industries are rapidly adopting Building Information Modeling (BIM) tools that allow not just the modeling of structures, but the processes that drive those structures during and after construction. Architects can model the indoor environmental quality of their building designs based on solar aspect and insulation, local microclimates, and r-values of windows. Engineers model the performances of storm and sanitary sewer networks, dams, bridges, and highways, all under simulated conditions. With the UK's mandate of BIM adoption in AEC by 2016, landscape architects will have to finally grapple with Information Modeling. Is BIM a silver bullet? Is it a cure for all of landscape architecture's problems, or is it the one thing that will finally kill landscape architecture off? That comes down to the question "who's BIM?"

2 What is BIM?

Building Information Modeling (BIM) describes software tools in use in the Architecture, Engineering, and Construction industries. This is more than a simple computer aided drafting (CAD) system. These tools allow the creation of smart components within building models which allow professionals to model the processes involved in construction and operation of buildings. BIM tools are represented by Autodesk Revit, Graphisoft ArchiCAD, Bentley Architecture and others.

These tools are built on a set of standards governing the interoperability of information within the design and construction workflow. This interoperability is one of the key features of BIM packages. It allows professionals working across disciplines the ability to collaborate on large projects. An oft quoted statistic states that the US National Institute of Standards and Technology have identified that "poor interoperability and data management costs the construction industry $15.8 Billion a year, or approximately 3-4% of the total industry" (TAYLOR & BERNSTEIN, 2009; SUERMANN, 2009; originally GALLAHER et al., 2004)

Streamlining the collaborative process with interoperable standards is just one advantage BIM tools offer. the ability to model building processes has been extended to perform energy efficiency simulations. This capability is essential, not merely for the improved performance of the resulting structure, but because it has supported the proliferation of sustainability standards like LEED.

And as BIM is a relatively recent development, it is still evolving and finding new uses. Researchers at The University of Texas at Austin have been exploring the integration of laser scanning and BIM models of construction sites to create a real time construction control application (TEZIER et al., 2005).

3 How Can Landscape Architects Use BIM?

The advantages to architects and engineers using BIM are considerable: maintaining consistency of vision across very large projects, ease of making changes or versioning, cost control and quality assurance, spec'ing. But how can landscape architects capitalize on the advantages of information modeling? What would a Landscape Information Model (LIM) look like?

3.1 Large Parks

Large parks are gaining ground, and are seen as one solution to the myriad problems caused by urbanization and disurbanization. In North America, these projects are becoming the marquee projects of landscape architecture, from Orange County Great Park outside Los Angeles, to Downsview Park in Toronto, Ontario, to Fresh Kills and Jamaica Bay in New York. This trend can be seen in Europe as well, with Duisburg-Nord Landscape Park in Germany and Cultuurpark Westergasfabriek in the Netherlands. Termed Landscape Urbanism in North America, these projects are predicated on designing landscape process at a large spatial scale, centering on dynamic ecological successional plans without fixed outcomes. This premise flies in the face of the traditional construction process. And yet, conceptually at least, these are exactly the sorts of tasks that BIM is suited to.

At Downsview Park in Toronto, Ontario, Canada, the need for a Landscape Information Model (LIM) is apparent. Originating as a design competition in 1999, from its inception the park was meant to be ecologically dynamic. The competition brief stated that, "The physical conditions of the site and its natural processes provide opportunities to create new as well as old ecologies. [...] The design of the Park is to provide both an inauguration and a framework for change over time. It should anticipate an adaptive management approach in which the effects of interventions are monitored, adjustments are made, and new directions and configurations emerge" (MERTINS, 2001; in NORTH, 2012). The winning competition entry, from OMA/Bruce Mau Design, met this challenge head on. However, as construction has begun, several factors both within and outside the design process have shifted the design "from a diagram signifying a process framework toward a static plan and hence eliminating the opportunity to test and reiterate the landscape-as-process theory" (NORTH, 2012)

While BIM cannot solve the political disputes involved in large scale civic work, it is suited to the challenges of large scale projects and of versioning. NORTH (2012) describes how the design of Downsview Park was broken up into discrete pieces handled by numerous subcontractors. A Landscape Information Model would have allowed these practitioners to work together on one model for the park, and allowed the project's principals more control over the multiple iterations that were produced. More importantly, the ability of BIM to model processes, when combined with versioning, offers a way forward for implementing dynamic landscapes within the traditional construction process. A LIM could offer landscape architects the ability to easily create sheet sets for each successional phase of a project, satisfying the needs of dynamic ecosystems and a linear construction process.

The failure at Downsview Park to realize the competition's call for change over time should be seen as a wakeup call. As NORTH (2012) states, "It is this critical aspect of time that leading landscape architects understand very clearly... but they are still lagging behind on solid design practices for orchestrating and developing such landscapes, especially once construction is complete." This is brinksmanship, plain and simple. If even "leading landscape architects" are unable to realize their designs, the hope for the rest of the profession does not look promising. The time for a Landscape Information Model has clearly arrived.

3.2 Smaller Scales

The advantages of BIM at Large Scales and for "leading landscape architects" is undeniable. But as Sipes notes, the cost not only of the software, but of the time lost to training may be quite high for smaller studios who work largely at residential scales (2008). What are the advantages of Landscape Information Modeling for smaller studios working at smaller scales?

LIM benefits smaller studios by helping to automate routine tasks and streamline administrative work. Vectorworks Landmark, which the publisher refers to as Site Information Modeling, offers tools for Grading and Plant Spec'ing. Autocad Civil 3D, part of Autodesk's Infrastructure Modeling Suite, offers very powerful grading tools. These programs also offer integrated libraries of materials and interoperable file formats to streamline the process of developing renderings.

These advantages could provide smaller studios with significant time savings, spending their billable hours on the work that matters. Small firms, however, do not have the luxury of being able to hire a BIM Manager, and so, "every team member must be committed to learning and implementing BIM" (SIPES, 2008).

3.3 Industry Advantages

Since 2005, the American Society of Landscape Architects (ASLA) has been developing a set of national standards for project sustainability called the Sustainable Sites Initiative. Modeled after the US Green Building Council's LEED system, Sustainable Sites awards credits for responsible site design decisions in categories including Water, Soil and Vegetation, Construction, Human Health and Well Being, among others. Still in the pilot stage, Sustainable Sites has the possibility to drive a new wave of sustainable landscape design and construction in much the same way as LEED has galvanized architecture. Landscape Information Modeling is organized to facilitate this process.

Architects use BIM like Autodesk Revit Architecture to count credits toward LEED certification of projects. Furthermore, BIM is able to easily recalculate the credits received by a project when the design is modified. This functionality is integral to the adoption of the Sustainable Sites Initiative credit system. Designers need the ability to know how each design element that is proposed affects the overall score if they are going to attempt certification of this type. Any implementation of the Sustainable Sites Initiative needs to consider Landscape Information Modeling if it is to succeed.

3.4 Current Opportunities

The benefits of maintaining consistency of vision across very large projects, ease of versioning, cost control, quality assurance, and spec'ing make a strong case for developing Landscape Information Models. Current tools and standards offer some of this functionality and could easily be extended for more functionality.

Geographic Information Systems (GIS) exist to represent discrete and continuous spatial data. These tools are often used by environmental scientists to represent relevant processes like migratory patterns, pollution dispersion, and to make statistical predictions of community interactions. There is vast potential to use these tools for representing site processes. Though interoperability between GIS software and BIM tools has improved greatly in recent years, this process remains very particular and requires training.

Existing file formats including LandXML and CityGML offer great opportunities for interoperability between software tools. LandXML has deep roots in the land development sector, and contains object classes of value to landscape architects. CityGML is much newer, but has been designed to store even more object classes including buildings, bridges, tunnels, and site furniture. While LandXML is writeable from many current software tools, CityGML is not yet available as a file format in many existing tools, and some work still exists before this file format is widely adopted by practitioners.

4 Obstacles to BIM Adoption

With all of these advantages Landscape Information Modeling seems poised to be the next big thing in landscape architecture, and yet it has not taken off. Why have landscape architects failed to adopt the tools necessary for flexibility in the design of large scale projects, increased productivity, and integration of sustainability standards? Many of the answers to this question do not withstand scrutiny.

4.1 Current BIM Tools Do Not Model Landscape Processes

Researchers exploring the topic of Information Modeling for Landscape Architecture often cite the fact that existing BIM packages do not model landscape processes. (PIETSCH et al., 2009; SIPES, 2008) This is not wholly true. Existing BIM tools like Autocad Civil 3D model many landscape processes including drainage and grading volumes. This toolset exists within a larger BIM suite from publisher Autodesk that promises interoperability with another of their BIM tools Revit. Ahmad and Aliyu note that the UK Landscape Institute has a list of BIM tools ready for use in landscape architecture including Vectorworks Landmark, Siteworks, Graphisoft ArchiCAD, and more (2012). Clearly, there exist already BIM tools which model landscape processes.

4.2 BIM Tools Are Too Expensive

Many Landscape firms are small outfits and would not be able to take advantage of the scale benefits of Landscape Information Modeling. "It is probably overkill to use BIM for residential design, planting design, or other projects that are still best completed with CAD" (SIPES 2008). BIM tools are expensive and have large learning curves. Yet, Sipes quotes Vectorworks in referring to themselves as "'BIM software for the smart-sized firm'", and goes further saying "[Vectorworks] is one of the most affordable BIM products on the market" (SIPES 2008). It seems that cost is not truly an obstacle. What then?

4.3 Cultural Resistance

Perhaps the biggest obstacle to BIM adoption is the attitudes and mores within the profession. Here's SIPES, "soon [we will] be able to integrate GIS and BIM data seamlessly. Data will be interchangeable, and landscape architects will not have to worry about how this data exchange works." (2008) First of all, GIS and BIM integration still requires finesse from practitioners, 5 years after Sipes' paper. But what is more troubling about this statement is SIPES' – and as his paper was published by ASLA, the profession's – passive approach to information technology; the notion that if we wait around long enough, someone will come and solve our problems for us.

To underline this point, an example. BIM is as much a set of standards as a piece of software. In the US, the National Building Information Model Standard (NBIMS) has been in development since 2005 by an interdisciplinary group of professionals, vendors, government agencies, and professional associations. And yet, even though NBIMS is embarking on Version 3 of the standard, the steering committee remains free from the presence of landscape architects in any capacity. This is stunning as much for landscape architects passive attitude as it is for the AEC industries' antipathy to site processes.

The Sustainable Sites Initiative is the closest that the American landscape profession has come to developing a shared set of standards for design. Yet unlike NBIMS, the Sustainable Sites Initiative is largely unilateral, being developed almost entirely by ASLA. this indicates that the programs reach will be limited by the sphere of influence of ASLA, which is already dwindling. The irony is that inclusion of site processes and Sustainable Sites credits in NBIMS would do much to counteract this decline.

5 Is BIM the Answer?

BIM has many advantages, and much of the resistance to BIM adoption is no longer valid. So the question then becomes, is a Landscape Information Model what landscape architects need?

Robert Aish is a senior software engineer at Autodesk and developed Generative Components for Bentley. He believes that designers need flexible tools which they can program themselves. AISH notes that, "Tools give possibilities, from these possibilities we discover advantages, advantages become a convenience, and convenience can too easily become a convention. Some tools, or the way some tools are used, make certain forms or processes 'convenient' and this can have a deeply conservative influence on design" (2011). OZIMEK & OZIMEK echo this sentiment when they note that "Autocad Civil 3D ... provides users with a great spectrum of solutions ...[yet] sometimes discourages students from experimenting and does not help develop their spatial imagination" (2011). Powerful software tools, in the hands of passive designers, lead to complacency and convention. We have already identified the US landscape architecture profession's generally passive attitude toward Information Modeling. What results should we expect, then, when landscape architects finally, inevitably adopt LIM?

Grasshopper, the generative modeling environment for Rhinoceros 3D is really quite startling in its flexibility and accessibility to beginners. The key to its power seems to be in the community that has grown around it. Grasshopper's designers have allowed the software to be extended by the community and have created a forum where those explorations can be shared and improved. The freedom to use and extend the tool for an individual designer's needs and the involvement of the software developers in the user community have created a robust technology that welcomes newcomers yet is powerful enough to be of value to experts. This is equally true of the creative coding environments openFrameworks and Processing. Is a BIM for landscape architects, developed by a for-profit company with a finite customer service budget the tool we need? Is there a way to couple LIM to an open community of experts and beginners engaged in the work of extending the tools?

WATSON (2010) discusses how there is a bottleneck in the digital education of Landscape Architecture students. He notes that increasingly students require more and not less instruction on digital tools, and that faculty are less and not more able to meet this challenge. FRICKER et al. (2010) note that until they had radically rethought what a landscape architecture curriculum could be, their students were not being adequately prepared for the complexities of the current design climate. If landscape architecture students are currently joining the profession underprepared, how will the development of new software substantially change anything?

BIM tools offer many advantages to the landscape architecture profession, yet the profession at large has been slow to adopt these tools. Many existing BIM packages could be used to model landscape processes, but overly complex software tools can lead to decreased creativity. Open source tools exist with robust user and developer communities that offer open ended design strategies and are more forgiving to beginners. Meanwhile, many students are joining the workforce without adequate exposure to existing tools. is the development of another BIM, specific to landscape architecture, the solution?

6 Conclusions and Further Research

6.1 Conclusions

BIM tools are powerful pieces of software with the potential to not only bring the landscape architecture profession into the present but to propel it into the future. These tools have the opportunity to address serious contemporary problems within the profession such as how to generate construction documents for dynamic designs and how to manage large scale projects which are increasingly the focus of "leading" firms. But the simple development of a new software tool will not in and of itself solve these contemporary problems.

Large firms should focus on adopting and extending existing BIM tools. Professional associations need to work with other entities within the AEC industries on developing shared standards. Software tools have the greatest success when they are supported by active user communities and when developers are involved in those communities. Educators need to expose students to these tools and to some basic programming to enable students with the agency to create their own tools.

6.2 Further Research

While the conclusions above are compelling, they are based on incomplete information. It would be illuminating to conduct surveys of landscape architecture firms working on large scale projects and see how they are or are not using BIM. Furthermore, it would be illuminating to survey landscape architecture and planning departments of large AEC firms to understand where they fit into the BIM workflow. The answers to these questions may provide a guide for how to develop a BIM for landscape architects.

Refrences

Ahmad, A. M. & Aliyu, A. A. (2012), *The Need for Landscape Information Modelling (LIM) in Landscape Architecture.* In: DLA 2012, Proc. of Digital Landscape Architecture Conference 2012, Bernberg, Germany. Hochschule Anhalt (Web), 13 Jan. 2013.

Aish, R. (2011), *Designing at T + N.* In: Architectural Design, 13 Jan. 2013, 20-27. Wiley Online Library (Web), John Wiley & Sons.

Fricker, P., Girot, C., Melsom. J. & Wemer, P. (2012), *From Reality to Virtuality and Back Again – Teaching Experience within a Postgraduate Study Program in Landscape Architecture.* In: DLA 2012, Proc. of Digital Landscape Architecture Conference 2012, Hochschule Anhalt, Bernberg, Germany. Hochschule Anhalt (Web), 13 Jan. 2013.

Gallaher, M. P. & Chapman, R. E. (2004), *Cost analysis of inadequate interoperability in the US capital facilities industry.* US Department of Commerce, Technology Administration, National Institute of Standards and Technology.

Mertins, D. (2001), *Downsview Park International Design Competition.* Case: Downsview Park Toronto, 24-33.

North, A. (2012), *Processing Downsview Park: transforming a theoretical diagram to master plan and construction reality.* In: Journal of Landscape Architecture, 7.1, 8-19.

Ozimek, A. & Ozimek, P. (2011), *The Idea of "Integrated Design" in Digital Techniques Teaching*. In: DLA 2011, Proc. of Digital Landscape Architecture 2011, Hochschule Anhalt, Bernberg, Germany. DLA 2011. Hochschule Anhalt (Web), 13 Jan. 2013.

Pietsch, M., Heins, M., Buhmann, E. & Schultze, C. (2009), *Object-based, Process-oriented, Conceptual Landscape Models – A Chance for Standardizing Landscape Planning Procedures in the Context of Road Planning Projects*. In: DLA 2009, Proc. of Digital Landscape Architecture 2009, Hochschule Anhalt, Bernberg, Germany. Hochschule Anhalt (Web), 13 Jan. 2013.

Sipes, J. L. (2008), *Integrating BIM Technology Into Landscape Architecture*. LATIS 2008, Technical Paper, American Society of Landscape Architects, Washington D.C.

Suermann, P. C. (2009), *Evaluating the Impact of Building Information Modeling (BIM) On Construction*. Diss. University of Florida, 2009. N. p.: n. p., n. d. Print.

Taylor, J. E. & Bernstein, P. G. (2009), *Paradigm trajectories of building information modeling practice in project networks*. In: Journal of Management in Engineering, 25.2, 69-76.

Teizer, J. et al. (2005), *Real-time 3D modeling for accelerated and safer construction using emerging technology*. In: Proceedings of the 1st International Conference on Construction Engineering and Management. 2005.

Watson, D. (2009), *Are Landscape Programs Meeting the Challenge of Educating the Second Generation of Digital Landscape Architects?* In: DLA 2009, Proc. of Digital Landscape Architecture 2009, Hochschule Anhalt, Bernberg, Germany. Hochschule Anhalt (Web), 13 Jan. 2013.

A Numerical Model of Solar Transmittance for Isolated Trees in an Urban Area

Eiko KUMAKURA, Kazuaki NAKAOHKUBO and Akira HOYANO

1 Introduction

Deciduous trees contribute to a more comfortable thermal environment by reducing the impact of solar radiation in summer and allowing more solar radiation in winter. These effects are influenced by tree geometry. Many studies on how leaves and branches affect the radiation interruption process (MONSHI & SAEKI, 2005; TALBOT & DUPRAZ, 2012) have been conducted; however, little is known about the difference in effects produced by the various tree species used for urban landscaping in Japan. This paper discusses the solar shading effects of an isolated tree and presents a numerical analysis using a computer graphics (CG) tree model.

2 Material and Methods

2.1 Model description

Simple solar shading models are useful as simulation tools to predict the thermal environment in outdoor living spaces. Our target users are landscape designers, architects, and urban planners who need to compare tree species when designing outdoor urban living spaces where various solar and reflected incident angle radiation beams must be considered. This paper defines ratio of direct solar radiation reaches on the tree shade by simulating of a ray-tracing method as solar transmittance. Simulation models of solar shading are related by the tree canopy structure and solar transmittance depending on tree geometries (KUMAKURA et al., 2011). A method for describing the solar transmittance of inhomogeneous canopy structures using Leaf Area Index, G-function, and a Clumping Index that uses Poisson distribution, has been proposed (NILSON, 1971). On the basis of on this methodology, Talbot et al. suggested a formula that also considers the Branch Area Index for an isolated tree, and we have based our model for predicting solar shading effects in summer and winter on this formula. Figure 1 shows our formula and the relationship between tree geometries and design simple parameters, such as tree species, crown shape and leaf density. We have added three parameters to the formula suggested by Talbot et al. to describe various tree crown shapes and changes in leaf density. The added parameters facilitate two new functions. One is the new overlapping coefficient function, which shows the relationship between leaves and branches and is dependent on branch structure and leaf density. The other is the new Clumping Index, which is dependent on crown shape and shows the complexity of the layered structure. From the computer graphics (CG) tree models, we determined the coefficient to calculate the solar transmittance by incident angle for each species.

2.2 CG tree model

De Reffye et al. found a correlation between the architectural model and measurements from actual trees. Their method could make CG tree model for various tree species (DE REFFYE et al., 1988, natFX, Bionatics). Therefore, we chose their model to represent actual tree geometry. Five tree species were constructed (*Ginkgo biloba*, *Zelkova serrata*, *Liquidambar styraciflua*, *Platanus orientalis*, *Prunus yedoensis*) using CG software. These species are all used in urban landscaping in Tokyo, and five models of each species were constructed using a random seed parameter to consider differences in leaf density.

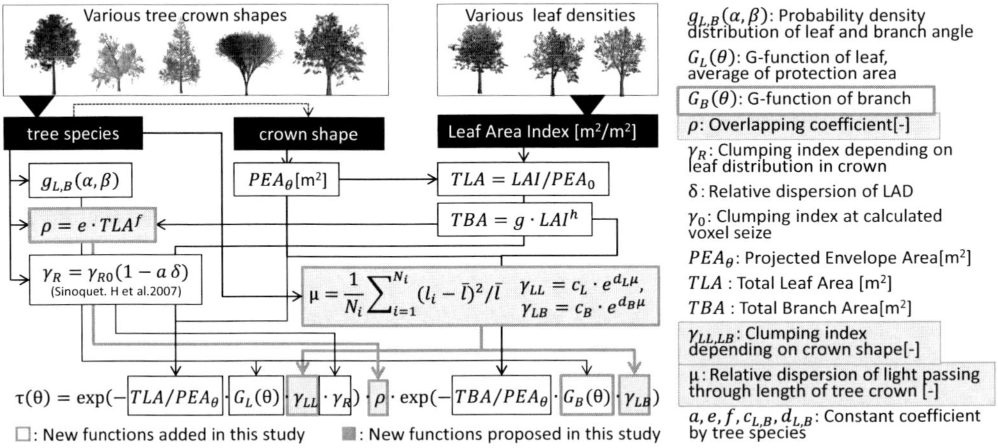

Fig. 1: Calculation process of solar transmittance by incident angle using CG tree

2.3 Modeling of solar transmittance

Using CG tree models, new parameters are formulated on the basis of numerical analysis using a regressive approach. Other parameters are obtained from a previously reported method that used CG geometry (SINOQUET et al., 2007; KUMAKURA et al., 2011). The new overlapping coefficient function is formulated in terms of total leaf area for each species. The new Clumping Index is formulated by the relative dispersion of light passing through the length of the tree crown. Using the new added parameters, our proposed solar transmittance formula more clearly shows the difference among tree species. We used the RMSE (root-mean square error) to check the accuracy by comparing projection images that were created by a ray-tracing method (KUMAKURA et al., 2011). Figure 2 shows the influence of the added parameters on the accuracy of solar transmittance for each species. *Ginkgo biloba*, which is characterized by low density and random distribution geometric features, is more influenced by branch angle G_B than leaf density. *Liquidambar styraciflua*, which is characterized by high density and clumping distribution geometric features, is more influenced by the Clumping Indexγ_{LL}, γ_R than the branch angle. *Prunus yedoensis,* which has a complex crown shape, is influenced by the new Clumping Index $\gamma_{LL,\ LB}$. *Platanus orientalis*, which has a large area per leaf, is influenced by the overlapping coefficient functionρ. In *Zelkova serrate,* the leaf layer and branch layer are clearly separated, and consequently this specie is influenced by both branch angle G_B and the new Clumping Index$\gamma_{LL,\ LB}$. These results show that our proposed model make it possible to compare seasonal change and leaf density in the summer season by species.

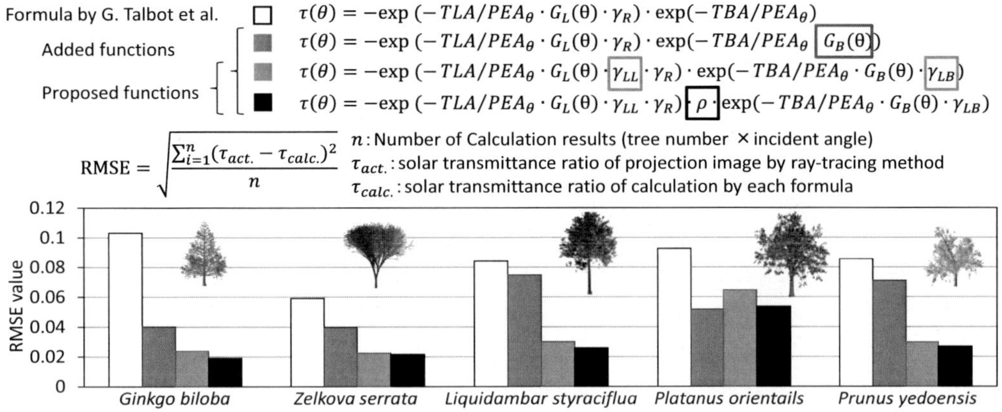

Fig. 2: Influence of added parameters on accuracy of the solar transmittance by species

3 Simulation Application

The tree shapes representing tree species were modeled using 3D CAD. The solar transmittance data by incident angle were calculated by the process shown in Figure 1. Using these solar shading models of tree crowns, we proposed an urban greenery plan that takes solar shading effects in an actual open space into consideration (Figure 3).

We used a numerical simulation system (ASAWA et al., 2008) to evaluate the effects of spatial forms and building materials on an outdoor thermal environment. Our calculations involved a comparison of the effects of these tree species on the thermal environment in summer and winter. Figure 3 shows a perspective depiction of the surface temperature distributions at 12:00 for both summer and winter. The surface temperatures in the shaded areas were calculated to be between 35 °C and 45 °C in summer and 22°C and 28 °C in winter. The values varied depending on tree crown shapes and the solar transmittance. The graph in Figure 3 shows a comparison of the differences of simulated incident solar radiation with and without a tree (0.6 m above the ground, i.e., sitting position under the tree). *Liquidambar styraciflua* and *Zelkova serrata* showed the highest incident solar radiation in summer and winter, respectively. These results show that this tree model can simulate the effects of tree species and leaf status on solar shading in both summer and winter.

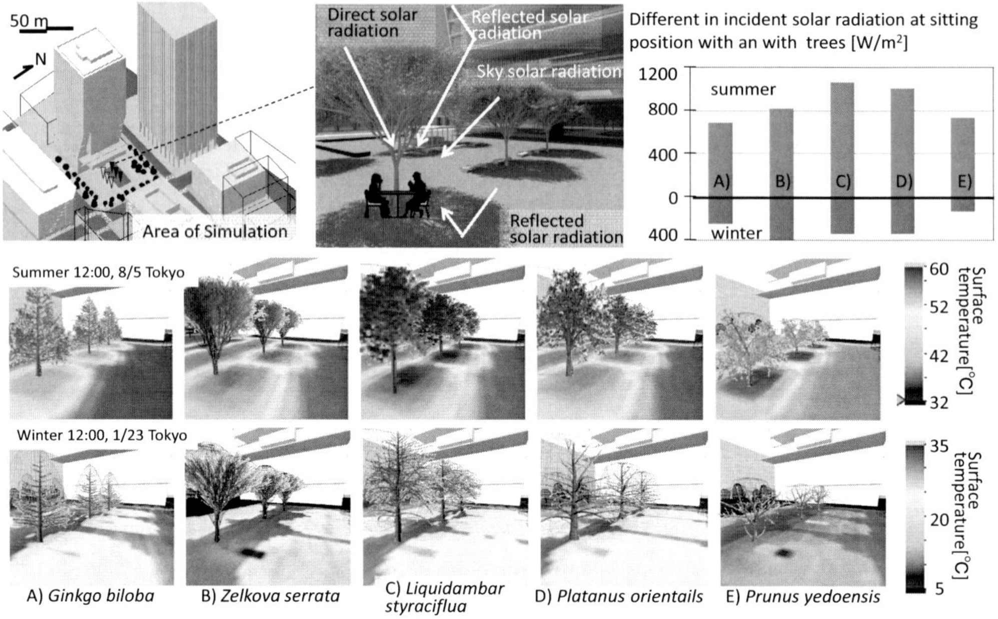

Fig. 3: Surface temperature distribution in the tree shadow

4 Conclusions

This paper has reported the development of a new numerical model that uses overlapping coefficient and improved Clumping Index functions to predict thermal effects caused by the solar shading of a tree in an urban area. On the basis of the results, the differences in solar shading effects of various isolated trees on the thermal environment in an actual urban area can be predicted using a 3D CAD compatible thermal environment simulator. The calculations provide some evidence that different tree species, represented by crown shape and solar transmittance, affect shaded surface temperatures in both summer and winter differently.

References

Asawa, T. et al. (2008), *Thermal design tool for outdoor spaces based on heat balance simulation using a 3D-CAD system.* In: Building and Environment, 43 (12), 2112-2123.

De Reffye, P. et al. (1988), *Plant Models Faithful to Botanical Structure and Development.* In: Computer Graphics, 22, 151-158.

Kumakura, E. (2011), *Numerical analysis of solar shading of a deciduous tree for seasonal change.* In: The 27th International Conference on Passive and Low Energy Architecture, Brussels, Belgium.

Monshi, M. & Saeki, T. (2005), *On the factor light in plant communities and its importance for matter production.* In: Annals of Botany, 95(3), 549-567.

Nilson, T. (1971), *A theoretical analysis of the frequency of gaps in plant stands.* In: Agricultural. Meteorology, 8, 25-38.

Sinoquet, H., Stephan, J., Sonohat, G., Lauri, P. E. & Monney, P. (2007), *Simple equations to estimate light interception by isolated trees from canopy structure features: assessment with three-dimensional digitized apple trees.* In: New Phytologist, 175, 94-106.

Talbot, G. & Dupraz, C. (2012), *Simple models for light competition within agroforestry discontinuous tree stands: are leaf clumsiness and light interception by woody parts relevant factors?* In: Agroforest Syst., 84, 101-106.

Visualization of 3D Landscapes

Who Is Afraid of "Light"?

Mintai KIM, Youngeun KANG and Shamsul ABU BAKAR

1 Introduction

This study investigates the effects of outdoor lighting on people's perceptions of nighttime landscapes. More specifically, it examines the relationship between lighting and specific physical features of nighttime landscapes in participants' reports of fear. To accomplish this, a combination of eye-tracking and self-reported data were collected. This study ultimately presents design guidelines intended to reduce fear levels experienced in the nighttime landscape.

Fear has been identified as a prominent factor in nighttime experiences of the urban environment (e.g., NASAR, 1998). Concerns regarding nighttime safety and security led to a more brightly lit nightscape. Although abundant lighting has been shown to increase individuals' perceptions of safety, the present research postulates that the effects of outdoor lighting on the perception of danger is a complex phenomenon needing more detailed study. Whereas under-lit nightscapes tend to induce fear, over-lit spaces can also cause fear because bright lighting can make it difficult to discern features and details in the nightscape.

Many studies have examined the relationships between outdoor lighting, crime, and fear (e.g., AUSTIN et. al., 2002; ZAHM, 2004; PAINTER & FARRINGTON, 2005; PEASE, 1999). Although some studies have found that lighting improvement did not help reduce crime (e.g., QUINET & NUNN, 1998), many others indicate that, in general, increased lighting reduces crime and fear (e.g., PAINTER, 1996). Researchers have also studied perceptions of nighttime safety (e.g., KNIGHT, 2010; THOMAS & BROMLEY, 2000). Approaching from a social sciences perspective, these studies used crime statistics from large-scale areas. These studies lacked site design scale understanding of lighting and crime and did not include human perception of fear in the situation. At a smaller scale, FISHER & NASAR (1992) studied the relationship between fear of crime and site features with respect to refuge, prospect, and escape characteristics. They concluded that an absence of refuge for potential criminals (i.e., the absence of hiding places) and high prospect for others (i.e., the ability to visually survey one's surroundings) reduce the fear of crime among study subjects. Their study and similar ones following it, however, did not examine the specific landscape elements that influence people's fear of crime, particularly at nighttime. Because of the nature of the data and the large study areas, the factors of nightscape design in small-scale places, such as types and brightness of lighting, and landscape elements, could not be accounted for.

Within the design disciplines, most of the lighting design literature focuses on lighting aesthetics and largely ignores questions of safety or the perception of safety (WILSON, 1984; MUNCHEN, 1998; RAINE, 2001; MOYER, 2005).

The present research studies subjects' reported levels of fear in response to the interaction of lighting position and level with the presence of specific physical elements in the landscape as viewed in photographs. This research studies participants' reported levels of fear directly corresponded to the interaction of lighting positions and the presence of specific physical elements in the landscape. The specific landscape elements addressed in this study are buildings, low, free-standing walls, tall and short trees, and shrubs. Additionally, the study examines the effects of the presence of a human figure in a nightscape, because the presence of a stranger in a nighttime landscape is suspected to elicit fear. The landscape elements above could increase or decrease level of fear. For example, FISHER & NASAR (1992) argued that although trees typically have restorative characteristics in daytime environments, they also provide "concealment, a limited prospect, and blocked escape routes [that] may increase fear at night" (FISHER & NASAR, 1993).

Although eye-tracking technology has been used rarely in landscape architecture research, it has been widely used in various fields such as market research, ergonomics, human behavior, psychology, education, neuroscience, and sports. LEE, TANG & TSAI (2005) studied color preference through eye tracking comparing combination colors. ITOH & FUKUDA (2002) studied eye movement in the extent of central and peripheral vision and use by young and elderly walkers. CUTRELL & GUAN (2007) used eye-tracking to examine eye-movement during Web browsing.

One of the significance of this study is that it used eye-tracking techniques to identify the specific elements in landscape photographs on which subjects' gaze rested and for how long in relation to fear. Unlike the above-mentioned studies that examined the overall image of a landscape, the use of an eye-tracking apparatus in this study provided a means of identifying the areas and objects within various nightscapes on which subjects focused their gaze while reporting that a landscape image elicited fear. By comparing the particular objects/areas of the photographs on which people focused, the sequence in which they focused on these objects/areas, and length of each focused gaze with reported levels of fear, specific physical elements of nighttime landscapes were identified.

2 Methods

The study described here tests two main assumptions about the sources of fear in the nighttime landscape:

Assumption 1: Physical elements in nightscapes, such as woods, shrubs, and man-made structures, will affect people's level of fear.

Assumption 2: If the other landscape elements are constant, the presence of an adult figure in the nightscape will increase fear.

2.1 Survey Instrument

In order to test these assumptions, a survey instrument containing nine digital photographs of three different nightscapes (three sets of three photos; see Figure 1). Photos in the first set (referred to as B1, B2, and B3) depicted a nightscape with a building in the background. The second set of photographs (T1, T2, and T3) depicted a nightscape with trees in the background. The third set (S1, S2, and S3) depicted a nightscape with landscape structures (i.e., retaining walls, sitting walls) in the background. Additionally, each set of photographs depicted three different conditions: the nightscape with no human presence, the nightscape with an adult male figure whose face was illuminated by site lighting, and the nightscape with an adult male figure lit from behind so that his facial features are not discernable. In the latter two conditions, the individual was centered in the frame, 3 meters from the camera. All photos were taken on the Virginia Tech campus in Blacksburg, Virginia.

Fig. 1: Images used for the experiment (three sets of three). The image set in the top row (B1-B3) has a building in the background, the middle set (T1-T3) has vegetation in the background, and the bottom set (S1-S3) has landscape structures in the background. Each image set presents one image without the presence of a person (left column), one with a light source behind the person and his face in shadow (middle column), and one with the person's face lit (right column).

2.2 Eye-Tracking Apparatus

This study used a video-based, pupil/corneal reflection eye-tracking apparatus that included an infrared eye movement camera and recording system (Red 250) manufactured by Senso-Motoric Instruments (SMI) of Germany (Figure 2). The infrared sensor was positioned directly below the monitor. As participants viewed a photograph on the monitor, the eye-tracking apparatus tracked and recorded points of gaze (or areas of interest; AOIs), length of gaze, and saccades (eye movements between focused gazes on AOIs). The system used Begaze2 software that is part of the Eye-Tracking system by SMI to record and analyze the collected data.

The research was conducted at the School of Visual Arts Perception and Usability Testing Laboratory at Virginia Tech in Blacksburg, Virginia. While the survey was conducted, only the researcher and research participant were present in the lab.

Fig. 2: Eye-tracking equipment in the Perception and Usability Testing Laboratory at Virginia Tech. An operator controls the foreground computer, and a subject sits in front of the screen in the back.

2.3 Participants

The participants consisted of 15 male and 11 female students and professors at Virginia Tech. Two men and one woman were eliminated from the study because of irregularities, leaving a total of 23 experiment participants of various disciplinary backgrounds.

2.4 Procedure

At the beginning of each session, participants were informed about the purpose of the study and the experimental procedure. Next, the eye tracker was calibrated to each participant's eyes by asking him or her to follow a red circle on the monitor with their eyes. In order to minimize distractions for participants, all lights in the lab were turned off during each session.

Participants were shown nine digital photographs (three different nightscapes with three different conditions; see 2.1 Survey Instrument) in a random order. Images appeared on the computer monitor one at a time. As participants viewed each image, they were asked by the researcher to rate (by typing into the computer) their level of fear on a 7-point Likert-type scale, where 1 is very fearful and 7 is very safe. No further elaboration on these instructions were given by the researcher. Participants were permitted to view each image for as long as they wanted before pressing the spacebar to rate their level of fear on the screen. Once they recorded this number, the monitor automatically advanced to the next, randomly selected image.

After rating their level of fear for each photograph, participants were asked to fill out a brief paper survey that recorded demographic information. Each session took about 30 minutes.

2.5 Measurement

To analyze participants' eye movements for each image, the investigator examined the eye-tracking data for the total duration of gaze for each image and each participant's fixation duration (ms), and saccade counts (see Table 1 for a summary of terms used in the paper). Total duration of gaze measured the total time a participant spent looking at an image before moving to the next image. Fixation duration measured the time spent looking at particular points or areas within a single image. Saccade counts accounted for eye movements between fixation points. Additionally each image was divided into areas of interest, such as a building, a structure, trees and grass, a concrete path, and the human figure, by tracing the edges of each of these elements in the Begaze2 software. Investigated next was the area of interest (AOI) coverage (%), AOI net dwell time (ms), AOI fixation count, and AOI revisits count.

To analyze fear and eye movement differences, an analysis of variance (ANOVA) was performed for the nine sets of images.

Table 1: Summary of terms used in the paper*

Terms	Description
Total duration	Total time that each participant spent viewing each image
Fixation duration	Total time of all fixations (no movement)
Saccade	Eye movements as the gaze travels from one point of fixation to another
Saccade count	Number of saccades
Scan path	Gaze positions and eye movement plotted on the stimulus image
Pupil size	Average size of a pupil
Heat map	Visual representation of time spent looking at different areas of a stimulus image
AOI coverage	The percentage of each AOI on an image
AOI net dwell time	Sum of all fixations and saccades within an AOI for all subjects
AOI fixation count	Number of fixations for selected subjects divided by the number of selected subjects
AOI revisits count	Number of revisits recorded when an AOI is revisited at least once.

*Definitions follow SMI (2012). All terms are per a participant for an image.

3 Results

3.1 Fear

The results comparing means of fear levels for the nine images showed that most participants tended to feel more fear while viewing nightscapes with structures (not buildings)

(Table 2). Image S2 found in Figure 1 (Structure type with a person's face in shadow) was the most feared image (mean: 2.70). On the other hand, the image rated safest among the nine images was T1 (a tree in the background, with no human figure present; mean: 4.43). An analysis of variance (ANOVA) was performed to confirm whether the level of fear was affected by the environment type (i.e., building, tree, and structure). Although the intensity of illumination, the size and height of the light fixtures were the same within each of the three photo sets, the results showed significant differences among the three environment types regarding the mean of fear levels ($F = 61.942$, $p < 0.05$).

To determine whether fear levels were affected by the presence of a human figure, another ANOVA was conducted. For the building type (B1, B2, B3), there were significant differences among the conditions: (1) without a human figure, (2) with a person whose facial features are lit, and (3) with a person whose facial features are in shadow ($F = 101.246$, $p < 0.000$). There were also significant differences in the condition for the Structure type (S1, S2, S3) ($F = 80.638$, $p < 0.000$). However, no significant differences were found among the images with the Tree type (T1, T2, T3). Overall, regardless of the location of the light source, images containing a person were feared more than images without a person.

Table 2: Participants' mean fear ratings for each photograph

	B1	B2	B3	T1	T2	T3	S1	S2	S3
Mean	4.43	2.96	3.13	3.83	3.78	4.00	3.83	2.70	3.70

Participants reported their level of fear in response to each image on a 7-point Likert-type scale where 1 = very fearful and 7= very safe.

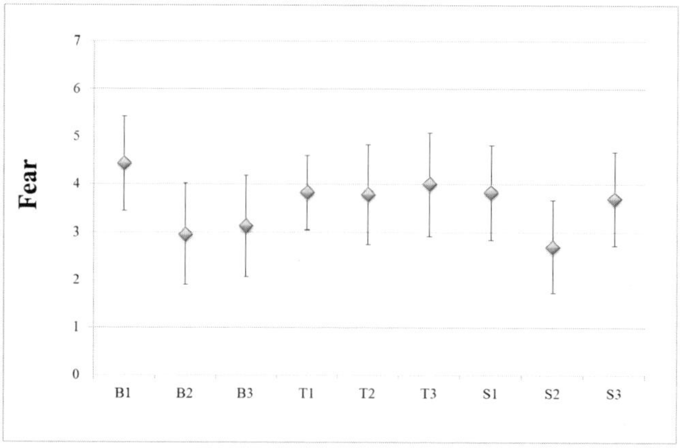

Fig. 3: Mean levels of fear (squares) and standard deviations (vertical lines) for the nine images

3.2 Eye Movement Results

Although the results above could be found with traditional visual preference study methods using slides, one goal of this research was to determine if an eye-tracking apparatus could be used to more accurately understand why subjects felt a particular level of fear. The eye-tracking results for the three environment types are summarized in Table 3.

For the eye movement results, the investigator analyzed subjects' eye movements using fixation duration, saccade count, scan path length, pupil size, and total duration. In general, when a human figure appeared in the images, subjects focused on the person's face more, regardless of whether his facial features were illuminated or obscured, whereas for the images without a person, subjects' foci were more scattered.

Table 3: Results of eye movement analysis

Image Back-ground	Type	Fixation duration (avg. in ms)	Saccade count (avg.)	Scan path length (px)	Total duration (avg. in ms)	Pupil size (px)	Fear rating (avg.)
Building (B)	No human present (B1)	222.34	10.70	1588.87	3211.17	15.96	4.43
	Adult male present (lit from behind) (B2)	284.43	10.09	1307.70	3993.91	16.50	2.96
	Adult male present (lit from the front) (B3)	304.63	12.61	1600.74	4804.74	16.65	3.13
Tree (T)	No human present (T1)	220.24	15.91	2192.22	5543.43	17.53	3.83
	Adult male present (lit from behind) (T2)	308.04	7.65	1092.96	2895.61	16.26	3.78
	Adult male present (lit from the front) (T3)	323.79	8.39	1070.91	3275.70	16.74	4.00
Structure (S)	No human present (S1)	198.20	16.61	2864.04	5312.35	16.54	3.83
	Adult male present (lit from behind) (S2)	276.90	7.30	1034.04	2443.22	16.56	2.70
	Adult male present (lit from the front) (S3)	259.55	9.39	1309.39	3107.13	16.07	3.70

For Tree and Structure type images, when a person was not present, subjects tended to spend more time looking at the landscape (total duration) when compared to other images with a person. For fixation duration, S3 had a higher number, suggesting that when a human figure appeared in a photo, subjects spent more time looking at the facial features of the person. Saccade count tends to have an inverse relationship with fixation duration because the participants spent more time looking at fixed objects than moving their gaze. Building type was an exception for an unknown reason.

Image Set B (Building)

After completing the overall analysis of eye movements, AOIs were analyzed in order to identify areas that received more attention from subjects (Table 4). The investigators first divided each image into AOIs, such as those shown in Figure 4.

Table 4: Notable factors of AOI for the Building type

Image		AOI coverage (%)	Net dwell time (avg. in ms)	Fixation count (avg.)	Revisits count (avg.)
Background only (B1)	Building	26.6%	1444.8	4.9	1.3
	Structures	4.2%	133.0	0.4	–
	Trees and grass	24.5%	250.2	1.0	–
	Concrete path	23.1%	1012.6	3.6	1.0
Adult male present (lit from behind) (B2)	Human	5.2%	2434.5	6.1	1.4
	Building	24.8%	515.0	1.3	–
	Structures	3.5%	195.4	0.7	–
	Trees and grass	26.8%	248.7	0.7	–
	Concrete path	18.8%	222.0	0.5	–
Adult male present (lit from the front) (B3)	Human	4.9%	2470.5	6.6	2.3
	Building	24.1%	458.1	1.5	0.1
	Structures	3.3%	47.9	0.0	–
	Trees and grass	25.2%	253.5	0.8	–
	Concrete path	19.6%	604.2	1.9	–

For the Building type, when a figure's face was illuminated (B3), subjects' gaze was more focused than when the figure's face was obscured by shadow for all three conditions. The heat maps in Figure 5 show accumulated time spent looking at different areas of the stimulus image for different conditions. Scan path maps in Figure 5 show gaze positions and eye movement path. While heat maps are good for showing areas of attention, scan path maps are good for depicting scattered eye movement.

The image without a human figure (B1) indicated that subjects were mostly interested in the building (fixation count: 4.9; dwell time: 1,444.8 ms) and the concrete path (fixation count: 3.6; dwell time: 1,012.6) when compared to other parts of the image. The building in the same image (B1) also recorded 1.3 revisits, which suggests that people wanted to return to the spot in which they were interested earlier. When a human figure was present in the photos (B2, B3), subjects were more interested in the figure's face than to landscape elements or structures.

Fig. 4: An example of area of interest (AOI) delineation and statistics for image T2. The image on the left was derived from the right-side image.

In addition, whether or not the facial details were visible showed significant impact on subjects' eye movements. When the figure was back-lit and their features were obscured (B2), the building was the second most interesting element (fixation count: 1.3) following the figure in the image. However, when the figure's face is illuminated (B3), it was found that the concrete path was the second most interesting factor to most subjects, after the figure. When a person is in an image (B2, B3), AOI coverage values were significantly smaller and more focused (B2: 5.2%; B3: 4.9%), while the net dwell time, fixation counts, and revisit counts were high. This means that a person's presence in an image is an important factor affecting subjects' evaluation of nightscape.

Fig. 5: Heat maps and scan paths for Building type

Image Set T (Trees)

For the image with a tree in the background (T1, T2, T3), subjects focused more on the figure when his face was illuminated (fixation duration: 323.79 ms) than on other elements on the image. The image with no human figure (T1) received the lowest fixation scores, as subjects showed more scattered scan path patterns (scan path length: 2192.22 px) (Figure 6, Table 5). The image also elicited the lowest fear levels of the three images (T1, T2, T3). The investigator suspects that when a person fears an element, they spend more time focusing on that object.

For an AOI analysis, the images were divided into areas of building, structure, tree and grass, concrete path, and human figure as appropriate. It was found that almost all subjects focused on the human figure. The net dwell times and fixation counts were the highest when a human figure was present (net dwell time of T2: 1601.2, fixation count of T2: 4.2, net dwell time of T3: 1833.3, fixation count of T3: 4.5).

The investigators have also found that when the figure's face was in shadow (T2), the second most focused upon area was the concrete path (fixation count: 1.4). When the figure's face is lit, the second most focused upon areas were trees and grass (fixation count: 1.3). Just like in the Building type, the location of the light relative to the figure's position made the difference in subjects' second most important areas of interest.

Fig. 6: Heat maps and scan path for Tree type

For the Structure type, as in the other two types, subjects spent more time on fixed areas when an image included a human with a back to the light (S2 and S3) than on other areas (fixation duration: 276.90) (Figure 7, Table 6). The lowest score of the fixation was recorded when the image was without a human (S1) as was in the other two types. Scan path length and total duration were the highest in the condition without a human (scan path length: 2864.04, total duration: 5312.35).

Table 5: Core contents of AOI

		AOI coverage (%)	Net dwell time (avg. ms)	Fixation count (avg.)	Revisits count (avg.)
Background (T1)	Building	1.5%	289.7	1.0	–
	Structures	23.2%	190.7	0.8	–
	Trees and grass	47.6%	1158.0	3.9	0.4
	Concrete path	23.2%	2093.3	7.2	2.0
Human with a back to the light (T2)	Human	4.9%	1601.2	4.2	1.2
	Building	0.9%	38.8	–	–
	Structures	2.1%	63.8	0.3	–
	Trees and grass	47.1%	378.2	1.1	–
	Concrete path	19.6%	419.6	1.4	–
Human facing the light (T3)	Human	5.3%	1833.3	4.5	1.5
	Building	0.9%	59.5	0.3	–
	Structures	3.1%	39.6	0.2	–
	Trees and grass	46.1%	427.5	1.3	–
	Concrete path	19.3%	59.5	0.3	–

Image Set S (Structure)

For an AOI analysis, the areas of the image were divided into building, structure, tree and grass, concrete path, and human (except S1). It was found that almost all subjects focused mainly on the human as in previous two cases. The net dwell time and fixation counts were highest when a human is added to the image (net dwell time of S2: 1184.0, fixation count of S2: 3.6; net dwell time of S3: 1713.3, fixation count of S3: 5.2). When a human is not present in the image, subjects focused on the concrete path than on other areas. When a human is present (S2 and S3), the second most focused area was the building in the background (fixation count of S2: 2.0; fixation count of S3: 2.3).

Fig. 7: Heat maps and scan path of the condition surrounded with structures

Table 6: Core contents of AOI in the condition surrounded by structures

		AOI coverage (%)	Net dwell time (avg. ms)	Fixation count (avg.)	Revisits count (avg.)
Background (S1)	Building	32.0%	2074.0	7.4	2.1
	Structures	2.8%	228.9	0.9	–
	Trees and grass	14.3%	235.0	0.7	–
	Concrete path	25.3%	1319.1	5.2	1.4
Human with a back to the light (S2)	Human	4.5%	1184.0	3.6	1.0
	Building	21.0%	546.6	2.0	–
	Structures	2.4%	92.8	0.3	–
	Trees and grass	22.5%	30.8	0.1	–
	Concrete path	23.1%	275.6	0.9	–
Human facing the light (S3)	Human	4.8%	1713.3	5.2	1.8
	Building	31.6%	666.9	2.3	0.8
	Structures	4.4%	60.1	0.3	–
	Trees and grass	10.7%	11.6	–	–
	Concrete path	21.7%	194.0	0.7	–

4 Discussion and Conclusion

Perceived fears of nighttime spaces have been studied previously (e.g., Fisher and Nasar, 1992). However, these studies did not investigate the relationship between fear and specific physical elements in the nightscape. This study builds on these prior studies by analyzing people's gaze into various parts of nightscape images.

This paper investigated the effect of elements in the nightscape and lighting on people's perception of nightscapes. Evaluations of nine different nighttime situations were conducted to verify assumptions that the degree of fear study participants reported when viewing nightscape photographs would correlate to their eye movement.

Through this study, the investigator learned that reported levels of fear were affected by the different situations in the photographs. Almost all participants felt more fear when a human figure appeared in an image, regardless of the background type. The results showed that participants felt more fear when the human figure's facial features were obscured by backlighting compared to the figure whose face was illuminated. This is likely because participants were not able to see the facial features of the individual and could not evaluate whether or not he was a threat. The investigator also could speculate that the higher the degree of fear, the shorter the distance of eye movement from the main points of interest of each image and the tendency of eye movement. In addition, when a human figure is added to an image, the intensity, frequency, and dwell time of focused gaze toward the human-being were increased greatly. This study suggests that to make the nightscape safe, it is important to light spaces evenly so that facial features are always visible.

A limitation of this study could be the relatively low number of participants. However, many other eye-tracking studies have had fewer number of subjects and still had meaningful results. Also because the experiment is done in a lab setting, the level of fear in a real outdoor environment could not be simulated. Wearable eye-tracking goggles might solve this problem. Future studies by the author plan to use these goggles.

The results of this study could help designers create safer nighttime environments through a growing understanding of the elements and conditions in nighttime landscapes that elicit fear. This knowledge will help add to nighttime design guidelines. Future studies might investigate other aspects of the perception of nighttime space to understand what people perceive as a preferred environment and not just fear.

References

Austin, D., Furr, L. & Spine, M. (2002), *The effects of neighborhood conditions on perceptions of safety*. In: Journal of Criminal Justice, 30 (5), 417-427.

Cutrell, E. & Guan, Z. (2007), *What are you looking for? An eye-tracking study of information usage in Web searches*. In: CHI 2007 Proceedings. San Jose, CA.

Fisher, B. S. & Nasar, J. L. (1992), *Fear of crime in relation to three exterior site features: Prospect, refuge and escape*. In: Environment and Behavior, 24, 35-65.

Itoh, N. & Fukuda, T. (2002), *Comparative study of eye movement in extent of central and peripheral vision and use by young and elderly walkers*. In: Perceptual and Motor Skills, 94 (3, Pt. 2), 1283-1291.

Knight, C. (2010), *Field surveys of the effect of lamp spectrum on the perception of safety and comfort at night*. In: Lighting Research & Technology, 42 (3), 313-329.

Lee, T. R., Tang, D. L. & Tsai, C. M. (2005), *Exploring color preference through eye tracking*. In: AIC Colour, 5, 333-336.

Moyer, J. L. (2005), *The landscape lighting book*. New York, NY, Wiley.

Munchen, C. (1998), *Landscape lighting design book*. ALS Landscape Design Institute.

Nasar, J. L. (1998), *The evaluative image of the city*. Thousand Oaks, CA, Sage.

Painter, K. (1996), *The influence of street lighting improvements on crime, fear and pedestrian street use, after dark*. In: Landscape and Urban Planning, 35 (2-3), 193-201.

Painter, K. & Farrington, D. (2005), *The financial benefits of improved street lighting, based on crime reduction*. In: Criminal Justice Policy Review, 16 (1), 41-163.

Pease, K. (1999), *A review of street lighting evaluations: Crime reduction effects*. In: Crime Prevention Studies, 10, 47-76.

Quinet, K. D. & Nunn, S. (1998), *Illuminating crime: The impact of street lighting on calls for police service*. In: Evaluation Review, 22, 751-779.

Raine, J. (2001), *Garden lighting*. Laurel Glen.

SMI (2012), *Begaze Version 3.2 Manual*. SMI.

Thomas, C. & Bromley, R. (2000), *City-centre revitalisation: Problems of fragmentation and fear in the evening and night-time city*. In: Urban Studies, 37 (3), 1403-1429.

Zahm, D. (2004), *Brighter is better. Or is it? The devil is in the details*. In: Criminology & Public Policy, 3 (3), 535-546.

Wilson, W. H. (1984), *How to design and install outdoor lighting*. San Francisco, CA, Ortho Books.

Quantitative and Qualitative Analyses of Landscape Views

Agnieszka OZIMEK and Piotr ŁABĘDŹ

1 Introduction

Evaluations of landscape visual resources and estimations of the acceptable change in its character usually base on human feelings. Both expert opinions and studies, taking into consideration public preference, are frequently criticised, because of their subjective nature (SMARDON 1996). The necessity of ***objective indicators*** for the assessment of scenic values has constituted a subject of academic debate since 1970's. Unwin emphasised that landscape measurement should serve as a tool of its resources inventory and evaluation (UNWIN, 1975). Robinson argued for ***numerical and quantitative values*** that would help to reduce the subjective character of landscape assessments and would guarantee similar results, independent from the individual judgement of the observer (ROBINSON et al., 1976). The requirement of ***quantitative measurement*** of the scenery elements impact on their perception was stressed by Buhyoff and Riesemann (BUHYOFF & RIESEMANN, 1976). A coherent framework for landscape classification and valuation was proposed by Cooper and Murray (COOPER & MURRAY, 1992). They distinguished geographical units possessing similar characteristics, regarding physical attributes. This system provided the environment that might support decision making in landscape planning and management.

The aim of this paper is to present the method that base on the cooperation between an expert in the field of landscape architecture and a programmer. While, in the initial phase of the research, the expert formulates the rating of the positive and the negative aspects of the given area, the programmer tries to find the visual equivalents of these issues. They can be measured and provide tools that help in the more objective landscape evaluation, in decision making, with reference to the choice of the best scenario of its development, or allow monitoring scenery alterations in time.

The most attractive viewpoints, from which several of the valuable resources of the terrain are perceptible, are determined basing on diagrams of visibility. The more positive elements can be seen, the more beautiful this view is. Reversely, the more negative objects, the less attractive the vantage point. These places are the subjects of the researches "in situ", because of evident inexactness of the visibility diagrams (MALOY & DEAN, 2001).

For the particular view the quantitative analysis can be executed, which answer the question: ***"how much?"*** The mean area of the entity in the image can precise the characteristic parameter of the typical element. If we consider several scenarios of development, we can estimate the scale of the new objects in this view and conclude, if they fit to the environment. Here, it is worth to notice, that they can be comparable in the view, especially with respect to their size, only under the condition, that they are located in the similar distance. Therefore, the segmentation of the image into sub-images containing distance-dependent zones is necessary (OZIMEK et al., 2012).The qualitative analysis of the view, which answers the question ***"what kind of?"***, is far more complex. In this case ***shape factors*** can be taken into regard.

When the particular view is recognized as valuable, its character should be maintained. Therefore, all the actions consisting in introducing new elements into this environment should be aimed at preservation of its general appearance. Shape factors may serve as tools for designation of existing forms features and afterwards, by the comparison, for ascertainment, if the proposed variant of development is acceptable.

2 Material and Methods

2.1 Input Data

The eye-level view from the tourist trail, presenting a terrain intended for building development in master plan, was chosen as a subject of the research. The photographs of the current state (Fig. 2 and Fig. 3) were supplemented with the images of houses, which could be erected in the vicinity. Three different scenarios were considered:

- a housing estate, with buildings similar in size and form to the existing ones (Fig. 4 and Fig. 5);
- one big edifice (a hotel or a rest house) – alike, in relation to its shape, but dominating over the neighbourhood (Fig. 6 and Fig. 7);
- an industrial building – contrasting with its environment (Fig. 8 and Fig. 9).

The images were binarized (converted into bi-level, black and white images) in order to distinguish the building development, which constituted the subject of analysis (GONZALES & WOODS, 2002). As to compare the elements in the similar distance, the fragment of panoramic view has been chosen, which depicts the group of buildings situated along one road and the adjacent areas. In the Fig. 1 it is distinguished with a red rectangle.

Fig. 1: The 180^0 panorama with the view of the lake – the analysed fragment marked with the red rectangle

Fig. 2: The fragment of panorama showing the existing development and the terrain of the potential investments

Fig. 3: The fragment of panorama showing the existing development – the bi-level image prepared for the shape factors calculation

Fig. 4: Scenario 1 – residential development similar to the existing one

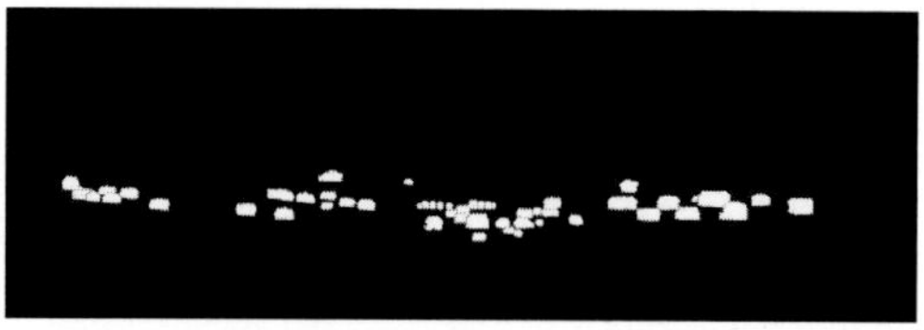

Fig. 5: Scenario 1 – residential development – the bi-level image prepared for the shape factors calculation

Fig. 6: Scenario 2 – a big rest house

Fig. 7: Scenario 2 – a big rest house – the binarized image

Fig. 8: Scenario 3 – an industrial building

Fig. 9: Scenario 3 – an industrial building – the binarized image

2.2 Modelling approach

In order to determine qualitative characteristics of objects, eleven shape coefficients were taken into consideration, which, for the most part, are applied in material and medical analyses (WOJNAR et al., 2002). In a binary image, with analysed objects marked white, every object is separated from each other by, at least, one pixel. In this manner, ***isolated regions*** were obtained, that can be investigated individually. Thus, any entity in the picture has no influence on any other, and the context in which that object exists, is not considered.

To maintain order and clarity, shape factors were numbered ascending from F_1 to F_{11}.

The first two of them are ***circularity coefficients***, calculated according to formulas:

$$F_1 = \sqrt{\frac{4 \cdot A}{\pi}} F_2 = \frac{P}{\pi}$$

where A stands for an area of the object, and P - it's perimeter. F_1specifies the diameter of a circle with the same area as the object, while F_2identifies the diameter of a circle with the same perimeter as the object.

$F_3$is called ***Malinowska factor***, computed basing on the object's area and perimeter. It achieves higher values for elongated objects.

$$F_3 = \frac{P}{\sqrt{4 \cdot \pi \cdot A}} - 1$$

In order to calculate ***Blair-Bliss coefficient*** (F_4) we have to determine a centroid of each object. This factor is defined by the equation:

$$F_4 = \frac{A}{\sqrt{2 \cdot \pi \cdot \sum_i r_i^2}}$$

where r_i is a distance between centroid and each pixel of the object and i a number of pixel.

Danielsson coefficient (F_5) requires specification of object's contour. We assume that contour and distance are determined in Euclidean space.

$$F_5 = \frac{A^3}{(\sum_i l_i)^2}$$

l_i – stands for a minimal distance between considered pixel and the contour of the object.

Another factor, demanding high time-consuming calculations, is ***Haralickcoefficient*** (F_6), for which we use the formula:

$$F_6 = \sqrt{\frac{(\sum_i d_i)^2}{n \cdot \sum_i d_i^2 - 1}}$$

where d_i is a distance between the centroid and each pixel of the contour of the object and n defines number of pixels in the contour.

F_7 is useful in determining ***circularity*** of the object, utilizing the equation:

$$F_7 = \frac{r_{min}}{r_{max}}$$

r_{min} means minimal distance between the centroid and the contour, while r_{max} determines analogical maximal distance.

F_8 helps to measure ***irregularity*** of the object:

$$F_8 = \frac{D_{max}}{P}$$

where D_{max} is maximal dimension of the object.

F_9 is ***modified Malinowskacoefficient.*** The more circular the object, the closer this value to 1.

$$F_9 = \frac{\sqrt{4 \cdot \pi \cdot A}}{P}$$

The next one is ***Ferret coeffictient*** (aspect ratio).

$$F_{10} = \frac{D_h}{D_v}$$

where D_h is maximal horizontal dimension and D_v is maximal vertical dimension (TADEUSIEWICZ & KOROHODA, 1997).

The eleventh factor (F_{11}) is used to calculate ***the extent*** understood as the ratio of pixels in the region to pixels in the total bounding box. All the computation were performed in Matlab with Image Processing Toolbox.

The shape factors were calculated for every object and, basing on them, the arithmetic means and standard deviations of all existing elements were computed. The results obtained for the particular quantities vastly differ in range. Some of them, like the extent coefficient (F_{11}), reach fractional values, while circularity (F_1, F_2) and Danielsson factors (F_5) fluctuate from about 10 to nearly 70.

A standard deviation estimated for every shape factor is an adequate tool for determination of geometrical features of the collection of existing buildings. If the coefficient has more coherent values, it characterizes the group of objects in a better way. On the contrary, when it is diverse, we assume that it is not specific for this set of elements. In order to compare the outcomes, the results computed for standard deviation were scaled by dividing them by the arithmetical mean, calculated for a given factor.

Three scenarios of development, mentioned above, were investigated. Every new element, planned in this surrounding, was analysed, with respect to its size and shape. If its' parameters diverge from the existing elements, the conclusion is that this object is contrasting. For the new group of buildings the mean area and average shape coefficients were computed and taken into account.

3 Results

The first part of a research consists in quantitative analysis. The mean area of building was calculated basing on the bi-level image of the current development (Fig. 3) and it equals 950.69 pixels. In the first scenario (Fig. 4 and 5), for the similar housing estate that was introduced, it reached 1083. The difference in size between the existing buildings and the proposed ones came to nearly 14%. In second case (Fig. 6 and 7) the area of the edifice (6065 pixels) was more than 6 times bigger (638%) than the arithmetic mean calculated for the existing objects. In the third variant (Fig. 8 and 9), the new element, which size reached a value of 9132 pixels, was nearly 10 times bigger than the present houses (961%).

In the initial phase of the qualitative analysis, results obtained in calculations of eleven shape factors were examined, with reference to their statistics. The coefficients, for which the values were dispersed, were eliminated, whereas these with the most coherent results were regarded characteristic for the analysed group of elements. Fig. 10 shows that for the factors F_4, F_6, F_8, F_9, F_{11} the results are the most similar, so they may be recognized as typical for this set of forms. They describe an object, concerning its irregularity (F_8), compactness (F_4, F_6, F_{11}), or circularity (F_4, F_6, F_9).

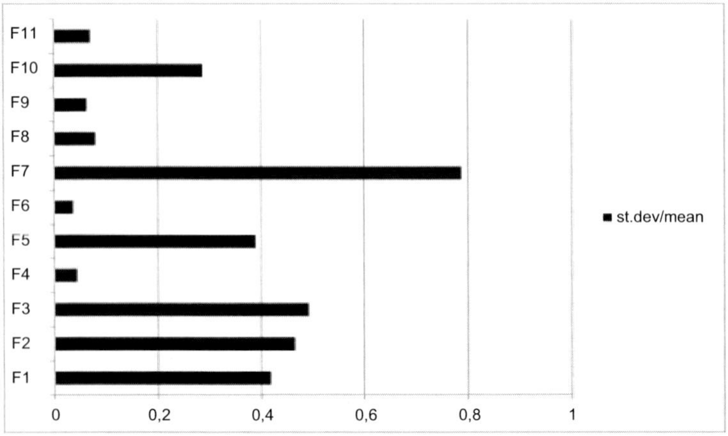

Fig. 10: A standard deviation of shape factors (numbered from 1 to 11) divided by the arithmetic mean of values obtained for the given coefficient (objects in present situation)

In the research conducted for three scenarios of development the selected shape factors of new buildings (or a mean values for the group of houses) were compared with values of the existing ones. The results of the proportional change are shown in the Fig. 11. It is worth to notice that the closer is this value to one, the smaller the difference.

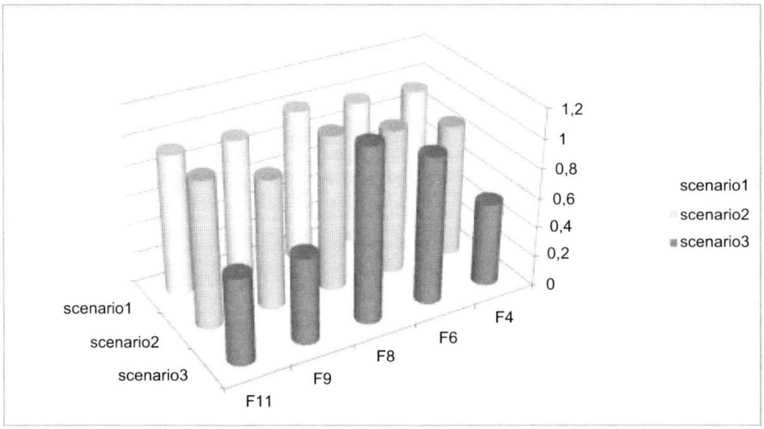

Fig. 11: A comparison between three scenarios of development and the current state

In the first situation, when the similar forms were introduced, the outcomes are in the range between 0.96 and 1.03. It is consistent with our intuition, suggesting that these houses possess features corresponding with the present ones. In the second case, the building was significantly larger than the average one, but its shape was not contrasting with the environment. These observations are reflected in change of shape coefficients, which reaches values from 0.88 to 1.05. On the contrary, in the third scenario, where the new factory visibly differs from the surrounding, the analysed shape factors clearly change (0.56 – 1.17). The biggest differences may be noticed for Blair-Bliss (F_4) and modified

Malinowska (F_9) coefficients, as well as for the extent (F_{11}), which decreases almost twice. These factors depend on the object compactness and circularity.

In every analysis of objects' geometrical parameters some imprecision appears, resulting from the fact that the elements, which are closer to the photo camera, obscure the distant ones, affecting their shape. The example of this situation is presented in Fig. 4 and 5. In order to estimate the extent of this error, another simulation was prepared. Fig. 12 and 13 shows the similar housing estate, but the layout of objects was modified, in order to separate the regions.

Fig. 12: Residential development with the change in buildings layout

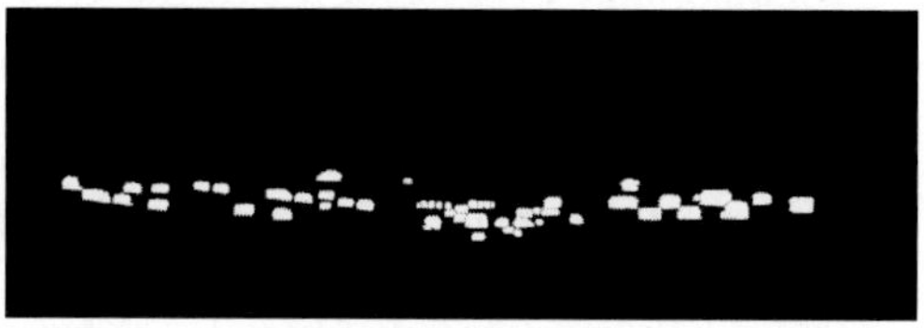

Fig. 13: Residential development – the bi-level image

Surprisingly enough, in this case, the differences in shape factors (in comparison to the scenario 1 – Fig. 4 and 5) are insignificant and they vary from 0.001 (for modified Malinowska coefficient – F_9) to 0.01(for the extent – F_{11}).

4 Conclusions and Outlook

For the research the evident examples have been chosen in order to compare the outcomes with the intuitive feelings. The results, as the numerical values, are objective in their character. They may constitute a criterion, which supports the expert in the process of decision making. The set of the factors, that characterize a particular group of elements, may differ; notwithstanding, it should be born in mind that some of the coefficients are by nature more variable than the other ones.

The presented approach is applicable, first and foremost, for open views in which the analysed objects are isolated, therefore, the elements do not interfere with each other. Thus, it is more suitable for rural and suburban terrains, than for the dense built-up areas. The investigation presented in this paper applied to the estimation of building development, however, it may be extended onto the other components of a view.

The restriction of this method results from the fact that, due to the perspective distortion, only objects located in the similar distance from the camera may be compared. The shape of the elements may be affected by inexactness of their detection or, in case when it is made manually, by the human factors. Image resolution may influence the results, as well.

For the valuable viewpoint observations can be made in the regular time spans and the changes can be investigated by the comparison of the successive images.

References

Buhyoff, G. J. & Riesenmann, M. F. (1979), *Experimental manipulation of dimensionality in landscape preference judgements: a quantitative validation.* In: Leisure Sciences, 2, 221-238.

Cooper, A. & Murray, R. (1992), *A structured method of landscape assessment and countryside management.* In: Applied Geography, 12, 319-338.

Gonzales, R. C. & Woods, R. E. (2002), *Digital Image Processing.* Prentice Hall.

Malina, W. & Smiatacz, M. (2005), *Metody cyfrowego przetwarzania obrazów* (*Methods of Digital Image Processing*). Oficyna Wydawnicza EXIT.

Maloy, M. A. & Dean, D. J. (2001), *An Accuracy Assessment of Various GIS-Based Viewshed Delineation Techniques.* In: Photogrammetric Engineering & Remote Sensing, 67 (11), 1293-1298.

Ozimek, A., Łabędź, P. & Michoń, K. (2012), *Discrimination of Distance-dependent Zones in Landscape Views.* In: Buhmann, E., Ervin, S. M. & Pietsch, M. (Eds.), Peer Reviewed Proceedings of Digital Landscape Architecture 2012 at Anhalt University of Applied Science. Berlin/Offenbach, Wichmann, 464-471.

Robinson, D. G. et al. (Eds.) (1976), *Landscape evaluation – the landscape evaluation research project 1970-1975.* University of Manchester.

Smardon, R., Palmer, J., Felleman, J. et al. (1986), *Foundations for Visual Project Analysis.* New York, A Wiley-Interscience Publication.

Tadeusiewicz, R. & Korohoda, P. (1997), *Komputerowa analiza i przetwarzanie obrazów (Computer image analysis and processing).* In: Wydawnictwo Fundacji Postępu Telekomunikacji, 256-269.

Unwin, K. I. (1975), *The relationship of observer and landscape in landscape evaluation.* In: Transactions of the Institute of British Geographers, 66, 130-133.

Wojnar, L., Kurzydłowski, K. J. & Szala, J. (2002), *Praktykaanalizyobrazu (Practice of Image Analysis).* Polskie Towarzystwo Stereologiczne, Kraków.

Digital Monuments and Augmented Reality – Study of the American Civil War Landscapes

Shamsul ABU BAKAR, Patrick MILLER, Brian KATEN, Troy ABEL, Mustafa K. M. SHARIFF, Peter SFORZA and Mintai KIM

1 Introduction

Because the Civil War is perhaps "the most compelling episode in American history" (SELLARS, 2005, p. 26), the preservation of American Civil War battlefields as National Military Parks has significantly benefited today's American society. They are known as places of reconciliation and remembrance, and contribute to the creation of a national identity. These places are in fact "sacred places" (LINENTHAL, 1991) that are visited by many who consider themselves pilgrims undertaking repeated journeys to a holy place (STODDARD, 1996).

The notion that a battlefield is a repository of memories cannot be argued. The landscape is not only stained with the nation's suffering but also offers the potential for healing. The battlefield is a place where "the combatants of old could come together, recall the shared horrors of war, and acknowledge their common actions as lessons for the nation" (Associates & Partners, 2003, p. 25). In addition, these annual reunions held on the battlefields served to "draw attention to the need to preserve the fields of battle as well as a symbolic act of healing sectional divisions" (ASSOCIATES & PARTNERS, 2003, p. 26).These battlefields long remained central places for reconciliation among veterans and their descendants and for keeping national memory alive.

1.1 Monuments in the Civil War Battlefields

Even before the war ended on April 9, 1865, Civil War battlefields received enormous attention from the soldiers, the public, and politicians who fought to preserve these historic landscapes. For instance, according to ABROE (as cited in ASSOCIATES & PARTNERS, 2003), during the course of war, active duty soldiers built three memorial structures as a remembrance to their fallen comrades. The first memorial was the Bartow Monument erected by the Eight Georgia Infantry in honor of their brigade commander who was mortally wounded during the Battle of Manassas in September 1861.Then in 1863, Union soldiers built two monuments: the Hazen Brigade Monument, which honored brigade members killed in the Battle of Stones River, and the Surrender Monument, which commemorated General Pemberton's surrender to the Union army after the Siege of Vicksburg (ASSOCIATES & PARTNERS, 2003).

Besides, SELLERS (2005) notes, the process of creating and marking battlefields with monuments not only serves as part of the commemorative efforts but also as an attempt to "impose order on the past on landscape of conflict and confusion" (SELLERS, 2005, p. 38). After the war ended, the veterans associations made numerous efforts to mark and accurately place monuments on the sites of significant events. The construction of monuments to represent specific combat units and their lines of battle has improved the ability of today's visitors to understand troop movements and avenues of approach. In fact, the important role of monuments in honoring and preserving the memory of important events was acknowledged more than a century ago when John Howard Wert noted that, "The more the field is decorated with these works of art, the more powerful becomes the impulse of the traveler and patriot to visit or revisit the field of glory" (as cited in ASSOCIATES & PARTNERS, 2003, p. 28).

1.2 Battlefield Monuments: Issues and Challenges in the 21st Century

Various government agencies, private associations, and veterans' organizations have been involved in the preservation of battlefields and the building of Civil War monuments. They organized annual reunions, erected battlefield monuments, and lobbied state legislatures to allocate funds for future monuments on different battlefields (HANSON & BLYTHE, 1999, p. 28). Among these organizations were the Gettysburg Battlefield Memorial Association (GBMA), the Society of the Army of Tennessee, the Southern Historical Society (SHS), and the Association of the Army of Northern Virginia (AANVA).

Despite these efforts, not every battlefield was successfully preserved and marked with monuments. Most of these battlefields remained on private land, which means access and site preservation became more difficult. Even worse, many historic battlefield shave been under continual threat from urban sprawl that has brought permanent changes to the physical landscape. Often these changes mean not only the loss of natural vegetation and the alteration of other natural features but also the loss of historical structures or important viewsheds related to the battlefield. For instance, MACMAHON (1991) has noted how landscape changes at the Gettysburg Battlefield impacted later historic interpretation:

> *"Through those motels and fried-chicken stands, Pickett's men charged. The first line faltered in the Burger King parking lot and regrouped next to the Tastee Freeze"* (MACMAHON, 1991, p. 16)

While monuments have long been used to convey the battle's memories, some scholars argued that they could potentially affect the historic landscape values. According to LINENTHAL (1991), mass erection of these commemorative monuments may simply degrade a battlefield's purified environment. The same concerns are echoed by BIRNBAUM (1993), who noted that contemporary monuments could potentially create a false sense of history when they are inserted in a historic landscape.

Besides, conveying historical lessons to today's tech-savvy visitors is more challenging. These visitors are highly dependent on computers, television screens, and movie theaters as a source of information (MOSS, 2008). According to MOSS (2008), this phenomenon has brought new patterns of learning, especially for "millennial learners" or "the net generation."It is very important to integrate visual technologies with historic landscape elements in order to bring on-site educational excitement. Although some of these technologies have been employed in interpretation and visitors centers, less effort has been made to bring these technologies closer to the actual ground. A combination of these elements could play a significant role in educating visitors in the near future.

Therefore, this study explores the potential of using digital technologies to convey historic memories in landscapes that are physically and visually inaccessible. Specifically, this study identifies the possibilities of using virtual 3D monuments as an alternative visual platform to commemorate historic events and facilitate users' ability to comprehend the horrific lessons.

2 Study Area

The Battle of Cedar Creek in Middletown Virginia, was fought on October 19, 1864. Major General Philips Sheridan commanded the Union troops, while the Confederate forces were led by Lieutenant General Jubal Early. This battle is considered unique by military historians because of the brilliant night maneuvers used by the Confederate forces to launch their surprise attack on the Union camps. Although the surprise attack was considered a victory for the Confederate forces, because of hunger and a lack of coordination, they were eventually defeated.

Cedar Creek and Belle Grove National Historical Park (CEBE) was officially created in 2002. It is partially situated in three different counties: Frederick, Shenandoah, and Warren in Virginia. The creek for which the battle is named served as a natural defensive feature, which both the Union and Confederate armies tried to maximize during the battle.

3 Methods of Study

The methods used in this study focus solely on the development of experimental 3D model layers used in an augmented reality platform and are divided into three stages. The first stage explains the process of developing a 3D model of battlefield monuments. The second stage involves developing augmented reality layers based on the battlefield's points of interest and integrating those layers into a mobile augmented reality browser known as LAYAR. In the third stage, several field tests were conducted by the researchers to review the quality of those models and to identify technical or user-related issues when viewing the model.

3.1 Developing Battlefield 3D Monuments

SketchUp 8 was chosen to create the 3D monuments because of its ease of use and simple interfaces that allow easy construction, manipulation, and visualization of 3D concepts. For this study, we constructed a 5-by 14-foot obelisk as an experimental model (Figure 1). The shape was chosen because it is one of the most common types of monuments used to commemorate soldiers at the Civil War battlefields. However, in order to be used in the LAYAR browser, the SketchUp (.skp) obelisk model file had to be converted to a3D object file known as (.l3d).

This process required two phases of file conversion. For the first phase, Bable3D online file translation service was used to convert the (.skp) file into a geometry definition file format (.obj). The new file format allowed geometrical characteristics such as the position of each vertex and texture coordinate vertex to be universally accepted by other 3D graphic applications. The (.obj) file was then imported to the LAYAR 3D model converter where the model vertices, faces, and material should be overviewed. It also allowed the editing process on the obelisk materials, textures, color, shininess, rotation, and scale. Once finalized, the model was then converted to the (.l3d) file (Figure 1).

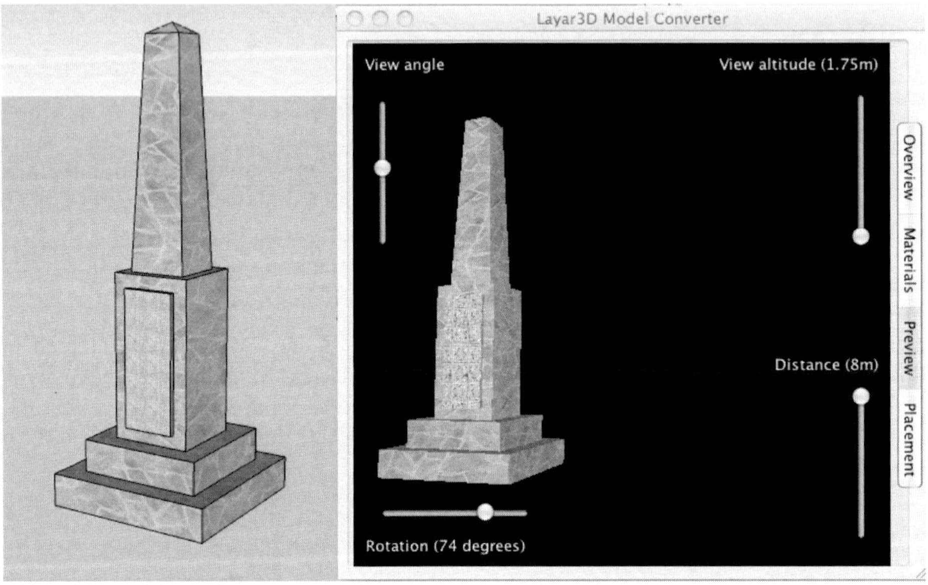

Fig. 1: Constructed 3D obelisk monument in SketchUp and it preview in LAYAR 3D Converter

3.2 Developing and Integrating the Layers

Using layer creation tools (LCT) known as Hoppala Augmentation, several points of interest (POI) identified from the Cedar Creek Battlefield events were used to create an augmented 3D monument model layer. Details were added to these POIs such as military unit descriptions, commanding officers, weblinks, sounds, and geo-referenced locations (Figure 2). Using the Hoppala content platform overlay uniform resource locator (URL), this layer was then linked to the mobile browser through the researcher's LAYAR developer account. The LAYAR browser generally allows developers to customize the augmented reality layer interface according to specific user preferences and launch the experimental layer in the LAYAR mobile application. For this study, the IPhone 3G with iOS mobile platform was used to visualize the final results.

Fig. 2: Layer creation tools (LCT) known as Hoppala Augmentation were used to augment several points of interest (POI) to create a monument layer using the 3D models (source: http://www.hoppala-agency.com)

4 Results

Several valuable outcomes were observed through the demonstration. First, the 3D obelisk model layer can be integrated and launched easily using both the Hoppala content management platform and the LAYAR mobile augmented reality browser. Next, the scale of the augmented model varies based on its location when viewed through the mobile phone's camera (Figure 3).The scale was found to be very useful in constructing "conceptual distance" between models and the user's location. Additionally, the Hoppala platform allows easy controls of 'z'[1] values for the models to be placed within the historic landform.

Meanwhile, integrated information linked with the model's pop-up interfaces enable users to have a better understanding and experience of the landscape's historic contexts (Figure 4). However, the materials and textures assigned to the obelisk model have lower image quality compared to the earlier model developed in SketchUp and the LAYAR 3D model converter. The reduced image quality may be caused by the size of smartphone screen, which limits pixelated details of the model's materials. In addition, it was found that the model's positions (x,y) have an acceptable 9 to 15ft level of accuracy in open areas. The

[1] Models' altitudes from ground level.

accuracy for the 'z' value however is more difficult to estimate, although the results show it is more credible.

Several difficulties were found when researchers used the application. For instance, the mobile electronic compass occasionally gave false orientations toward POIs. On sunny days, the user often faced difficulty seeing the models' details due to reflections and glare on the mobile's screen. Finally, users should also be aware of safety issues when using this application close to moving traffic.

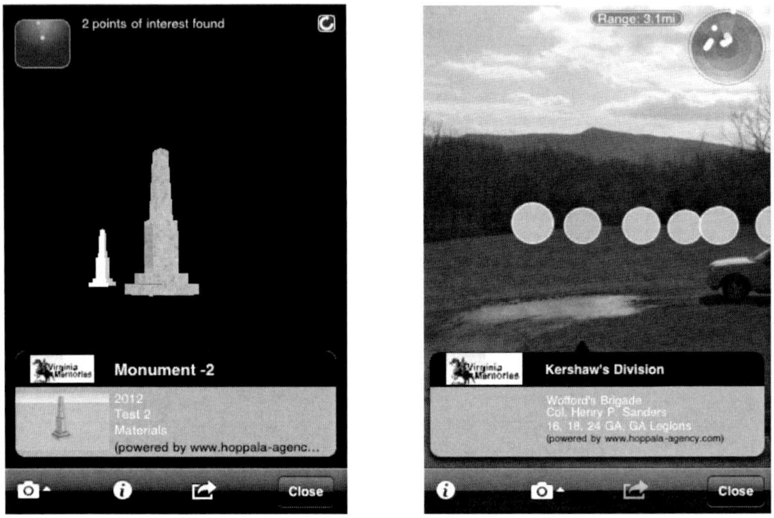

Fig. 3: 3D models provides better 'conceptual distance' between POIs than 2D symbols

Fig. 4: The POIs 'pop up' interface provide additional historic contexts and experience

Fig. 5: 3D Monuments were simulated within a degraded landscape

5 Conclusion

This paper explores the potential of using 3D models within physically and visually inaccessible historic landscapes to augment the memory of past events. Based on our observation, instead of using typical 2D objects to represent the POIs, the 3D models have shown its potential to facilitate user imagination to commemorate relevant lessons learned from the battle. We also believe that the new technique used to visualize 3D monuments could attract and provoke user interest to better experience the battlefield.

This technology can also be used to make the invisible visible. Important historic structures that have been removed or destroyed in the historic landscape such as bridges, fences, and stonewalls can be easily constructed. The use of new technologies to visualize historic events can certainly act as a contemporary communicator of social remembrance. In fact, based on this approach, the historic integrity of the battlefield can be enhanced and new ways of experiencing the historic landscape can be promoted. We speculate that 3D models can be an important visual representation tool in understanding historic human conflicts.

In addition, this technology allows users to easily access the historical battlefield and maximizes their opportunity to explore different points of interest. Considering the rapid adoption of smartphones, particularly in the younger demographics, AR can be a valuable learning tool. Some computer scientists, for instance, believed that in the future, an increasing percentage of information would be visual rather than written (STALEY, 2003).

Finally, this study also offers opportunities to further investigate the potentials and limitations of using 3D monuments in the Civil War landscapes. For instance, future studies may involve use of eye-tracking technology to determine usability of the application or user perception of the quality of 3D monuments. These will lead to better understandings of the potentials and limitations of 3D monuments within an augmented reality environment to commemorate lost historic events.

References

Associates, J. M. & Partners, J. (2003), *Design of the commemorative military park: contextual documentation & resource evaluation.* Phase 1: Contextual documentation essay. Unpublished work. Charlottesville, Virginia.

Babel 3D, *Translate and visualize 3D files*. Retrieved from https://www.babel3d.com.

Birnbaum, C. A. (1993), *Treatment for historic landscape*. In: Cultural Resource Management, (20), 21-22.

Hanson, Jill K. & Blythe, R. W. (1999), *Chickamauga and Chattanooga National Military Park: Historic Resource Study*. Atlanta, Department of the Interior/National Park Service.

Hoppla Augmentation. Retrieved from http://www.hoppala-agency.com.

Linenthal, E. T. (1991), *Sacred ground: Americans and their battlefields*. University of Illinois Press, Urbana.

McMahon, E. T. (1991), *Saving Our Sense of Place.* The Environmental Forum.

Moss, M. H. (2008), *Toward the visualization of history: the past as image*. Lanham, Lexington Books.

Sellars, R. W. (2005), *Pilgrim Places: Civil War battlefields, historic preservation, and America's first national military parks, 1863-1900.* In: CRM: The Journal of Heritage Stewardship, 2 (Winter 2005).

Staley, D. J. (2003), *Computers, visualization, and history: how new technology will transform our understanding of the past*. Armonk, N.Y., M. E. Sharpe.

Stoddard, R. H. (1997), *Defining and classifying pilgrimages*. In: R. H. Stoddard & A. Morinis (Eds.), Sacred Places, Sacred Spaces: The Geography of Pilgrimages. Baton Rouge, Geoscience Publications, Dept. of Geography and Anthropology, Louisiana State University, 41-60.

Modeling of 3D Landscapes

Analysis of Sequential Landscape Experiences

Ian BISHOP

1 Introduction

Much landscape change – for example the construction of wind farms, or logging of native forests – has negative impacts on visual quality of surrounding areas, and this impact has become the focus of public protest. Many land management agencies therefore consider visual assessment of landscape change a vital tool for exploring public responses to alternative landscape futures. Most landscape assessment work to date has relied on public reactions to edited photographs showing static views of landscape change. More recently, researchers have used a combination of geographic information system (GIS) based visual analysis and public responses gathered from original or manipulated photographs to estimate visual impacts of change in a larger landscape (BISHOP & HULSE, 1994; GERMINO et al., 2001; MOLLER, 2006). Both of these approaches have been criticized for ignoring critical aspects of landscape experience (HULL & STEWART, 1992). Of particular concern is the dependence of both approaches on public evaluation of individual views. While people often pause to admire a single view, large landscapes cannot be experienced at a single point, at a single time. More usually, large landscapes are revealed gradually or sequentially over time as a person moves through the landscape.

Indeed, much of our landscape experience comes while moving: cross-country skiing, cycling, hiking, driving, travelling by train or flying. More distance is covered driving than through any other mode of movement. In Victoria, Australia each vehicle averaged 14,500 km in 2010. Someone insulated from the landscape within a speeding capsule may not be as aware of his or her environment as the skier or the hiker (OKU & FUKAMACHI, 2006). Nevertheless this paper focuses on extended movement, such as travelled by car, in order to test some parametric approaches to landscape evaluation in this dynamic context.

Few studies have sought to understand how people evaluate landscapes experience sequentially. HULL & STEWART (1992) considered evaluations of views within a 'trip', but provides little insight to the relationship between individual views and evaluation of the trip as a whole. STEINITZ (1990) analyzed the driving experience around the Loop Road in Acadia National Park based on view analysis and a visual quality model. While the processing was based on point evaluations, Steinitz commented that "the rhythm of positive views along the clockwise direction is more frequent ..." (p. 238) indicating a sense that the sequence was as important as the set-by-step conditions.

To gather data on sequential experiences, QIN et al. (2008) created 2 minute animated drives using Visual Nature Studio (3dnature.com). They then recorded the perceived scenic beauty progressively using an electronic slider devices controlled by the viewer. The drive sequences included elements of interest such as billboards, a barn, lakes and different levels of enclosure. In that research they also asked for an overall assessment of the drives. However, they did not report the relationships between the sequential ratings and the overall rating.

Determining this relationship is one of the key questions relating to sequential experiences. It seems unlikely, although possible, that the overall scenic assessment of a journey, for most people, will be a simple average (time-weighted) of the value of the views along the way. There are other possibilities which also do not draw on the ordering within the sequence of views: the overall perceived quality might be approximated by the best view scene on the journey, or the lowest rated view, but this also seems unlikely. No matter how pleasing the view of a wide beach leading to an azure sea, it may start to loose its lustre if unchanging over a long distance. As Schoenberg wrote, in a musical context: "Two impulses struggle with each other within man: the demand for repetition of pleasant stimuli, and the opposing desire for variety, for change, for a new stimulus". (SCHOENBERG, 1911, quote on p48 of 1978 translation)

In the particular context of landscape change, such as created by forest harvesting or energy infrastructure, we need to know whether a short exposure to very high/low quality landscape raises/lowers the overall assessment disproportionately. There may exist the sequential equivalents of meta-variable such as complexity and coherence. What kinds of transitions are expected, unexpected, confusing, overwhelming, soothing, exciting?

The many parameters that may be used to quantify the sequential experience should include those that are normally evaluated in studies of static views and, in addition, those that are specific to moving views. ZIPF (1972) argued that many things in nature and society have a 1/k relationship meaning that the kth ranked item in a set (such as a list of city populations) occurs with 1/k of the frequency of the first ranked item (the second biggest city is half the size of the biggest etc). VOSS & CLARKE (1978) measured physical variable, including fluctuations in loudness and pitch, for several types of music including classical, jazz, blues, and rock. They found that the pitch and loudness fluctuations did approximate Zipf's distribution. MANARIS et al. (2003) sought to develop 'fitness functions for pleasant music' based on Zipf-distributions. They concluded: "If nothing else, since Zipf distributions appear to be a necessary, but not sufficient condition for aesthetically pleasing music, such fitness functions could minimally serve as an automatic filtering mechanism to prune unpromising musical samples." (p. 530) They suggested that Zipf analysis was not sufficient because two pieces of music could both fit the Zipf distribution perfectly but one could be quite monotonous because all the highest ranked notes occurred in one block (as if, for example, all occurrences of 'the' in a book came at the beginning). To go beyond the global balance identified by the Zipf coefficient, they suggested order could be incorporated in the analysis by also estimating the fractal dimension (a statistical measure of complexity) of a musical piece. They suggested that the box-counting method, as applied to the paintings of Jackson Pollock by TAYLOR et al. (1999), could also be applied to music. Here I suggest it might also be applied to sequential landscapes experiences.

2 Method

2.1 Landscape Elements

The following were identified as potential parameters of interest along road segments in Victoria, Australia: road orientation, relative elevation, openness, distance to horizon, and presence of specific elements (individual trees, forest, water, beach, shore rocks, snow, steep slopes, rock outcrops, green grass, brown grass, crop, bare earth, wind turbines, buildings, farm animals, wildlife). Some of these are well recognized in the literature of landscape aesthetics, others are less well accepted as features of interest but included because of their potential contribution to the unfolding landscape. Additional variables of interest during a journey include changes in road orientation and elevation. These were also recorded along with a very subjective impression of 'openness', an estimate of the distance to the furthest horizon (no including views directly along the road itsclf) and the extent to which high slopes were present in the view.

2.2 Sample Road Segments

One of the most popular tourist drives in Australia is the Great Ocean Road (GOR) running along the Victorian coast for 243 km from Torquay to Port Campbell. This is too long for this exploratory analysis and so a 16 km long section between Lorne and Apollo Bay was chosen (Figure 1).

To contrast with the GOR, I chose a section of road that I travel often and find pleasant but uninspiring. This is part of the Princess Highway (PHW) running east from Melbourne for around 400 km before turning north towards Sydney. Again a 16 km section was chosen between Nar Nar Goon and Drouin (Figure 2).

(a)

Fig. 1:
(a) The chosen section of the Great Ocean Road, Victoria, Australia
(b) A panorama illustrating the kind of landscape typical on this road segment

(b)

(a)

(b)

Fig. 2:
(a) The chosen section of the Princess Highway, Victoria, Australia
(b) A panorama illustrating the kind of landscape typical on this road segment

2.3 Data Extraction

The sample road segments were selected as travel routes in Google Maps using the names of towns beyond the two ends. After dragging the start marker to my chosen start point I activated Street View at that point. By rotating the view I could indentify the presence or absence of each of the landscape elements. When an element was present I assigned it a prominence, within the full 360-degree arc, on a scale of 1-10. I then dragged the end mark to a point 0.5 km along the route from the start point and repeated the process. By repeatedly dragging the end marker I could assess the landscape every 0.5 km over the 16 km segments. One possibly pertinent variable I could not extract in this way was the elevation. For this I switched to Google Earth and chose the corresponding point using the satellite images in the two products.

Street views have been captured more frequently than every 0.5 km and so a more fine-grained analysis would have been possible. However this seemed to be a useful initial distance at which to test the process. It did however created some ambiguity in selecting the Street View since Google Maps only gave the route length to the nearest 0.5 km. An alternative approach would have been to select locations where the view included elements of special interest, such as water bodies.

It would also have been an option to only use the part of the view in one direction of movement and so use, for example, an arc of 100 degrees on either side of the direction of motion. However I used 360 degrees because landscape elements behind the car would have typically been seen in front a few hundred meters earlier. This also made the assessment direction independent.

2.4 Creation of Indices

While each entity was recorded individually, they also form compatible groups with similar expected influences. These were combined into interest indices as follows:

- road_interest (variation in direction and elevation) = ABS(change_of_orientation (degrees)/10 + change_of_elevation (meters)/5)/3. (Changes are since last view point).
- view_extent = openness + furthest_horizon (meters)/10000.
- terrain_interest = slope + beach + shore_rocks + rock_outcrops
- vegetation_interest = individual trees + native forest - plantations_and_exotics
- agricultural_interest = crops_and_horticulture + green_pasture + brown_pasture/2 - bare_earth + farm_dam*2 + domestic_animals*2
- water_interest = sign_of_creek/2 + visible creek + river + lake + sea/ocean + snow
- infrastructure_impact = enhancing_buildings - ugly_buildings - transmission_pylons*2 - transmission_poles

Each index was transformed to an integer to allow frequency ranking and fractal analysis. Each was also normalized, so as to give an approximate range of 0 to 10, over the two road segments. However the indices would need adjustment if different roads with higher scores on particular indices were included in the analysis.

2.5 Analysis

For the Zipf distribution analysis, the frequency of occurrence of each index score, from 0 to 10, was recorded and then ordered by frequency rank. The $\log_{10}$ of each rank and each frequency were calculated and a simple linear regression run for each index. The coefficient (slope) of the regression and the R-squared value, showing goodness of fit to the regression line, were recorded for comparison.

The fractal dimension of each index distribution was estimated using the R statistical software (v2.15.2 for MacOS X) in conjunction with the Fractaldim package (v0.8-1). Both can be found at cran.r-project.org. Fractaldim gives a number of different techniques for estimating the fractal dimension of 1-dimensional time series. In this case, in the absence of a particular rationale for a more sophisticated method, the simple box-counting method was used. Dimensions were first calculated for each segment independently. This could be misleading because then a different interval would be used, on each segment, for the side of the boxes representing the index scores. This gave a high fractal dimension for some sequences with low variation in their index. To create a valid comparison it was necessary to give the sequences of index scores for each segment the same score range. This required introduction of a dummy maximum value into the sequences since the box dimensions could not be controlled independently.

2.6 Sensitivity to Landscape Change

To explore the effect of repeated instances of negative changes in landscape quality such as by forest harvesting or wind energy installations a different approach was necessary. We know quite a bit about the effect of these elements on individual views but next to nothing about their sequential effects. In order to begin exploration of the effect and the influence of different spacings, I made changes to the scores within the two road segments. To check this properly an experimental distance greater than 16 km may be necessary to allow a gradual increase in prominence, but in this case I look the liberty of assuming that the installations could appear suddenly and disappear just as suddenly. Three conditions were tested, (a) 10 consecutive points (5 km) were deemed to have a prominent wind farm, (b) 10 points in 5 equally spaced groups of 2, and (c) 10 individual equally spaced points. At these points the sum of the index values was reduced to zero. That is, the high infrastructure impact (II) negated any other positive index values. This is almost certainly too extreme but serves to test the effect on fractal dimension.

3 Results

Table 1 shows the index scores for each 0.5 km along each road segment, and summary statistics for the segments. A very clear difference between the two roads is the high level of water interest along the GOR and the complete absence of water features at the sample points along the PHW. In general the scenic value of the GOR is higher and so the segment as a whole has higher total and averages then the PHW segment. This tells us nothing about the effect of sequence however.

Table 1: Illustrative index values for the two 16 km road segments. The index values for the first 6 km (at 0.5 km intervals) of each segment and the summary values for the whole segment are shown.

Dist. (km)	Road Interest (RI)		View Extent (VE)		Terrain Interest (TI)		Veg. Interest (VI)		Agriculture Interest (AI)		Water Interest (WI)		Infrastructure Impact (II)	
	PHW	GOR	PHW	GOR	PHW	GOR	PHW	GOR	PHW	GOR	PHW	GOR	PHW	GOR
0	1	1	0	9	1	3	6	3	1	2	0	7	1	-2
	1	0	0	4	0	5	4	1	2	0	0	6	0	0
1	1	2	6	6	2	6	1	3	2	0	0	7	1	-1
	1	7	2	4	1	9	6	2	1	0	0	6	0	0
2	0	9	1	5	0	10	0	3	7	0	0	10	0	-1
	0	10	0	7	0	7	4	1	0	1	0	7	0	0
3	2	6	5	4	1	7	3	3	2	1	0	6	0	-1
	1	1	4	7	4	6	4	4	6	0	0	6	2	-1
4	0	1	1	9	0	7	4	3	2	2	0	7	-1	-2
	0	1	2	8	0	5	4	1	4	3	0	8	2	-3
5	0	1	2	8	1	3	4	1	3	1	0	6	1	-3
	0	1	7	1	1	0	0	0	4	0	0	0	-7	0
6	0	3	8	7	2	4	3	3	3	3	0	5	-3	-3
Sum	15	98	123	179	35	181	113	106	68	23	0	182	-56	-29
Mean	0.45	2.97	3.73	5.42	1.06	5.48	3.42	3.21	2.06	0.70	0.00	5.52	-1.7	-0.88
SDev	0.67	2.82	3.16	2.24	1.22	2.60	1.80	2.07	1.56	1.24	0.00	2.32	2.82	1.17

Table 2 shows the Zipf statistics for each index on the two segments. On the GOR the coefficients range from -0.69 (view extent) to -1.82 (agriculture interest) indicating that the variation in view extent and terrain interest are less than 'ideal' while the variation in agriculture and infrastructure interest is more than 'ideal' (where ideal is a coefficient of -1). Water interest and road interest have coefficients very close to -1 and reflect the sense that these are the dominant reasons why people choose the GOR as a driving experience. On the PHW, there is no water interest recorded and so no Zipf statistic. Among the other indices, view extent drops away only slowly with a wide range of values whereas road interest has only few values and so drops more quickly than 'ideal'. The latter reflects the general straightness and relative flatness of the PHW whereas the former reflects a fairly even mixture of view bounded by vegetation and open views to distant mountain ranges.

Table 2: Log(Rank) versus log(Frequency of Indices) give Zipf slope coefficients and R-squared estimates of fit

A. Princess Highway

	log(RI)	log(VE)	log(TI)	log(VI)	log(AI)	log(WI)	log(II)
Coefficient	-1.712	-0.616	-0.994	-0.871	-1.432	n/a	-1.134
R-squared	0.797	0.679	0.851	0.972	0.966	n/a	0.896

B. Great Ocean Road

	log(RI)	log(VE)	log(TI)	log(VI)	log(AI)	log(WI)	log(II)
Coefficient	-1.084	-0.691	-0.759	-1.157	-1.815	-1.067	-1.574
R-squared	0.959	0.935	0.785	0.78	0.986	0.898	0.902

Table 3: Local and global fractal dimensions (FD) of road segments based on index variability

	PHW		GOR	
	local FD	global FD	local FD	global FD
Road Interest	1.67	1.36	1.65	1.65
View Extent	1.71	1.58	1.78	1.62
Terrain Interest	1.62	1.53	1.76	1.76
Vegetation Interest	1.62	1.55	1.58	1.58
Agriculture Interest	1.63	1.63	1.66	1.58
Water Interest	1.00	1.00	1.83	1.83
Infrastructure Interest	1.36	1.36	1.82	1.64
TOTAL Interest	1.54	1.42	1.63	1.63

Table 3 shows the fractal dimensions of each of the indices on the two road segments, both 'local' and 'global' dimensions are shown the former using box counting based on the range of local index values and the latter using the full range of values found across the two road segments, giving a better direct comparison. Except for agricultural interest, the dimensions on the GOR are higher than on the PHW with water interest and terrain interest having the highest dimensions.

Each of the three sets (group of 10, 5 groups of 2, and 10 individual views) of wind turbines added to the segments by reduction of the index sum to zero at 10 points naturally had the effect of reducing the total and mean scores of the whole road segments. The degree of change depends on whether the insertions are at high scoring or low scoring points, but the effect is independent of the sequential pattern. Similarly, changes in the Zipf distributions are not order dependent. It is the fractal dimension that is prone to different effects. New fractal dimensions were calculated for the sum of the indices (representing overall visual quality). As with the total and mean, it was quickly apparent that the new fractal dimensions were dependent on which actual points were used as the start point for insertions. Therefore I used six replicates of the three test conditions each with a different starting point for the turbine-affected views. When these were averaged a clear trend emerged. On both road segments the 10 consecutive points with turbines gave the result with the lowest fractal dimension (PH $\mu = 1.38$, $\sigma = 0.087$, GOR $\mu = 1.50$, $\sigma = 0.087$), the case of 5 groups of 2 was next (PH $\mu = 1.66$, $\sigma = 0.046$, GOR $\mu = 1.70$, $\sigma = 0.064$) and the case of 10 individually affected points had the highest fractal dimension at each segment (PH $\mu = 1.69$, $\sigma = 0.082$, GOR $\mu = 1.84$, $\sigma = 0.052$). This can be compared with the global dimensions in the absence of the turbine insertions (PH 1.42 and GOR 1.63). Thus, 10 consecutive affected points somewhat reduced the fractal dimension of each road segment but the other configurations increased the fractal dimension in each case.

4 Discussion

At this stage we have no empirical data on liking for the road segments from which to judge these results. A small survey has been conducted on four different 16 km segments on the PHW. However neither the process nor results are reported here because the survey was very exploratory. Indeed, undertaking the analysis reported here on the feature sequences along road segments will help in designing a better survey.

So, what can we determine from the analysis so far? The major outcome is a set of more precise questions about sequential experience and the statistics that may be used to characterize that experience. These include:

- What is the relative importance of individual views and the sequential aspects of the experience?
- Is a Zipf distribution with a slope of -1 really 'ideal'?
- A higher fractal dimension suggests a more interesting range of experiences, and this is backed up by the higher dimension on the tourist favored GOR, but is this really so?
- If it is so, is it preferable to break intrusive elements, such as wind farms, into smaller groups, which create a higher fractal dimension, than to localize them in a single larger group?
- The fact that the more distributed turbines increased fractal dimension over the situation with no wind infrastructure suggests that using the index sum as the basis for calculation of fractal dimension is too blunt an instrument. Can an index such as fractal dimension, which has promise, be made more sensitive?
- Even though the analysis included infrastructure (houses, farm buildings, electrical pylons and poles etc) there was no clear way to score these as part of an index. It would be become still more problematic if the roads included (or might include in the future) elements like wind turbines, which provoke very mixed responses.

- What is the effect of landscape familiarity on these relationships? In high quality landscapes in or adjacent to key tourism areas (e.g. GOR) many of those making the journey will be first time visitors. On a major intercity highway (e.g. PH) many people will have made the journey before and the making the journey for reasons other than enjoyment.

References

Bishop, I. D. & Hulse, D. W. (1994), *Predicting scenic beauty using mapping data and geographic information systems.* In: Landscape and Urban Planning, 30, 59-70.

Germino, M. J., Reiners, W. A., Blasko, B. J., McLeod, D. & Bastian, C. T. (2001), *Estimating visual properties of Rocky Mountain landscapes using GIS.* In: Landscape and Urban Planning, 53, 71-84.

Hull, R. B. IV & Stewart, W. P. (1992), *Experience patterns: the dynamic nature of a recreational experience.* In: Journal of Leisure Research, 24, 240-252.

Moller, B. (2006), *Changing windpower landscapes: regional assessment of visual impact on land use and population in northern Jutland Denmark.* In: Applied Energy, 83, 477-494.

Oku, H. & Fukamachi, K. (2006), *The differences in scenic perception of forest visitors through their attributes and recreational activity.* In: Landscape and Urban Planning, 75, 34-42.

Manaris, B., Vaughan, D., Wagner, C., Romero, J. & Davis, R. B. (2003), *Evolutionary music and the Zipf-Mandelbrot law: Developing fitness functions for pleasant music.* In: Lecture Notes in Computer Science, 2611, 522-534.

Qin, X-c, Meitner, M. J., Chamberlain, B. C. & Zhang, X-n (2008), *Estimating visual quality of scenic highways using GIS and landscape visualizations.* ESRI Users Conference, San Diego CA, August 2008.

Rodrigues, M., Montañés, C. & Fueyo, N. (2010), *A method for the assessment of the visual impact caused by the large-scale deployment of renewable-energy facilities.* In: Environmental Impact Assessment Review, 30, 240-246.

Schoenberg, A. (1978), *Theory of Harmony.* Berkeley, University of California Press (translated by R. E. Carter from 1911 original).

Steinitz, C. F. (1990), *Towards a sustainable landscape with high visual preference and high ecological integrity: the Loop Road in Acadia National Park, U.S.A.* In: Landscape and Urban Planning, 19, 213-250.

Taylor, R. P., Micolich, A. P. & Jonas, D. (1999), *Fractal Analysis Of Pollock's Drip Paintings.* In: Nature, 399, 422.

Voss, R. F. & Clarke, J. (1978), *1/f Noise in Music: Music from 1/f Noise.* In: Journal of Acoustical Society of America, 63, 258-263.

Zipf, G. K. (1972), *Human Behavior and the Principle of Least Effort.* New York, Addison-Wesley (original publication Hafner Publishing Company, 1949).

Mapping and Modeling of Village Landscape Character

Sevgi GÖRMÜŞ, Dicle OĞUZ and Serhat CENGİZ

Abstract

Landscape character analysis is essential for the planning of rural and protected areas. Since Turkey joined the European Landscape Convention in 2003, landscape character analysis has become important for national planning laws and regulations. Therefore mapping and modeling instruments become more important for defining, monitoring and protecting landscape characteristics.

In this study, context of landscape character analysis perspective, land survey and literature studies have been realized. Obtained data have been evaluated using mapping and modeling programs (like Geographic Information System, AutoCAD, NetCAD, 3Dmax, Global Mapper, Lumion). As a result of modeling and analysis, settlement type analysis, interactions analysis between land use and forest area, landscape diversity analysis and population analysis, landscape character of villages which are on the periphery of Bartın-Kastamonu Küre Mountains National Park have been identified. The effect of identified landscape character type on the natural area has been questioned in the context of landscape fragmentation.

Keywords: Landscape Character, Village Character, Bartın-Kastamonu Küre Mountains National Park, Kapısuyu Basin

1 Introduction

Landscape Character Assessment is a well-established tool for systematically identifying, classifying and describing the landscape, recognising it as a continuous system that does not adhere to administrative boundaries. By identifying, features and elements of the landscape (and their combination and expression), the essence or special character of a particular place can be revealed, explored and understood (ANONYMOUS, 2007).

Due to properties of landscape character analysis, landscape planning will be integrated into social and economic planning.

Important effects of rural settlement areas on natural areas is landscape fragmentation. The rural feauture in Turkey has a quite complex structure with its socio-economic and cultural characteristics, therefore the outcome of these identify current problems, setting goals for solutions in both politics and in develpoment of these rural areas which have great importance. The unique feautures of these rural areas (removed from urban areas, messy settlements, physical conditions) lead to many problems and prevent these areas from progressing in terms of socio-economics (KELES, 2006a; KELES, 2006b).

The legislation that the villages in Turkey are bound to is the Village Act (LAW NO. 442) issued on March 18th, 1924. According to this act, the village was defined as "a local government whose population does not exceed more than 2000 and also has movable or immovable properties and takes decisions with its units about given duties, based on the Village Act. As the proposals related to spatial layout of these villages have not been applied by this law, there have been many problems with spatial layouts.

One of the most common problems in many developing countries and Turkey is chaotic spatial dispersions in numerical and functional terms of cities, towns and villages that make up the settlement system (MARIN, 2010). The rural settlements are known to be dispersed in a large number and a wide range within the settlement system in Turkey (MARIN, 2010). State Planning Organization (2000) states that there were a total of 75,631 rural settlements (35,014 villages and 40,617 units bound to these villages) in 2000.

Villages in Turkey have been classified according to their relationship with their spatial layout and the natural resource values. Classification that is done according to its spatial order are paid attention to by considering the villages' collective, dispersed or less dispersed settlements. As for the relationship between villages and natural resource values there is a classification that is done according to their proximity to the forest areas. In this classification, villages are regarded as off-forest, in-forest and edge-forest. TÜRKDOĞAN (2006) classified the villages and defined each class number in Turkey according to their distribution and relationship with forests. (Tab. 1 and Tab. 2)

The high rate of units in rural settlements and the presence of low-population and dispersed units (MARIN 2010) leads to landscape fragmentation and habitat loss as well as economic and social problems. The most significant rural settlements that lead to landscape and habitat fragmentation are in-forest and edge-forest villages. These villages constitute the poorest part of the rural areas (AÇIKSÖZ VE GÖRMÜŞ 2010). These villages take form according to their physical conditions and natural environment and their unique way of life. The differences can be observed among the regions, even in the same region or from one village to another, in rich culture of Anatolia in Turkey. The settlements are shaped by the topographic structure, cultural and physical environmental conditions (EMINAĞAOĞLU & ÇEVIK 2007). The villages that are shaped and get their identity within principles of experienced construction and layout, and in general, (EMINAĞAOĞLU & ÇEVIK 2007) are in interaction with surrounding forest areas, cultivated areas, roads and landscape around them.

Table 1: Spatial arrangement of villages in Turkey

Type of Village	Number of villages	Population
Cumulative village	25.453 (71.9%)	13.160.279
Sparse village	5.467 (5.5%)	3.339.917
Less sparse	4.192 (11. 8%)	2.189.975

Table 2: Proximity of village to forest

Position	Number of villages	Population
Off-forest	19.746 (55.8%)	11.118.111
In-forest	5.093 (14.4%)	2.355.067
Edge-forest	7.225 (20.4%)	3.828.227

In this study, which is based on the basis of the requirements of the European Landscape Convention's biological diversity conservation, landscape management and landscape determination, the identity of villages, their effects on the forests and protected areas they are in or on the edge of are discussed within landscape character analysis technique. In this study it is thought that the classification defined as landscape character analysis in village-scale will be useful for rural development, rural landscape planning, forest management, landscape management and nature conservation policies and strategies.

2 Material and Method

Kastamonu-Bartın Küre Mountains National Park covering an area of 37.000 hectares is one of nine hot spots. The Ministry of Forest launched a project in 1998 titled "Management of National Parks and Preserved Areas: Conservation of Bio-diversity and Rural Development" with the financial support of UNDP (United Nations Development Program) and FAO (Food and Agriculture Organization). Thanks to the "Küre Mountains Draft Development Plan" prepared within the framework of the project, a "planning zone" around the national park was planned aiming at reducing the threats caused by the immediate surrounding so as to secure "Kastamonu-Bartın Küre Mountains National Park" covering an area of 37.000 hectares and biodiversity within the national park, for the first time in Turkey (National park conservation zone is 37.000 ha and planning zone is 80.000 ha, total area is 117.000 ha). The core area (national park) is delineated by a range of cliffs and canyons that include pristine and semi pristine natural mixed deciduous and coniferous forest. The global significance of the Küre Mountains' biodiversity has been highlighted by its inclusion in the WWF's list of European forest hotspots for conservation. The site is considered to represent the best remaining example of deciduous and and coniferous forest. The global significance of the Küre Mountains' biodiversity has been highlighted by its inclusion in the WWF's list of European forest hotspots for conservation. The site is considered to represent the best remaining example of deciduous and coniferous forest of the North Anatolia ecoregion as well as being the best remaining example of the highly endangered karstic mountain areas of the 'Black Sea Humid Forests' ecotype (WWF, 2001). The Küre mountains hosts 40 out of the 132 mammals in Turkey, including large mammal species, such as gray wolf, brown bear, Eurasian lynx, red deer, roe deer and wild boar. The park and its buffer zone have been identified as one of the 122 Important Plant Areas (IPA), and also one of the 305 Key Biodiversity Areas (KBA) in Turkey (ANONYMOUS, 1999; WWF, 2001; UNDP, 2008; BANN, 2010).

Kastamonu-Bartın Küre Mountains National Park (KMNP) consists of two zones: KMNP Conservation Zone and KMNP Planning Zone. KMNP Conservation Zone doesn't include settlements and agricultural areas having protection status and has been admitted as a "natural area". Rural areas that includes settlement and agricultural areas in KMNP Planning Zone don't have "protection status". The study area Kapısuyu Basin is composed of KMNP Conservation Zone and KMNP Planning Zone. Forests in Kapısuyu basin are divided into two groups: production forests and natural forests. Because of these different features, transition has occured between natural and cultural landscapes in Kapısuyu basin.

The study has been realized in 13 villages which are classified as forest villages according to number 6831 Turkey Forest Law (Fig. 1).

2.1 Obtaining land cover/Land use map

The data bases of working field (Rapideye satellite images, present layouts of village, Forest Management Plans, Land Use Map, Digital Elevation layouts) have been digitized by Remote Sensing (RS) and Geographic Information Systems (GIS) techniques. ERDAS 8.7 software has been utilised for the classification of Raster data (satellite imagery) and ArcGIS 9.2 software has been used for the vector data digitizing and analysis. The accuracy control of spatial data obtained from RS and GIS has been provided with GPS (The Global Positioning System) based on the projection of coordinates of with UTM (Universal Transverse Mercator) WGS84 (World Geodetic System) in the field.

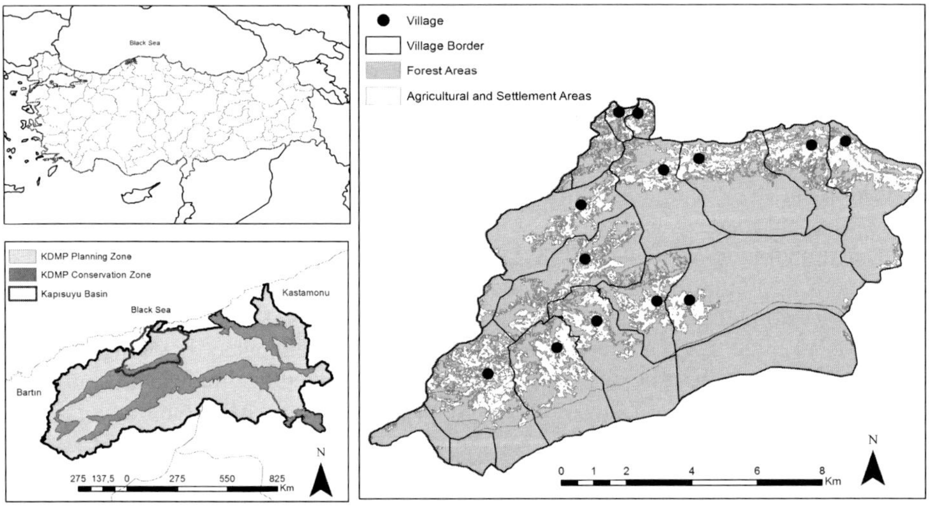

Fig. 1: Location of study area

2.2 Identifying village landscape character variables

To determine the type of landscape character of each village a series of following analyses have been carried out: The settlement pattern analysis, analysis of interaction land use and forest areas, analysis of landscape diversity, population analysis (Fig. 2).

The settlement pattern analysis: Villages have been classified according to topographic position (GERAY, 1985; KURTKAN BILGISEVEN, 1988; ROBERTS, 2003; TÜRKDOĞAN, 2006), land arrengement (KURTKAN BILGISEVEN, 1988; TÜRKDOĞAN, 2006; ROBERTS, 2003; GÖRMÜŞ, 2012) and settlement patterns (ROBERTS, 2003; GÖRMÜŞ 2012) (Tab. 3).

Analysis of the interaction land use and Forest areas: Land use map and patches and the edge density of these patches on the land cover have been determined by using landscape metrics (MCGARIGAL, 2002). Weighted average of all values has been calculated. The forest-agricultural land interaction of the villages of which edge intensity is over the average value have been considered as negative; and those of which edge intensity is below the average value have been considered as positive (Tab. 3).

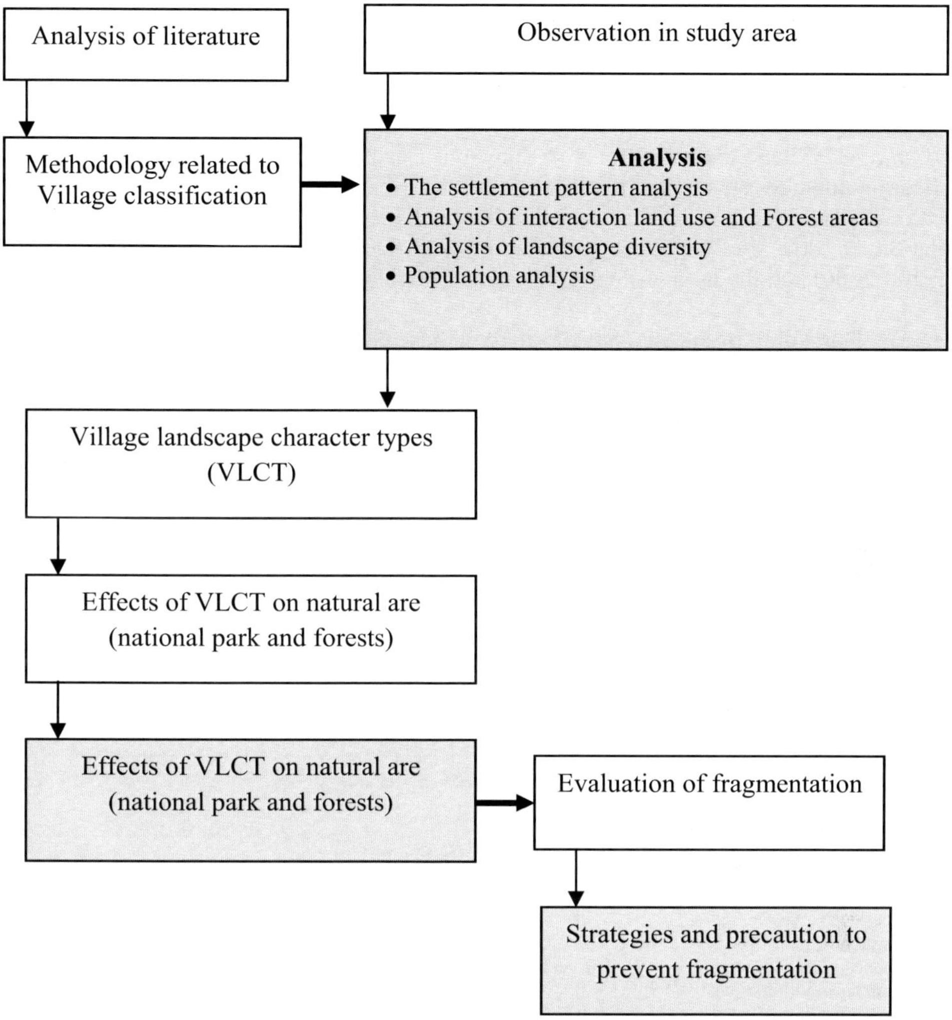

Fig. 2: A flowchart describing the process of identify of village landscape character

Analysis of landscape diversity: After borders of the villages have been assigned to the of land use and land cover maps, landscape diversity in each village has been obtained by the number of stains and Shannon Diversity Index (SDI) (MCGARIGAL, 2002) calculations. Weighted averages of values of all the villages have been calculated. The landscape diversity of he villages of which SDI value is over the average value have been considered as high; and those of which SDI value is below the average value have been considered as low (Tab. 3).

Population analysis: Main variables of the migration rate, the necessity of village-organization and the size of the village have been used in this analysis (Tab. 3).

Immigration rate: Migration rate has been determined according to the censuses between the years 1960-2010.

Necessity of local village-organization unit (According to the latest census): The necessity of village-organization is based on population data in the Village Act. According to the act, to establish and organization in a village, the population should not be less than 150 and more than 2,000.

Size of village: Village classification for Kapısuyu Basin has been expressed in line with classification method developed by Mitkovic et al. (2002) which is based on current population data. According to this classification, villages separate as very small villages (population: 0-100), small villages (population: 100-500), medium-sized villages (population: 500-2000) and the large village settlements (population greater than 2000).

Table 3: Desription of the analysis of village landscape character

The Settlement pattern analysis	Variables	Source
Topographic position	Plain village (Pl)	Geray, 1985; Kurtkan Bilgiseven, 1988; Türkdoğan, 2006; Roberts, 2003; Görmüş, 2012
	Valley foot village (Vf)	
	Valley slope village(Vs)	
	Mountain village(Mo)	
	Mountain slope village(Ms)	
Land arrangement	Nucleition village(N)	Türkdoğan, 2006; Roberts, 2003; Görmüş, 2012
	Dispersion/scattered village(D)	
	Line village(L)	
Settlement patterns	Random (Rm)	Türkdoğan, 2006; Roberts, 2003; Görmüş, 2012
	Regular(Rr)	
	Clustered(Cd)	
	Mixed (linear random (Lr), clustered random (Cr), clustered (Cl) lineer)	
Analysis of interaction land use and Forest areas Using landscape metrics: Edge Density Patch Area,	Villages that have low pressure on forest areas (F+) Villages that have high pressure on forest areas (F-)	Görmüş, 2012
Landscape diversity analysis Using landscape metrics: Patch Number in village Landscape Richness in village Shannon Diversity Index in village	Villages that have low landscape diversity (Ld+) Villages that have high landscape diversity (Ld-)	Görmüş, 2012
Population Analysis Immigration rate	Immigration rate (1960-2010) Villages with a pozitive immigration rate (I+) Villages with a negative immigration rate (I-)	Görmüş, 2012
Necessary of local village goverment unit (Accordingin to lastest cencus)	Villages that can be established local village goverment unit (G+) Villages that can not be established local village goverment unit (G-)	Turkey Village Act (Law No. 442); Görmüş, 2012
Size of village (Accordingin to lastest cencus)	Small village (Sv) Medium village(Mv Large village(Lv)	Mitkovic et al., 2002; Görmüş, 2012

2.3 Calculating and visualizing variables

By using mapping and modeling programs (like Geographic Information System, AutoCAD, NetCAD, 3Dmax, Global Mapper, Lumion) landscape character of villages which are on the periphery of Bartın-Kastamonu Küre Mountains National Park, is identified. Using cadastral, slope, elevation, topographic maps and social-cultural maps are realized analysis. Landscape character variables have been quantified and visualized by means of the programs mentioned below (Tab. 4, Fig. 3).

Table 4: Variables and programs

Analysis	The program used
Settlement Analysis	Arc view/Surface Analysis, Global Mapper Cadastral Data/AutoCAD/Netcad/3D Max
Analysis of interaction land use and Forest areas	Arc view/Patch Analysis
Analysis of landscape diversity	Arc view/Patch Analysis
Population analysis	Legal statistical data/SPSS

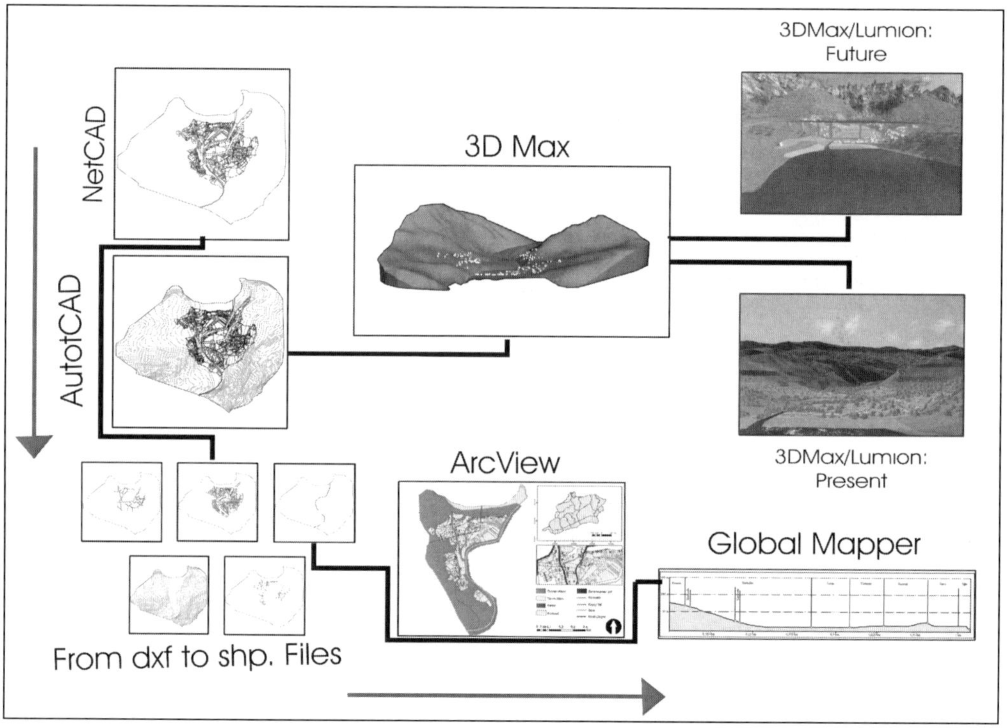

Fig. 3: Process of digital data

3 Result

Obtained data from analysis are classified in order to present landscape character of villages. To determine the type of landscape character of each village a series of following analyses have been carried out: the settlement pattern analysis, the interaction land use with forest areas analysis, analysis of landscape diversity and population analysis.

As a result of these analyzes, landscape character types that primarily cause fragmentation of the landscape have been obtained. Obtained some of village landscape character types are as follows:

Vf_ N_Cr_F$^-$_Ld$^+$_I$^+$_G$^+$Mv (Başköy): Nucleation at the foot of valley, clustered in random order, highly repressive to the forest area, high landscape diversity, the sum of migration rate is positive, medium-sized village settlement (Fig. 4).

Ms_N_Cr_ F$^-$_Ld$^-$_I$^-$_G $^+$Sv (Kaleköy): In a mountain slope, nucleation, random clustered, highly repressive to the forest area, low landscape diversity, the sum of migration rate is negative, a small village settlement where a local village goverment unit can be established (Fig. 4).

Vs_D_Lr_F$^-$_Ld$^-$_I$^-$_G$^-$SV (Nanepınarı): Scattered in the valley slope, linear random, highly repressive to the forest area, low landscape diversity, the sum of migration rate is negative, a very small village settlement where a local village goverment unit cannot be established (Fig. 4).

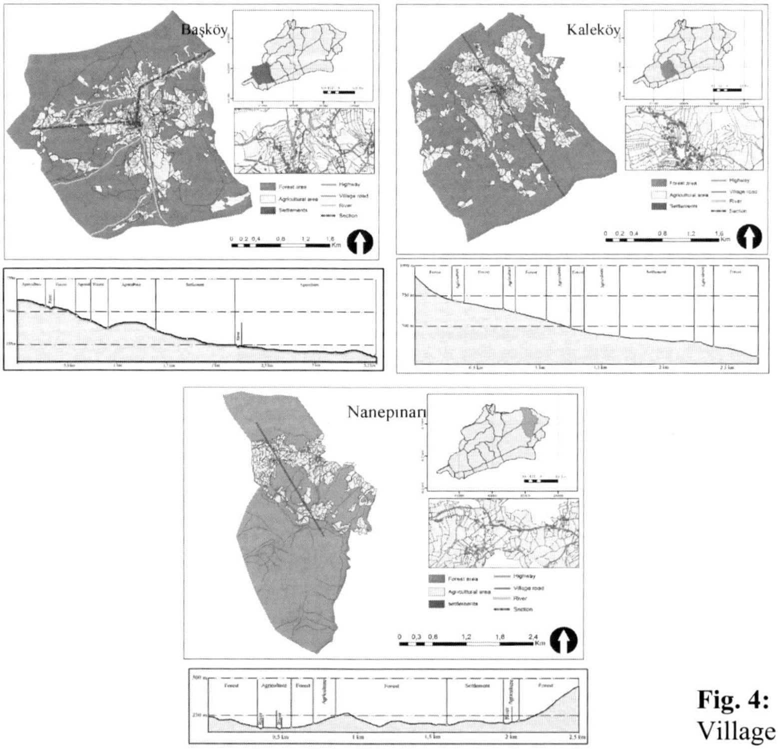

Fig. 4:
Village maps

4 Conclusion

Landscape fragmentation affects biodiversity in spatial and temporal scales. The decrease of landscape types and species causes an increase in the rate of fragmentation in landscape. As increase of patchiness and decrease of connectivity restrict the movement of populations (SHUKLA, 2002). Landscape fragmentation affects the pattern-process relationship. By splitting large stains in the landscape into small ones, fragmentation leads to loss of original habitat, reduction of habitat stains and increase of isolation between habitat stains (BOTEQUILHA LEITÃO & AHERN, 2002).

That agricultural land parcels are irregular in shape and position may lead to an increase in edge habitat types. However, this causes a decrease in biodiversity. According to Odum and Barrett (2008), species richness between two different land cover is defined as the edge effect. Special arrangements have been advised in order to increase the number of species between the field and forest. As it causes reduction in species diversity, precise boundary or sudden and sharp edge between the two habitats is not desirable. On the other hand, more irregular edges means reduction of biological diversity.

There are direct and inverse proportions among landscape diversity of villages, migration rate and land use with its interaction with forest area. When these variables are compared to analysis of the settlement pattern, the following conclusions are reached.

The concept of diversity of landscape does not only include natural landscape features but also includes cultural landscape elements. For this reason, "landscape diversity" concept includes both positive and negative effects. While landscape diversity is high in the villages that are at slope of the valley, that of villages at the foot of the mountain is low. There is not a significant difference between landscape diversity and forest areas interaction in the villages that have "cluster-randomized" and "linear-randomized" layout. Landscape diversity of the villages of which field order is "line" type is higher and their negative impact to the forest areas is lower. "Messy" nature of villages' landscape diversity is low but their impact to the forest area is the least.

In this study, the most important criterion that determines villages' impact to the forest area and landscape diversity has been found to be location. Location of land and housing layout take shape according to topographic scheme. For this reason, the location of villages around the national park and their negative effects arising from the location should be especially evaluated.

Acknowledgement: This paper is made under "Determination of Landscape Character type in Preserved Areas in the Case of Kastamonu-Bartın Küre Mountains National Park (110Y270) supported by The Scientific and Technological Research Council of Turkey (TÜBİTAK).

References

Anonymous (1999), *Kure Daglari National Park, Development Plan.* Prepared by the Ministry of Forestry (General Directorate of National Parks and Game-Wildlife) in collaboration with UNEP and FAO.

Anonymous (2000), *Eight Five Year Development Plan.* Ankara, Republic of Turkey, Ministry of development,

Anonymous (2007), *Exmoor Landscape Character Assessment. Commissioned by Exmoor National Park Authority.* Prepared by Emma Jane Preece. Available from Exmoor National Park Authority Exmoor House, Dulverton, Somerset, TA22 9HL. www.exmoor-nationalpark.gov.uk.

Bann, C. (2010), *Enhancing Forest Protected Areas System of Turkey. Developing a Business Plan for Küre Mountains National Park and its Buffer Zone.* Final Report.

Botequilha Leitao, A. & Ahern, J. (2002), *Applying Landscape Ecological Concepts and Metrics in Sustainable Landscape Planning.* In: Landscape and Urban Planning.

Eminağaoğlu, Z. & ve Çevik, S. (2005), *Evaluation Of Design And Planning Policies Related to Rural Settlements in Regional Scale.* In: Journal of the Chamber of City Planners, Planning, 2, 72-81.

Eminağaoğlu Z. & ve Çevik S. (2007), *Design Policies and Means Related to Rural Settlements.* In: J. Fac. Eng. Arch. Gazi Univ., 22 (1), 157-162.

Geray, C. (1985), *Settlement arrangement of Villages in Turkey.* Village sociology. İstanbul, Bogazici university Publisher.

Görmüş S. (2012), *Landscape Character Analysıs For Protected Areas Case study: Kastamonu-Bartın Küre Mountaıns Natıonal Park.* Ph. D. Thesis, Ankara university, Graduate school of natural and applied sciences, Department of landscape architecture, Ankara.

Keleş, R. (2006a), *Urbanization policy.* Ankara, İmge Bookstore.

Keleş, R. (2006b), *Governance and politics.* İstanbul, Cem Publishing.

Kurtkan Bilgiseven, A. (1988), *Village sociology.* İstanbul, Filiz Bookstore.

Lêon, Y. (2005), *Presidential Addres Rural Development in Europe: A Research Frontier for Agricultural Economists.* In: European Review of Agricultural Economics, 32, 3.

Marin M. C. (2010), *An Analysis Effects of Small and Medium Sized Cities on Rural and Balanced Regional Development.* Project of The Scientific and Technological Research Council of Turkey (TÜBİTAK), PN: (107K520). Kahramanmaraş.

McGarigal, K. (2002), *Landscape Pattern Metrics.* In: Encyclopedia of Environmentrics. A. El-Shaarawi and W. W. Piegrorsch (Eds.). Sussex, England, John Wiley and Sons, 1135-1142.

Odum, E. P. & Barrett, G. W. (2008), *Fundamentals of Ecology* (Ed. Kani Işık). Ankara, Palme Publishing.

Roberts, B. K. (1996), *Landscapes of Settlement. Prehistory to the Present*, London, Routledge.

Shukla, S. (2002), *How to Reduce the Adverse Impacts of Landscape Fragmentation on Biodiversity.* http://deathstar.rutgers.edu/advgeo/2002papers/Shukla.htm (available date: 04.12.2012.)

Türkdoğan, O. (2006), *Village sociology in Turkey.* İstanbul, IQ Culture and art Publisher.

UNDP (2008), *Project Document GEF MSP PIMS 1988: Enhancing Coverage and Management Effectiveness of the Subsystem of Forest Protected Areas in Turkey's National System of Protected Areas.*

WWF (2001), *Mediterranean Forest: A New Conservation Strategy.*

Humansensorik und Kartierung von Emotionen in der räumlichen Planung

Human Sensory Assessment and the Cartography of Human Emotions in Spatial Planning (English summary)

Peter ZEILE, Jan-Philipp EXNER, Benjamin S. BERGNER und Bernd STREICH

Summary

The objective measurement of human's well-being in urban areas or in natural environment shows enormous potentials for environmental planning. Actually, there is no known method, which makes it possible to get a "neutral" view of people's perception in space and the associated emotions concerning the built and non-built-up environment. Human sensory assessment, defined as joint analysis of real-time vital data and geo data, offers a solution. Psycho-physiological monitoring techniques collect and measure vital data like skin conductivity and skin temperature over time. Specific changes in this vital data indicate mental load synonymic stress. Stress in this context is the emotional construct of fear and anger. According to affective sciences, these negative emotions are identified when the skin conductance increases and the skin temperature decreases (the phenomenon of "cold sweat"). In combination with geo-position of the test people, it is possible to identify local stress hotspots. Supporting this new approach, other tools like the RADAR Sensing smartphone app or RADAR Infrastructure deliver real-time heat-maps of stress experienced by a multitude of people. Another tool is the new developed GeoVisualizer, which displays the vital data and the resulting stress spots on a digital globe interface. Three projects, in which human sensory assessment methods are used, will be presented in this contribution. The projects are the identification of bad accessibility for handicapped people in a city, the orientation of hikers on a forest trail in the Palatinate Forest and also the impact of noise in recreation areas.

1 Einleitung

Die Frage, ob es für planende Disziplinen möglich und notwendig ist, das Wohlbefinden eines Menschen in seinem Umfeld objektiv zu messen, wird immer wieder in Fachkreisen gestellt. Entwurfsmaxime sollte stets die Verbesserung der menschlichen Umwelt im Einklang aller Belange der Planung sein. Ein Planungsgrundsatz, der, so könnte der objektive Beobachter meinen, in der Vergangenheit bei vielen Neuplanungen außer Acht gelassen wurde. Aber wie ist das Wohlbefinden überhaupt messbar? Wie kann der Mensch als Messfühler seiner Umwelt in ebendiese miteinbezogen werden, ohne einer Beeinflussung vonseiten Dritter zu unterliegen? Eine Lösung dafür bietet der Ansatz, den Menschen als Sensor seiner Umwelt zu verstehen. Seine Reaktionen auf ein bestimmtes Ereignis werden mithilfe von Methoden der Humansensorik und im speziellen Fall des sogenannten psychophysiologischen Monitorings aufgezeichnet, ausgewertet und präsentiert. Vereinfacht formuliert steht hiermit eine Methode zu Verfügung, die es erlaubt, Emotionen (im Raum) messen zu können. Anwendungsfälle sind beispielsweise die Messung der Barrierefreiheit von Infrastrukturen, die Orientierung und Leitwirkung auf Wegen (z. B. Wanderungen oder touristische Routen in einer Stadt) oder die menschliche Reaktion auf Angsträume oder Lärm im urbanen Kontext. Der Beitrag stellt ein Teilergebnis des von der DFG geförderten Projektes „Städtebauliche Methodenentwicklung mit GeoWeb und Mobile Computing“ dar.

2 Stand der Forschung

Das folgende Kapitel beschreibt, inwiefern Emotionen über humansensorische Parameter messbar sind und welche Forschungsansätze in dieser Sparte bis dato durchgeführt wurden.

2.1 Raumsensorik

Zur Verwendung humansensorischer Daten in der räumlichen Planung ist es zuerst notwendig, Systeme für die Aufnahme, Speicherung und Analyse dieser Daten zu entwickeln. Die Art der Distribution der Daten ist dabei zweitrangig: Sie können sowohl in analoger Form über eine Publikation als auch über Web 2.0 oder diverse Geowebtechniken geschehen (ZEILE, 2011). Beispiele hierfür sind die Urban Sensing Systeme (CAMBELL, 2006, LANE et al., 2008), bei denen die Nutzer Informationen sammeln und gleichzeitig mit der Community teilen. Dies ist möglich, da die Vision des Ubiquitous Computings (WEISER, 1991) durch die allgemeine Datenverfügbarkeit und -Zugänglichkeit mittlerweile Realität geworden ist. Sensornetzwerke sind somit nicht mehr statisch, sondern dynamische Systeme von vernetzten Knoten. Sie sind mobil und immer dort verfügbar, „wo sich das Leben abspielt“ (WETTER, 2009, 6). Auch können urbane Daten für planerische Zwecke nahezu in Echtzeit verarbeitet werden (RESCH et al., 2012). Crowdsourcing-Ansätze, die die sogenannte „Weisheit der Massen“ (O'REILLY, 2003) bündeln, helfen hierbei genauso wie das Konzept, den Menschen als Sensors zu betrachten (GOODCHILD, 2007 und ZEILE et al., 2009).

Dem „Multisensor“ Mensch fällt dabei also eine große Bedeutung bei der Erfassung seiner Umwelt als aktiver Sensor zu. In Kombination mit den Sensoren des Smartphones ergibt sich so ein neues System aus intelligenten und smarten Sensoren (ZEILE, 2011). Neben klassischen Messparametern wie der Temperatur können indirekte Messgrößen wie das persönliche Wohlempfinden – ermittelt durch ein Scoring der Hautleitfähigkeit und der Hauttemperatur – vor Ort gesammelt werden und durch die Sende- und Empfangsmöglichkeit mobiler Kommunikationsgeräte einem größeren Netzwerk oder Community zu Verfügung gestellt werden.

2.2 Humansensorik und Emotionsmessung

Doch wie können menschliche Emotionen in einem objektivierbaren Ansatz aufgenommen und analysiert werden? Ein möglicher Ansatzpunkt sind die „Affektiven Wissenschaften“, die die Auswirkungen von Emotionen auf das menschliche Verhalten und die Gesellschaft erforschen (AFFEKTIVE WISSENSCHAFTEN, 2011). Damit könnte die Methoden wichtige Informationsgrundlagen für die räumliche Planung liefern. Zur Messung und Feststellung unbeeinflusster Emotionen sind zwei Parameter notwendig (BERGNER et al., 2011):

- Die Aufnahme von (affektiven) Emotionen muss in Echtzeit und zu jedem Zeit-punkt gewährleistet sein, um bestimmte Vorortsituationen identifizieren zu können.
- Die Messung darf den Betroffenen weder behindern noch stören, um die Erhebung kognitiv beeinflusster Emotionen zu verhindern.

Emotionale Reaktionen können durch die Veränderung in der Aktivität des autonomen Nervensystems identifiziert werden (KREIBIG, 2010). Dies belegen Forschungsergebnisse aus der Emotionsforschung und den affektiven Wissenschaften: „Diese nervlichen Reaktionen schlagen sich in spezifischen, physiologischen Parametern (wie Hautleitfähigkeit und Hauttemperatur) nieder, klassifiziert als Vitaldaten der Betroffenen. In Verbindung mit der Lokalisierbarkeit mithilfe von GPS-Sensoren ist nun möglich, diese Daten zu verorten. Die dazu zugehörige Methode wird als psychophysiologisches Monitoring bezeichnet. Mithilfe neuer Technologien im Bereich des „Ambulatorischen Assessments", unterstützt durch „wearable computers", ist eine Erhebung der personenbezogenen Vitaldaten in Echtzeit möglich" (EXNER et al., 2012, nach KREIBIG, 2010). Technisch ist diese Messung mit dem „BMS Smartband" (PAPASTEFANOU, 2009) möglich, einem Sensorarmband, das physiologischer Parameter aufnimmt. Durch die Messung dieser physiologischen Parameter kann die mentale Belastung, also der Stress, ermittelt werden. Stress wird in diesem Zusammenhang als das emotionale Konstrukt der höchst negativen Emotionen „Ärger" und „Angst" verstanden (BERGNER et al., 2011). Sind diese Stresssituationen noch im Raum verortbar, eröffnen sich neue Perspektiven des Raumverständnisses und dementsprechend auch für die räumliche Planung. Erste Ansätze in diesem Kontext liefern die Arbeiten im Rahmen des Biomapping von Christian Nold (NOLD, 2009), die Emomap-Studie (ZEILE, 2010) und das EmBa-GIS (Emotionales Barriere-GIS) (BERGNER et al., 2011).

3 Methoden

Mit den nachfolgenden Methoden ist es möglich, Daten des Sensors „Mensch" zu erfassen und anschließend eine Analyse durchzuführen. Zusätzlich werden Techniken vorgestellt, mit denen eine einfache und verständliche Visualisierung möglich ist. Dementsprechend ist der nächste Abschnitt unterteilt in die grundlegende Methode des psychophysiologischen Monitorings und weitere Methoden der Analyse und Visualisierung unterteilt. Das Ergebnis dieser zusammengefügten Bausteine stellt die Humansensorik in der räumlichen Planung dar.

3.1 Psychophysiologisches Monitoring

Die Emotionen eines Menschen sind durch das psychophysiologische Monitoring extrahierbar. Unter Monitoring wird die systematische Beobachtung eines Phänomens über eine Zeitachse verstanden. Dabei kann im Sinne der räumlichen Planung in deduktive und induktive Monitoringsysteme unterschieden werden (STREICH, 2011). Die Messung geschieht in Echtzeit und sammelt Daten der sich verändernden Körperreaktionen (Körperphysiologie) eines Probanden. Eindeutig identifizierbar sind negative Emotionen, bei denen aggregiert auch von „Stress“ gesprochen werden kann. „In Übereinstimmung mit Emotionsforschern liegt eine negative Erfahrung dann vor, wenn die elektrodermale Aktivität zunimmt und kurz danach die Hauttemperatur abnimmt“ (BERGNER et al., 2011, 435). Erfährt der Proband nun eine negative Emotion, wie Ärger oder Angst, nimmt bei ihm die Hautleitfähigkeit (Differenz zwischen Schweißproduktion und Schweißabsorption der Haut) zu und die Hauttemperatur in den Extremitäten ab. Ein bekanntes Beispiel ist die Situation einer extremen mentalen Belastungssituation, der auch als der „kalte Angstschweiß“ bekannt ist (BERGNER & ZEILE, 2012). Für eine bessere Identifizierung des Stressereignisses wird für die Verlaufskurven die “Erste Ableitung“ gebildet, um die jeweilige Steigung oder Gefälle zu ermitteln.

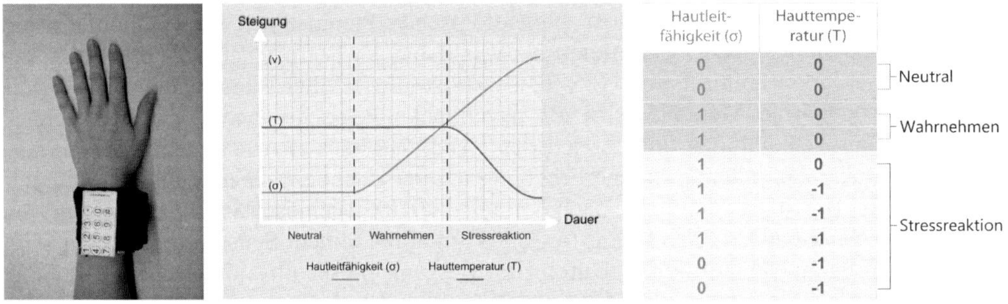

Abb. 1: Links: BMS-Smartband. Mitte: idealtypischer Kurvenverlauf bei einer Stressreaktion. Rechts: Ableitung der Kurvenverläufe; bei Auftreten des markierten Musters kann auf eine Stressreaktion geschlossen werden.

Das anschließende Scoring folgt nach dem Muster, dass die Zunahme der Hautleitfähigkeit den Scoring-Wert „+1“ erhält, die Abnahme der Hauttemperatur wird mit „–1“ klassifiziert (BERGNER et al., 2011). Bei Auftreten dieses Musters (s. Abb. 1) liegt eine negative Emotion vor. GPS-Daten zur Positionsbestimmung können über die Laufzeit des GPS-Signals mit den Messwerten zeitlich synchronisiert und dementsprechend auch verortet werden. Die Daten liegen anschließend als Tabelle im CSV-Format zur Weiterverarbeitung bereit.

3.2 Datensammlung und Visualisierung

Der wichtigste Schritt für die Auswertung der Emotionsdaten der BMS-Smartbänder ist die schon vorher erwähnte Auswertung der Stresssignale und die Verortung der Messwerte. Dabei werden die Messparameter Hautleitfähigkeit (elektrodermal) und Hauttemperatur (kardiovaskulär) mit der Geo-Position über einen GPS-Logger oder über ein Smartphone gekoppelt. Die erhobenen Vitaldaten Hautleitfähigkeit und Hauttemperatur unterlaufen danach einer statistischen Bereinigung. Dies dient der Eliminierung von Messartefakten und der Glättung der Daten. Erst danach kann die Auswertung durchgeführt werden. Zusätzliche Methoden können zur Verifizierung und retrospektiven Analyse der einzelnen Läufe noch nach Bedarf hinzugefügt werden. Dies sind im Folgenden: Fragebögen vor dem Lauf zur Ermittlung der allgemeinen Verfassung der Probanden und retrospektiv zur Erkennung von subjektiven Stresssituationen, die anschließend mit den ermittelten Daten verglichen werden. Alle Läufe werden zusätzlich aus der Ego-Perspektive per Kamera (wie z. B. der GoPro) mitgeschnitten, um Blickrichtung und Wahrnehmung sowie die Wechselwirkung zwischen Proband und Umwelt zu erkennen. Zusätzlich steht auch die Smartphone App „RADAR Sensing“ zur Verfügung, bei denen die Probanden GPS-gestützt positive oder negative Eindrücke direkt verorten können (vgl. hierzu ZEILE et al., 2012), die anschließend direkt über die RADAR-Infrastruktur (MEMMEL & GROSS, 2012) in Form von Heatmaps/Density Maps ausgewertet werden können.

Weitere Visualisierungsmöglichkeiten sind in einer einfachen Variante der GPS-Visualizer (SCHNEIDER, 2011) oder der neu entwickelte GeoVisualizer (vgl. hierzu z. B. MICHEL et al., 2013). Auch eine filmbasierte individuelle Auswertung ist möglich (EXNER et al., 2012). „Das Gesamtergebnis [also die Humansensorik in der räumlichen Planung] repräsentiert somit einen Methodenmix aus deduktiv und induktiv erhobenen Raumdaten sowie aus empirischen Individualdaten“ (EXNER et al., 2012). Der Vorteil des Ansatzes ist, dass er ähnlich des Web 2.0 Gedankens mashup-fähig ist, d. h., einzelne Bausteine können immer wieder in neuen Kombinationen, angepasst auf die Fragestellung, fungieren.

4 Techniken für die Humansensorik

Im folgenden Kapitel werden die Grundzüge der jeweiligen Techniken kurz erläutert. Jede für sich liefert ein gewünschtes Teilergebnis im Zusammenhang der Humansensorik und der räumlichen Planung. Die Vision besteht zweifelsohne, diese noch solitären Techniken in Zukunft unter dem Dach einer Applikation zu bündeln.

4.1 Geoanalytische Visualisierung

Ein Online-Tool für die Visualisierung der Emotionsdaten im zeitlichen Verlauf ist der GPS-Visualizer. Er kann GPS-Daten aufgeschlüsselt nach Punkten, Tracks sowie die Höhe des Signals als KML-Datei für Google Earth visualisieren. Zur Visualisierung der Emotionsdaten muss dazu lediglich die Tabellenspalte „Höhe“ für den Verlauf der Emotionskurve in der CSV-Datei angepasst werden. Für jeden Lauf muss die Prozedur auch manuell wiederholt werden. Für eine erste Visualisierung ist die Technik jedoch sehr empfehlenswert.

Abb. 2: Bilder aus der Echtzeitumgebung des GeoVizualiser eines Testlaufes zum Thema „Wandern und Orientierung“. Der rote Kopf des Wanderers deutet auf einen Stresspunkt im Datensatz hin (MICHEL et al., 2013a).

Eine Automatisierung dieses Vorgangs in Kombination mit weiteren Geodiensten bietet der GeoVisualizer. Dabei wird auf aktuelle Visualisierungskonzepte und -methoden aus dem Bereich Computergrafik (der Informationsvisualisierung) zurückgegriffen, die nun durch eine einfache Benutzeroberfläche auch von Nicht-Visualisierungs-Experten genutzt werden kann. Eine aussagekräftige Repräsentation ist möglich, um einen schnellen Eindruck der zugrundeliegenden georeferenzierten Daten zu bekommen (MICHEL et al., 2013a).

Der GeoVisualizer basiert auf dem NASA World Wind SDK und greift auf die 3D Virtual Globe API zurück. Weiterhin kommt die Java Web Start Technologie zum Einsatz, die auch offline nutzbar ist. Das System besitzt eine flexible und erweiterbare Anwendungsarchitektur zur Darstellung von georeferenzierten 3D-Daten im Zeitverlauf. Animationen als auch die interaktive Exploration und Analyse der Daten sind dadurch möglich. Weitere Features sind 3D-Stereodarstellung sowie der „Vizualisation Wizard VizWiz“, mit dem die Darstellung der Kurven angepasst werden kann (STEFFEN et al., 2013b).

Die manuelle Auswertung der Emotionsdaten sowie die retrospektive Ermittlung der Filmabschnitte, in denen eine Emotion aufgetreten ist, stellte bisher den schwierigsten Bereich der vorgestellten Techniken dar. Der Zeitaufwand war für die Grundlagenforschung noch zu rechtfertigen. Eine Anwendung mit einer repräsentativen Probandenmenge ist in einem kurzen Zeitrahmen jedoch noch nicht möglich gewesen. Mithilfe der RADAR-Plattform, die auf die ALOE-Infrastruktur (MEMMEL & SCHIRRU, 2007) aufbaut, konnten folgende Aufgaben beschleunigt werden: (1) Automatische Synchronisation der Datensätze von GPS-Tracker, Kamera und Smartband, (2) Identifizierung der Stress Phasen eines GPS-Tracks und (3) die automatische Extraktion von Film-Snippets zum jeweiligen Stresspunkt. Die identifizierten Stresspunkte werden automatisch in einer Geo-Datenbank gespeichert und anschließend als aggregierte Heatmap visualisiert.

Mithilfe von Perl-Skripten, die auf jedem gängigen Betriebssystem lauffähig sind, ist eine web-basierte Verarbeitung der Daten möglich. Die Daten werden automatisch bereinigt und synchronisiert. Fehlen auf der Route Signale des GPS-Trackers, so hilft das Skript "GPS Interpolation“, das automatisch fehlenden Positionsdaten unter Berücksichtigung der letzten und nächsten bekannten Position und Zeitstempel einer identifizierten Lücke interpoliert.

Abb. 3: Links: Identifizierte Stresspunkte eines individuellen Laufes. Rechts: die Heatmap aller Läufe. Ort der Untersuchung: Alexandria, Ägypten.

Nach der Aufbereitung der Daten wird der „StressPhaseIdentifier“ angewandt, um die in Kap. 3.1 erläuterten Stressphasen in den physiologischen Daten zu identifizieren. Zusätzlich können für jeden Track alle identifizierten Stressphasen, statistischen Informationen sowie die durchschnittliche Länge einer Stressphase als Übersicht dargestellt werden. Zu jeder Stressphase extrahiert das Skript auch automatisch die entsprechende Filmsequenz mittels der „ffmpeg-Bibliothek“. Abschließend können alle Läufe mit den identifizierten Stressphasen durch die "GenerateHeatmapInput" Komponente zu einer Dichteberechnung der Häufung der Stresspunkte zusammengefasst und visualisiert werden. Grundlage hierfür ist eine JSON-Darstellung zur Generierung der Heatmaps im RADAR-System (Abb. 3, rechts) verwendet werden.

4.2 Videobasierte Auswertung

Die filmbasierte Auswertung ist eine qualitative Analyse, mit der die jeweiligen stressauslösenden Elemente retrospektiv abgeglichen werden. Durch die Zusammenführung der Messdaten in einem Video erfolgt wird die qualitative Auswertung der zurückgelegten Wegstrecke erleichtert. Die Wechselwirkungen zwischen Stressreaktion und Situation können so schneller analysiert und nachvollzogen werden. Qualitativ kann so geklärt werden (EXNER et al., 2012): Wann und wo eine Stressreaktion eingetreten ist? Wie (stark) eine Stressreaktion eingetreten ist? Welche Faktoren diese Stressreaktion möglicherweise ausgelöst haben?

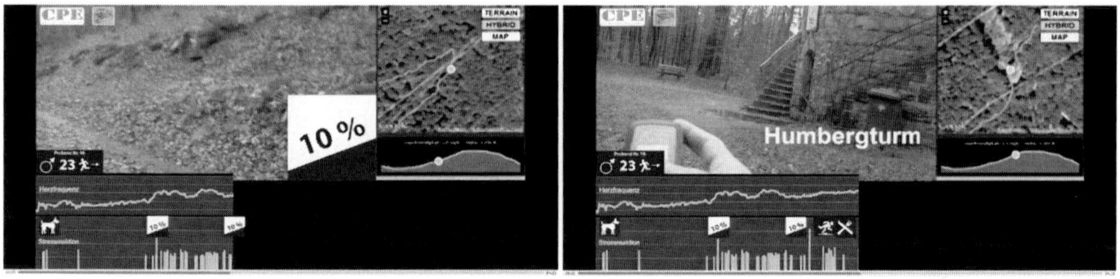

Abb. 4: Videoanalyse (EXNER et al., 2012, Screendesign: D. Broschart)

Wie in Abbildung 4 zu sehen ist, werden so gleichzeitig Herzfrequenz, Hautleitfähigkeit und Stressreaktionen mit GPS- und Videosignal dargestellt. Bestimmte Situationen (schlechte Wegoberfläche, Orientierungsprobleme oder Ähnliches) sind so besser zu erkennen. Ein Vorteil dieses Arbeitsschrittes ist, dass die Probanden ein „personal feedback“ bekommen, wie sie die Situation erlebt haben. Dies ist insofern wichtig, da die Teilnehmer einer solchen Studie dadurch besser zu motivieren sind. Nachteilig bei dieser Technik ist die lange Nachbereitungszeit.

4.3 Taggen mit RADAR Sensing

Zur Realisierung des Taggings (digitales Markieren eines Ortes auf einer Karte) und der damit zusammenhängenden Bewertung von speziellen urbanen Attributen wie bspw. Stadtgestalt oder Verkehr in „gut“ und „schlecht“, wurde eine bestimmte Instanz der RADAR-Infrastruktur – das RADAR-SENSING – entwickelt. Die RADAR SENSING App läuft auf android-basierten Geräten, und verbindet sich mit dem RADAR-Sensing-Backend, wo alle Informationen wie Benutzerdaten, die Beiträge der Benutzer, etc. in einer entsprechenden Datenbank gespeichert werden. Als Austauschformat wurde wiederum JSON gewählt. Nach einer Registrierung im RADAR-SENSING Web Frontend und der Installation der RADAR SENSING App können die User positive und negative Bewertungen für eine Reihe von vordefinierten Kategorien abgeben. Optional kann auch eine freie Kategorie eingegeben werden. Daraufhin lokalisiert die App die Position des Benutzers mithilfe von mehreren Sensoren (vorzugsweise GPS) und zeigt diese Position auf einer Karte an. Falls eine Lageungenauigkeit besteht, haben die Benutzer die Möglichkeit, diese Position zu ändern. Interessant für die planenden Disziplinen ist hierbei wiederum die interaktive Heatmap der markierten Punkte. Der Benutzer kann dort wählen, welche Attribute für die Erzeugung der Heatmap berücksichtig werden. Weiterhin ist es möglich, sich auch nur eine gewisse Zeitspanne anzeigen zu lassen oder die Generierung der Heatmap in Echtzeit zuzulassen.

5 Projekte

Nachfolgend werden einige Projekte aufgezeigt, die das Thema der Humansensorik und der dadurch erstellten Karten (umgangssprachlich „emomaps“) im Einsatz für die räumliche Planung näher erläutern.

5.1 Analyse von Barrierefreiheit

Das Thema „Barrierefreiheit“ erlangt im Rahmen von Neuplanungen eine immer stärkere Bedeutung. Menschen sollen unabhängig von einer eventuell vorhandenen Behinderung oder Mobilitätseinschränkung Gegenstände, Medien und Infrastruktur uneingeschränkt nutzen können. Bei der Gestaltung des öffentlichen Raumes verschiedene Probleme der „Accessibility" ergeben. Eine neue und objektive Form der Messung der Barrierefreiheit von Infrastrukturen liegt im oben beschriebenen Ansatz des „emotional mappings“ oder des „Menschen als Messfühlers“ an. Ein Experiment zur Barrierefreiheit wurde in Kirchheimbolanden, mit einer Elektro-Rollstuhlfahrerin, einer Sehbehinderten, Kinderwagenfahrer und einer „non-handicapped“ Referenzperson durchgeführt (vgl. zu dieser Studie ausführlicher BERGNER & ZEILE, 2012).

Nach einer städtebaulichen Aufwertung wurde eine bestimmte Route nochmals auf ihre Barrierefreiheit hin überprüft: Zum einen wurde eine Untersuchung nach DIN 18024 „Barrierefreies Bauen“ und DIN 32984 „Bodenindikatoren im öffentlichen Raum“ als Referenzvergleich durchgeführt, zum anderen konnten Stresspunkte der Barrierefreiheit im öffentlichen Raum ermittelt werden.

Innerhalb des Stadtraumes konnten so eine Reihe von Hotspots identifiziert werden und wurden anschließend nochmals vor Ort anhand der DIN-Vorschriften überprüft, um abzugleichen, ob entweder ein Planungsfehler vorliegt, die Barriere eventuell temporär oder allein persönlicher Natur war. Beispielsweise war ein Sektor im Südwesten des Untersuchungsraums eine Fußgängerzone, die durch ihr unterschiedliches Kopfsteinpflaster und temporären Barrieren wie Außenaufstellern des Einzelhandels, in der die Probanden mehr oder weniger stark beeinflusst wurden. Je nach Art der Behinderung lassen sich individuelle Stresskarten erstellen (Abb. 5). Die Ergebnisse wurden anschließend öffentlich präsentiert und diskutiert.

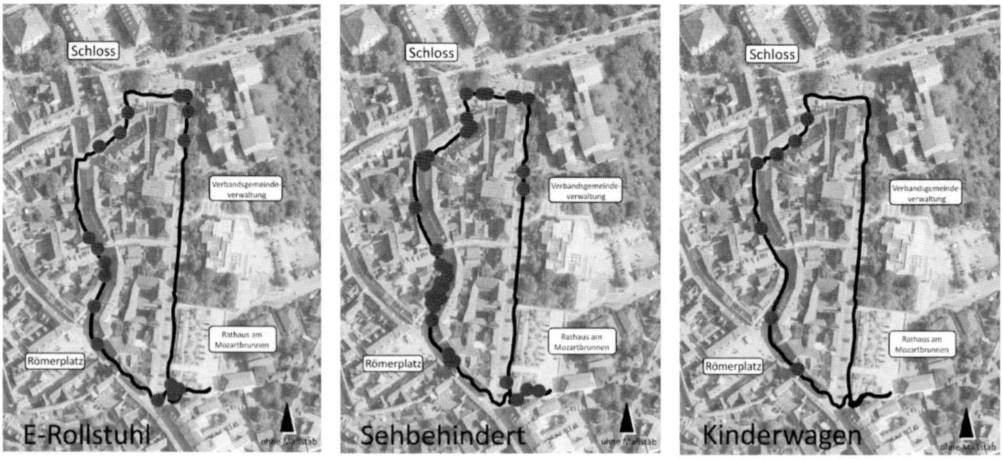

Abb. 5: Ermittelte Stresspunkte von E-Rollstuhlfahrerin, Sehbehinderte, Mutter mit Kinderwagen (BERGNER & ZEILE, 2012)

5.2 Orientierung auf Wanderwegen

In diesem Projekt sollten sowohl der Erholungswert, die Orientierung als auch allgemein die Qualität von Wanderwegen, hier im Speziellen eines „Premium Wanderweges“, untersucht werden. 22 Studierende sollten anhand einer offiziellen Karte eine ihnen unbekannte Route durch den Pfälzer Wald erwandern. Der Weg ist gekennzeichnet durch lichten und dunklen Wald, verschiedene Steigungs- und Gefällesituationen, Wegbreiten, Bodenbeschaffenheiten sowie einen markanten Aussichtspunkt und gilt damit als „abwechslungsreich“ laut den Qualitätskriterien „Qualitätsweg Wanderbares Deutschland“ (vgl. EXNER et al., 2012).

Abb. 6: Physisches Belastungs- und Erregungsniveau eines Probanden, erstellt mit GPS-Viualizer (rechts), Dichtekarte Stresssignale und markierte fehlende Wegemarkierungen mit blauem Pfeil identifiziert (EXNER et al., 2012)

Für die Analyse wurde hier eine kartenbasierte Darstellung der Stresspunkte der Testpersonen herangezogen. Unter Zuhilfenahme aller Läufe kann eine Dichtekarte aller Stressreaktionen erstellt werden. Die Abbildungen 6 zeigen den Hautleitfähigkeitslevel einer Probandin sowie auch eine Dichtekarte aller ausgelösten Stresssignale während des circa einstündigen Parcours. Die Hautleitfähigkeitskurve (Abb. 6 links) beschreibt das physische Belastungs- und Erregungsniveau der Probandin. Die Dichtekarte (Abb. 6 rechts) beschreibt die Überlagerung aller gemessenen Stresssignale aller Probanden auf dem Parcours. Gleichzeitig können die einzelnen Läufe entweder in Echtzeit mit dem GPS-Visualizer (Abb. 6 links) oder mit der retrospektiven Videoanalyse (Abb. 3) noch auf die speziellen Eigenheiten hin untersucht werden. Schlecht markierte Weggabelungen (Abb. 6. rechts; markiert mit Pfeil) und Hindernisse lösten die meisten negativen Reaktionen aus. Innerhalb des Experimentes konnte anhand der Daten beobachtet werden, wie sich ein Proband verläuft, was Stress auslöste (für mehr Ergebnisse s. EXNER et al., 2012).

5.3 Wirkung von Lärm

Das Ziel dieser Studie (BERGNER et al., 2012) war es, Rückzugs- und Erholungsgebiete vor dem Hintergrund der EU-Umgebungslärmrichtlinie mithilfe von psychophysiologischen Messungen zu identifizieren. Die Hypothese dahinter war, dass in einer Stadt qualitative Rückzugsgebiete mit einem hohen Erholungswert auch durch qualitativ hochwertige Gestaltung entstehen, obwohl der Lärm die in der Richtlinie festgelegten Grenzwerte übersteigt.

Folgende Parameter wurden in der Studie verglichen: Eine prognostizierte Lärmbelastung (quantitative Modellierung mit Immissionsschutzsoftware), die tatsächliche Lärmbelastung (quantitative und qualitative Messung mit Lärmpegel-Messgerät), die emotionale Belastung mittels Smartband, subjektive Eindrücke über die RADAR SENSING App sowie das subjektive Empfinden über einen ICBEN-Fragebogens (International commission on biological effects of noise), der die Wahrnehmung der Lärmbelastung beschreibt. Alle Probanden wurden während der Messung zusätzlich mit Kamera aufgezeichnet. Erster Schritt war eine Freiraumanalyse der Stadt Kaiserslautern, um potenzielle Bereiche mit hoher Eignung für ruhige Erholung zu identifizieren.

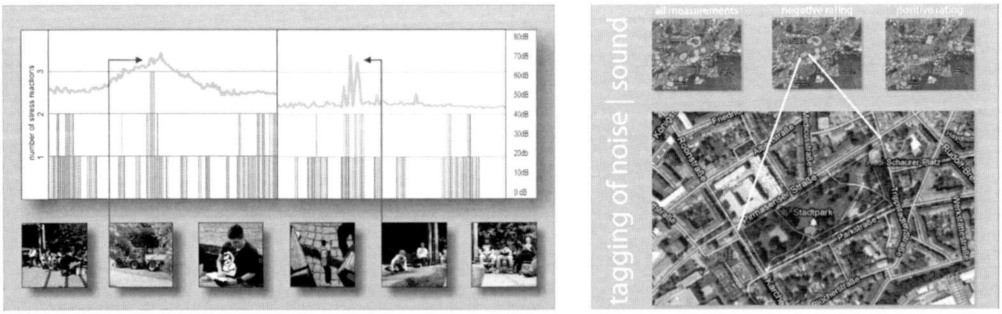

Abb. 7: Qualitative Analyse der Geräusche an der Messstation „Volkspark" (BERGNER et al., 2012) und getaggte Eindrücke der Probanden zum Thema Lärm innerhalb der Stadt (ZEILE et al., 2012)

In der Phase der Datenerhebung mit dem SMART-Band wurde zusätzlich der Lärm vor Ort gemessen. Interessant hierbei war, dass nicht eine dauerhaft kontinuierliche hohe Lärmbelastung als störend empfunden wurde, vielmehr waren punktuelle Lärmereignisse problematisch. Bei einer relativ hohen Grundbelastung war z. B. eher ein nochmals verstärkendes Geräusch (Rasenmäher) Stressauslöser (Näheres zur Studie bei BERGNER et al., 2012). Gleichzeitig lieferte die app-gestützte Analyse über die RADAR SENSING App weitere wichtige Hinweise, welche Punkte innerhalb der Stadt die Studienteilnehmer positiv als auch negativ im Kontext der selbstbewerteten Lärmbelastung identifiziert haben.

6 Fazit und weiterer Forschungsbedarf

Der hier vorgestellte Ansatz des Einsatzes der Humansensorik in der räumlichen Planung mit den zugehörigen Techniken und den realisierten Projekten zeigt deutlich auf, welche Potenziale in diesem zwar noch experimentellen, aber menschzentrierten Ansatz liegen. Die Einsatzbereiche sind vielfältig und bis dato nicht abschließend formuliert. Gerade im Bereich der Landschaftsarchitektur, bei der Aufwertung von ungeordneten räumlichen Situationen kann dieser Ansatz eine Erfolgskontrolle in Form eines Vorher-Nachher Vergleichs darstellen (räumliches Monitoring).

Stärken besitzt die Humansensorik vor allem in den Bereichen der Überprüfung von Leitsystemen und der Orientierung im Raum, aber auch im Bereich der Verkehrsplanung. Die Experimente zum Thema Barrierefreiheit zeigen auf, dass die erlebten (und gefühlten) Beeinträchtigungen im Stadtraum besonders gut zu messen und zu identifizieren sind. Schwierige Kreuzungssituationen mit dem Begegnungsfall Kfz-Mensch konnten so nachgewiesen werden. Schwächen des Ansatzes liegen bis dato immer noch in der noch ausbaufähigen Auswertungsprozessen. Jedoch zeigen die oben beschriebenen technischen Lösungswege in die richtige Richtung. Die Bearbeitungsdauer konnte bei gleichzeitiger Produktion einer größeren Quantität und Qualität der Daten so schon erheblich verkürzt werden. Für validere Ergebnisse und aufgrund der komplexen Wechselwirkungen zwischen Raum und Mensch, besteht die Notwendigkeit die Teilnehmerzahl für weitere Studien die Teilnehmerzahl deutlich zu erhöhen. Das zukünftige Ziel wird sein, Datensätze effektiver zu analysieren und mittelfristig die Ergebnisse in Echtzeit zu verarbeiten und zu visualisieren. Die Bausteine dazu sind definiert, deren Zusammenführung bestimmt die nächsten Arbeitsschritte.

Literatur und Internetquellen

Affekte Wissenschaften (2011), http://www.affective-sciences.org/themes-common/public_img-files/pdf/NCCR_FLYER_D.pdf.

Bergner, B. S., Zeile, P., Papastefanou, G. & Rech, W. (2011), *Emotionales Barriere-GIS als neues Instrument zur Identifikation und Optimierung stadträumlicher Barrieren.* In: Strobl, J., Blaschke, T. & Griesebner, G. (Hrsg.), Angewandte Geoinformatik 2011. Berlin/Offenbach, Wichmann Verlag, 430-439.

Bergner, B., Exner, J.-P., Zeile, P. & Rumberg, M. (2012), *Sensing the City – How to identify Recreational Benefits of Urban Green Areas with the Help of Sensor Technology.* In: Schrenk, M., Popovich, V. & Zeile, P. (Eds.), Proceedings REAL CORP 2012. Wien.

Bergner, B. & Zeile, P. (2012), *Ist Barrierefreiheit messbar?* In: PlanerIn – Thema Gesundheit, 2/2012.

Campell, A. T., Eisenmann, S. B., Lane, N. D., Miluzzo, E. & Peterson R. A. (2006), *People-centric urban sensing.* In: Proceedings of the 2nd annual international workshop on Wireless internet. Boston, Massachusetts: ACM, 18.

Exner, J.-P., Bergner, B. S, Zeile, P. & Broschart, D. (2012), *Humansensorik in der räumlichen Planung.* In: Strobl, J., Blaschke, T. & Griesebner, G. (Hrsg.), Angewandte Geoinformatik 2012. Berlin/Offenbach, Wichmann Verlag, 690-699.

Goodchild, M. F. (2007), *Citizens as Sensors: the World of Volunteered Geography.* In: Geo-Journal, 69 (4), 211-221.

Kreibig, S. D. (2010), *Autonomic nervous system activity in emotion: A review.* In: Biological Psychology, 84 (3), 394-421.

Lane, N. D, Eisenman, S. B., Musolesi, M., Miluzzo, E. & Campbell, A. T. (2008), *Urban sensing systems: opportunistic or participatory?* In: Proceedings 9th workshop on Mobile computing systems and applications. Napa Valley, CA: ACM, 2008, 11-16.

Memmel, M. & Groß, F. (2011), *RADAR – Potentials for Supporting Urban Development with a Social Geocontent Hub.* In: Schrenk, M., Popovich, V. V. & Zeile, P. (Eds.), Proceedings REAL CORP 2011, Schwechat, 777-784. ISBN 978-3-9503110-1-3.

Memmel, M. & Schirru, R. (2007), *ALOE – A Socially Aware Learning Resource and Metadata Hub.* In: Wolpers, M., Klamma, R. & Duval, E. (Eds.), Proceedings of the EC-TEL 2007 Poster Session. CEUR workshop proceedings.

Michel, F., Steffen, D., Bergner, B., Exner, J.-P. & Zeile, P. (2013), *A New Approach in the Visualization of Georeferenced Sensor Data in Spatial Planning.* In: Schrenk, M., Popovich, V. V., Zeile, P. & Elisei, P. (Eds.), Proceedings REAL CORP 2013.

Michel, F., Steffen, D., Bergner, B., Exner, J.-P. & Zeile, P. (2013), *GeoVisualizer – Ein neuer Ansatz zur Visualisierung von Sensordaten im raumplanerischen Kontext.* AGIT 2013.

Nold, C. (2009), *Emotional Cartography – Technologies of the Self.* Published under Creative Commons License. ISBN 978-0-9557623-1-4S.

O'Reilly, T. (2005), *What is the Web 2.0? Design Patterns and Business Models for the Next Generation of Software.*

Papastefanou, G. (2009), *Ambulatorisches Assessment: eine Methode (auch) für die Empirische Sozialforschung.* In: Weichbold, M., Bacher, J. & Wolf, C. (Hrsg.), Umfrageforschung: Herausforderungen und Grenzen. Österreichische Zeitschrift für Soziologie, Sonderheft, 9. Wiesbaden, VS Verl. für Sozialwiss., 443-469.

Resch, B., Zipf, A., Breuss-Schneeweis, P., Beinat, E. & Boher, M. (2012), *Towards the Live City – Paving the Way to Real-time Urbanism.* In: International Journal on Advances in Intelligent Systems, 5 (3&4), 470-482.

Schneider, A. (2011), *GPS-Visualizer.* www.gpsvisualizer.com.

Steffen, D., Michel, F., Bergner, B., Exner, J.-P. & Zeile, P. (2013), GeoVisualizer – Ein neuer Ansatz zur Visualisierung von Sensordaten im raumplanerischen Kontext. AGIT 2013.

Streich, B. (2011), *Stadtplanung in der Wissensgesellschaft: ein Handbuch*. 2. Auflage. Wiesbaden, VS Verlag.

Weiser, M. (1991), *The Computer for the 21st Century*. Scientific American, 265.

Wetter, M. (2009), *Sensorbasierte Datenerfassung im Dienst der Gesellschaft*. Fachseminar Verteilte Systeme, ETH Zurich's Distributed Systems Group, Zürich. http://www.vs.inf.ethz.ch/edu/FS2009/DS/berichte/ds2009_matthiaswetter.pdf (Zugriff: 26.7.2010).

Zeile, P. (2010), *Echtzeitplanung – Die Fortentwicklung der Simulations- und Visualisierungsmethoden für die städtebauliche Gestaltungsplanung*. Dissertation, TU Kaiserslautern.

Zeile, P. (2011), *Städtebauliche Methodenentwicklung mit GeoWeb und Mobile Computing – Untersuchung über die Fortentwicklung des städtebaulichen und raumplanerischen Methodenrepertoires angestoßen durch technologische Neuerungen im Internet*. Weblog des Forschungsprojektes. TU Kaiserslautern, Fachgebiet CPE, Prof. Streich, Kaiserslautern, 2011. http://geoweb.arubi.uni-kl.de/ (Zugriff: 5.3.2013).

Zeile, P., Exner, J.-P. & Streich, B. (2009), *Humans as Sensors? The Measurement of Physiological Data in Urban Areas and Its Possible Use for Urban Planning*. 11th Inter-national Conference on Computers in Urban Planning and Urban Management (CUPUM), Hong Kong.

Zeile, P., Memmel, M. & Exner, J.-P. (2012), *A new urban sensing and monitoring approach: Tagging the city with the RADAR SENSING App*. In: Schrenk, M., Popovich, V. V. & Zeile, P. (Eds.), Proceedings REAL CORP 2012. Wien.

Geodesign Applications

Cross-Case Comparison of Four Climate-Related Geodesign Processes over Time

Olaf SCHROTH, Ellen POND and Stephen SHEPPARD

1 Introduction

In the presented ***multiple-case study***, we analyzed collaboration processes over three years in four landscape-related planning processes in Canada that address climate change using geodesign tools including environmental modeling and 3D landscape visualization (POND et al., 2012). The case study sites were located in Greater Toronto, Metro Vancouver, the Calgary/Elbow River watershed, and Clyde River, Nunavut. The multiple-case study shows that the underlying Local Climate Change Visioning (LCCV) framework (POND et al., 2010) was successfully applied to local climate change adaptation and mitigation and gives insights in the adaptation of a planning framework to local circumstances over time.

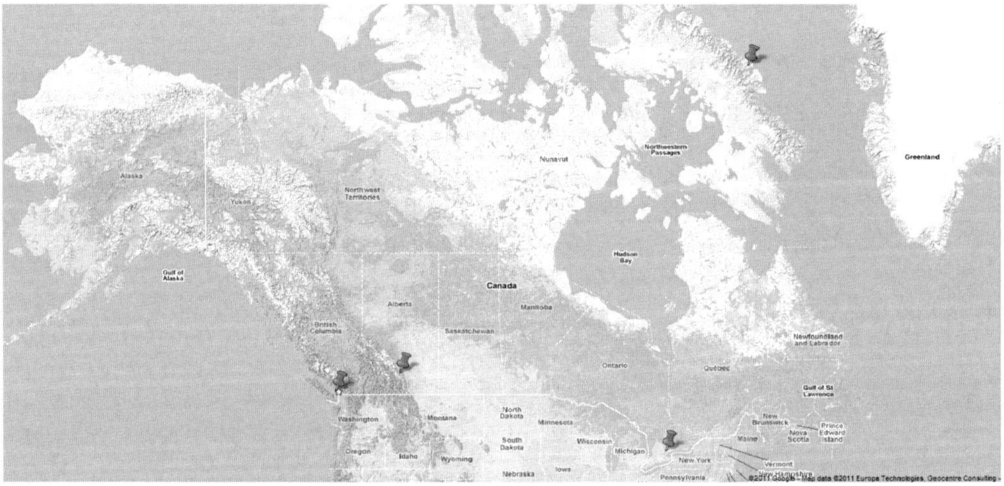

Fig. 1: Locations of the four case study sites (source: Google Maps)

Although the term ***geodesign*** was not used in the case study areas, it well describes the application of geospatial tools in the analyzed design process, which is inherently the idea of geodesign. The terms' origins go back to MCHARG (1969) and his concept of layers. According to SCHWARZ VON-RAUMER & STOKMAN (2012), KUNZMANN (1993) was then the first to use the actual term geodesign in the context of European spatial development. Later, the term was introduced by ESRI with the release of ArcSketch in 2010. The idea of geodesign basically refers to the interdisciplinary collaboration of public stakeholders, natural sciences, design professionals and geospatial information experts (STEINITZ, 2012). Although geodesign could be both, an output and a process, or rather a noun or a verb, Steinitz puts the emphasis on geodesign as a verb. Steinitz further develops a geodesign framework and goes through its iterations, i.e. A) scoping the study, b) designing the study and c) carrying out the study. In the cross-case study presented here, we looked at the third iteration, i.e. the carrying out of a geodesign project. With regard to geodesign case studies, he asks the questions what, where and when regarding the implementation phase and looks at tools, techniques and methods; representation; process; evaluation; change; impact; decision-making; the level of spatial-analytic complexity; ways of designing and visualization and communication.

2 Methods: Local Climate Change Visioning (LCCV)

The case studies in this paper all looked at landscape-related climate change impacts and mitigation and/or adaptation responses on local (e.g. hamlet) to regional (e.g. watershed) scales. All four planning processes were informed by the LCCV framework (Fig. 2., POND et al., 2010), cases differ with regard to landscape types (urban, suburban, rural, Arctic), scale, the nature of climate change impacts, and adaptation and/or mitigation measures. All case studies were part of the GEOIDE network and applied geospatial information systems (GIS).

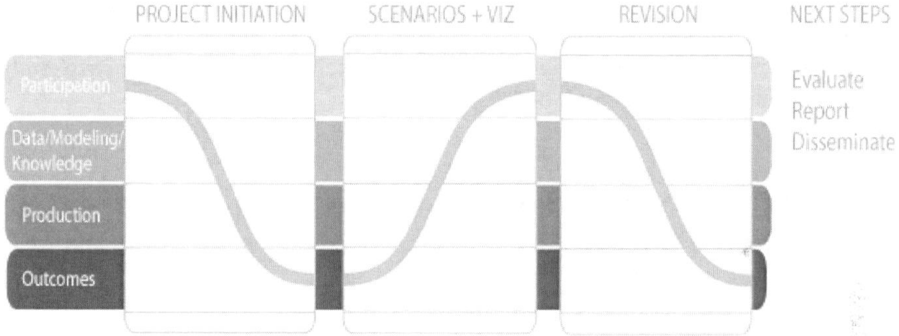

Fig. 2: LCCV process diagram

Although none of the projects used a complete geodesign approach as defined by ERVIN (2012) or STEINITZ (2012), several important geodesign tools and technologies were applied: geodatabases with base data, objects and constraints; collaboration tools; regional environmental models; site-level design-oriented approaches; diagrams; and 3D landscape simulations for contentious issues at hamlet scales (Fig. 3).

Fig. 3: Use of photomaps, 3D models and indicators in the Clyde River case

For the evaluation, the study builds on the case study method (FRANCIS, 2001), qualitative data collection methods, and STEINITZ (2012) geodesign framework for cross-case comparison. On May 18, 2010, researchers from all four case study sites met at the University of Toronto and discussed the collaboration process, the use of geospatial tools, and products and outputs in a workshop. Key part of the workshop was a mapping exercise, in which the researchers drew process diagrams for each case exploring and documenting the collaboration process among researchers, students and local stakeholders over time. Later in 2011, we conducted phone interviews with local stakeholders from three case study areas to evaluate the outputs and outcome of the project.

3 Results of the Comparison of Process Methods through Flowcharts and Timelines

The cross-case comparison exemplifies the trend towards climate change mainstreaming, i.e. the integration of climate change issues into everyday landscape architecture and planning. The cases address traditional landscape related themes (water, infrastructure, housing development, and green roofs), balancing quality of life, design, and mitigation/adaptation. The additional long-term evaluation of the planning outcomes provides further insights into local capacity building with regard to mitigation and adaptation. Key factors are local champions and change agents, who can be supported through the right choice of geodesign tools (please see POND et al. 2012 for a discussion of geodesign tools and collaboration).

Key outcome of the May workshop was a set of collaboratively drawn process diagrams for each case. The diagrams illustrate the different adoptions of the LCCV process and how modeling and visualization tools were integrated. The following diagrams are unitized simplifications of the original sketches.

3.1 Toronto case study: Design-driven non-linear process on neighbourhood to building scales leading to best practise example

In Toronto, the overall project on the city scale (Toronto A)early branched into multiple design-driven projects (Toronto B) on scales of individual lots and buildings (Fig. 4). Adaptation to urban heat islands was a key consideration in the design process and the design interventions targeted the areas where the highest surface heat had been predicted from thermal images (STEINITZ 2012: "anticipatory design"). The branching process is particularly interesting because it allowed working around institutional barriers and opened new opportunities such as implementation of demonstration projects with business partners. In this regard, the Toronto case has proven to be the most flexible among the four cases, although such an approach could easily become fragmented. Hence, Toronto was the only case with an actual demonstration project being implemented on the ground.

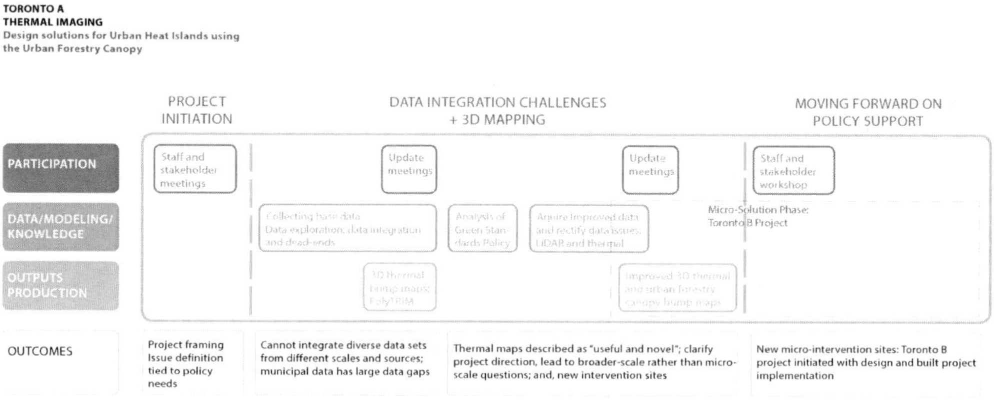

Fig. 4: Toronto process flowchart

3.2 Delta: Stakeholder-driven process on city scale leading to policy recommendations

In Delta, a participatory modeling approach was implemented together with City staff (Fig. 5). After a shared delineation of scenarios and indicators, modeling and staff meetings took place iteratively. Outputs were produced throughout the whole process and iteratively improved. SALTER & ROBINSON (2010) describe such a way of designing, informed through participatory meetings, as "facilitated modeling approach"; STEINITZ (2012, 57) as "participatory design approach". Again, key drivers were local policies and the stakeholder needs. Participation in the Delta process was particularly successful because the design team could build on six years of shared work with the community and support through champions in local administration.

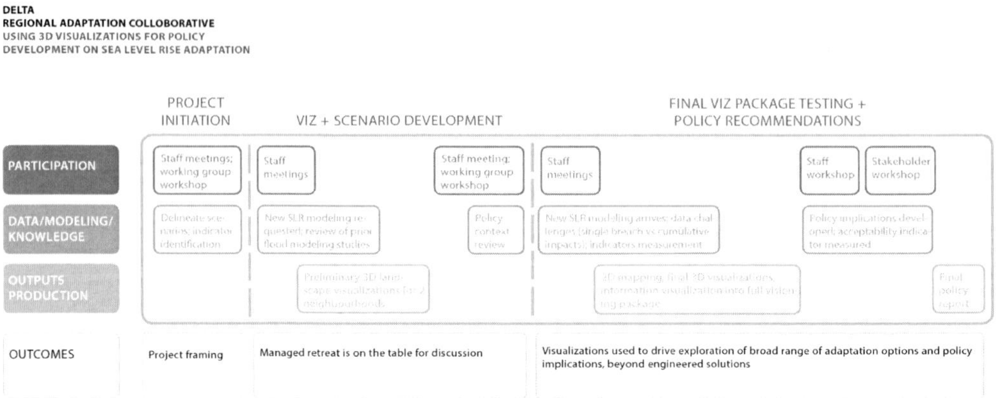

Fig. 5: Delta process flowchart

3.3 Clyde River: Community-driven bottom-up process on Hamlet-scale leading to policy recommendations

The Clyde River process was implemented by partly the same researchers as the Delta case and is therefore similar to Delta. In comparison to Delta, even more emphasis was put on community participation as can be seen in Fig. 6 (community meetings instead of stakeholder workshops). The strong community orientation lead to an early re-framing of the project goals; away from climate change mitigation/adaptation towards the exploration of new housing sites and the design of more resilient and more liveable housing types that respond not only to climate change but also to current housing shortages and social problems caused by overcrowding.

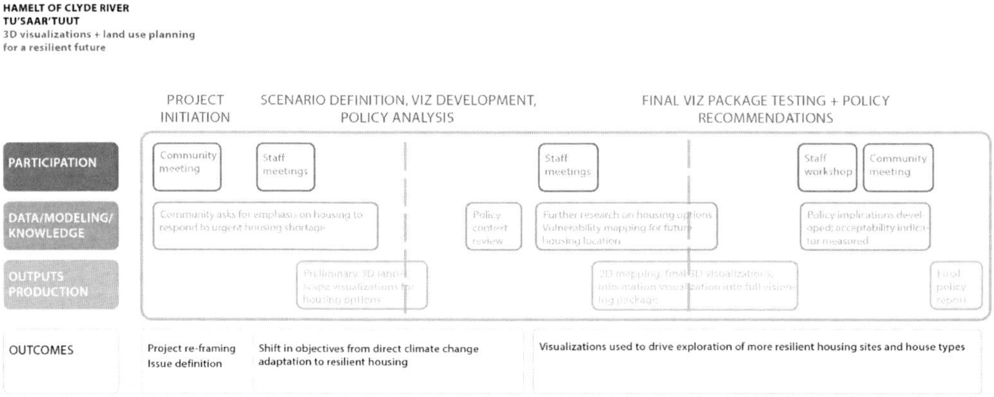

Fig. 6: Clyde River process flowchart

3.4 Calgary: Model-/stakeholder-driven process on watershed scale leading to regional strategies

The Calgary case covered the whole Elbow River watershed on a regional scale. Therefore, a very structured model-driven approach was chosen with an early stakeholder workshop for the collaborative definition of shared goals and structured stakeholder interviews to inform the modeling process (WIJESEKARA et al., 2011). In terms of STEINITZ (2012), the methods resembled mixed sequential and agent-based change models (called "hybrid modelling", Fig. 7.) with participatory elements. In comparison to the other three case studies, the Calgary study covered the largest area (watershed) and worked on the highest level.

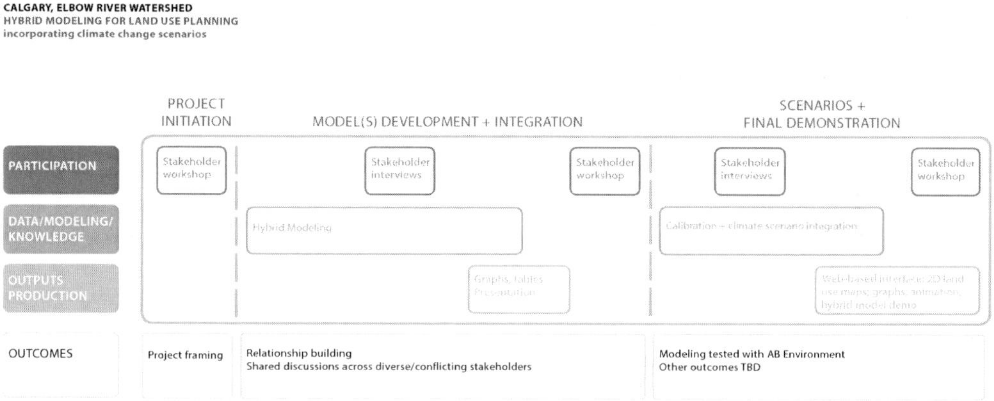

Fig.7: Calgary Elbow River process flowchart

3.5 Cross-case comparison with regard to spatial-analytic complexity, process, design and decision-making

The cross-case comparison provides examples and anecdotal evidence how to adapt a geodesign framework (e.g. STEINITZ 2012 for an overall geodesign framework), here particularly the LCCV framework (POND et al., 2010), to individual sites and contexts. With regard to **spatial-analytic complexity**, the Calgary case covered the biggest area applying the most sophisticated hybrid models whereas Delta and Clyde River focused on participatory processes on Hamlet and city scale and Toronto finally scaled down from neighbourhood to building scale emphasizing the anticipatory design approach.

In order to adapt the LCCV framework, not all **processes** ran linear: while the high-level intervention in Calgary was still the most linear, the participatory processes in Delta and Calgary had to respond to stakeholder needs and changed directions. Finally, the design process in Toronto branched out into multiple sub-projects. Similarly, the **design** process varies from model-driven to open and anticipatory. Topics ("what") changed in Toronto and Clyde River due to policy changes and community/stakeholder needs and focused even on different sites than previously anticipated ("where"). The example shows that geodesign has to respond to community/stakeholder input and policy change.

Finally, all cases follow different approaches of **impacting the decision-making process**: While Calgary addresses the high-level long-term planning of land use and water management on a regional scale (top-down approach, but with stakeholder participation in the modelling); Delta and Clyde River impact change through participatory bottom-up approaches. Toronto finally attempts a third way: inspiring change through the implementation of best practise examples.

4 Conclusions: Case by Case Application of the Visioning Process

The cross-case comparison clearly shows that a geodesign framework such as the LCCV guidance manual cannot be replicated without case-by-case adaptation and changes. Some of the cases in this study change scope and area size during the process and in respond to stakeholder needs and policy change, i.e. there is no "one size fits all" approach to local climate change visioning. However, there are replicable principles: a) the level of spatial-analytical complexity has to be adapted to scope and scale of the planning task at hand; and b) the process has to be flexible enough to respond to external drivers such as policy change, stakeholder needs and data issues (to name the three most important in these case studies).

Although not designed as a geodesign tool, the LCCV guidance manual (Pond et al., 2010) can provide process approaches, modeling and visualization tools that provide solutions for a range of scales and scopes in geodesign, particularly to climate-related questions. This cross-case comparison and previous work by Pond et al. (2012) have shown that hybrid modeling approaches with stakeholder participation work best on a regional scale for a watershed, whereas more community-driven approaches facilitate small working groups in neighbourhoods and towns and the design-driven process successfully engaged business partners and led to implementation. The cases show three ways of fostering change: Working on long-term regional plans; bottom-up community participation; and change through best practise examples. Only the latter has resulted in a visible change yet but it is the nature of plans that there outcome will take more time the larger area and scope are. Therefore, it is too early for a final assessment of the outcomes of community participation and regional plans at this point.

References

Ervin, S. M. (2012), *A System for GeoDesign*. In: Buhmann, E., Ervin, S. M. & Pietsch, M. (Eds.), Peer Reviewed Proceedings Digital Landscape Architecture 2012 at Anhalt University of Applied Sciences. Berlin/Offenbach, Wichmann, 158-167.

Francis, M (2001), *A Case Study Method For Landscape Architecture*. In: Landscape Journal, 20 (1), 15-29.

Kunzmann, K. R. (1993), *Geodesign: Chance oder Gefahr?* In: Planungskartographie und Geodesign. Hrsg.: Bundesforschungsanstalt für Landeskunde und Raumordnung. Informationen zur Raumentwicklung, 7/1993, 389-396.

McHarg, I. L. (1969), *Design with Nature*. New York, Wiley.

Pond, E., Schroth, O., Sheppard, S., Feick, R., Marceau, D., Danahy, J., Burch, S. (2012), *Collaborative Processes and Geo-Spatial Tools in Support of Local Climate Change Visioning and Planning*. In: Chrisman & Wachowicz (Eds.), The Added Value of Scientific Networking: Examples Contributed by GEOIDE Network Participant 1998-2012. Quebec City, GEOIDE Network.

Pond, E., Schroth, O., Sheppard, S. R. J., Muir-Owen, S., Liepa, I., Campbell, C., Flanders, D. & Tatebe, K. (2010), *Local Climate Change Visioning and Landscape Visualizations: Guidance Manual*. Vancouver, Collaborative for Advanced Landscape Planning.

Salter, J., Robinson, J. & Wiek, A. (2010), *Participatory methods of integrated assessment-a review*. Wiley Interdisciplinary Reviews: Climate Change, 1 (5), 697-17.

Steinitz, C. (2012), *A Framework for Geodesign: Changing Geography by Design*. Redmond, ESRI.

Schwarz-v.Raumer, H. G. & Stokman, A. (2011), *GeoDesign-Approximations of a catchphrase*. In: Buhmann, E., Ervin, S. M. & Pietsch, M. (Eds.), Peer Reviewed Proceedings Digital Landscape Architecture 2012 at Anhalt University of Applied Sciences. Berlin/Offenbach, Wichmann, 106-115.

Wijesekara, G. N., Gupta, A., Valeo, C., Hasbani, J.-G., Qiao, Y., Delaney, P. & Marceau, D. J. (2011), *Assessing the impact of future land-use changes on hydrological processes in the Elbow River watershed in southern Alberta, Canada*. In: Journal of Hydrology, 412-413, 220-232.

Hypergraphic – Superstudio – A Manifesto for Digital Landscape Architecture

Paul CURETON

Abstract

Hypergraphic – **Superstudio** is a marriage of discrete terminologies aimed at forming a manifesto and pedagogic grouping. Such formation arises from what is to be argued as a representational crisis in Digital Landscape Architecture. Drawing (Digital & Analogue) involves an imaginative and speculative activity – of seeing possibilities (CORNER, 1992). It is theoretical, abstract and also tactile – it arguably mediates between the conceptual and productive (AMOROSO & HARGREAVES, 2012). Given the role that drawing plays within wider information design, paradoxically it has received little attention and research in Landscape Architecture. Given this lack of critical attention, it is worth forming new methods and new insights in this area leading towards a transdisciplinary digital landscape – for effective information systems.

Such unification of disciplines (NICOLESCU, 2002) is not without context and is applicable to Digital Landscape. **Hypergraphics** (or Hypergraphy) as process, was developed by the Lettrist movement in the 1950s, Lettrist's worked towards a critical synthesis of visuals and writing, from poetry to cinema. Hypergraphics, a form of post-writing, supplements the means of expression through an ensemble of communicative signs (ISIDORE, 1964; FORD, 2005). Lettrism continued its experimental activity in the form of Psycho – Geography (GUY DEBORD 1955) and its criticism of the undertows of the then Parisian urbanism. Adopting such design methods arguably provides efficient communication of complex landscape sites. Founded by Natalini & Toraldo di Francia in 1966, **Superstudio** concerned itself with such political questions such as 'What is Architecture?' (RINGEN, 2004). **Superstudio** sought a form of architectural practice focused upon conceptual, cultural and theoretical criticism of the built environment. By extension, this paper is focused on the development of a Landscape Architecture Superstudio interested in the theoretical, the conception and criticism of built landscape in digital form. Such a manifesto is built marshalling specific representational cases from Sketchpad (SUTHERLAND 1963), Geographic Markup Language (GOLDFARB, MOSHER, LORIE, 1969) and Habitat (LUCAS FILM, 1986) in which each computational invention marked radical changes to the digital realm and evidences a continuing requirement for invention.

While each of these terms developed out of particular cultural contexts, as Harvey states, "at the representational level the emergence of spatial concepts is inextricably bound up with the structure of the culture in which spatial concepts are being developed" (HARVEY, 1969, p. 194). Given Harvey's assertion a manifesto for Digital Landscape is focused on the development of a Landscape Architecture **Hypergraphic** – **Superstudio** interested in both critical ideational activity and effective digital communication, recognising the cultural and research milieu in which production took place.

1 Introduction

"(Studying under Gropius), I soon appreciated the basic idea of the Bauhaus that the arts were not segmented." (SAFFA, 1991)
Lawrence Halprin

The push and pull between professional bodies, universities, and design firms with varying interests in the environment, competition and public perception has affected the position of landscape today. This effect has led Landscape to something of a peculiar position. This position arguably has led to a paradigm which is leading to a strange restriction on the ability of landscape design to respond in meaningful ways to changing climate, cultural and economic conditions. To solve, or at least move beyond the paradigm, we need to look outside of the discipline itself, or realistically pull back the basis – roots and segments- in which the discipline has constructed itself today. Thus my justification of a manifesto is not just for the sake of breaking the shelf on the manifestos of urban design, but to talk about an intention and motive to make landscape architecture representation more focused and considered in both academic and professional practice (CORNER, 1992). It is a manifesto, as I believe there is a sense of urgency in its task. Part of the realisation of the manifesto is in the area of digital landscape and fabrication.

That is to say, the process by which landscape design emerges and is constructed 'becomes' through drawing. We primarily design using drawing which then moves into something else, and provide questions and possibilities on solutions for urban issues and threatened environment. We need to focus research and the communication that the drawing continues to play, to analyse the characteristics of the Rhizome "a map not a tracing" (DELEUZE & GUATTARI, 1988, p. 12) perhaps equivalence to what Kundera calls a "poetic mediation on existence" (KUNDERA, 2005, p. 35).

Representation as part of the design process is an area in which to locate these roots. Both concept and ideational activity at the early design stage, but also the signifier in which the public constructs the image of place from built production – "Referential art is, by its very nature, reduced from its referents"(EVANS, 2000). Common constructed representational debates focus on analogue to digital transition, building a debate about the fidelity of digital production and the artistry and directness of hand drawing (MITCHELL, 1989). But this is arguably an irrelevant moot point. The movement or transition between representation and production is the real issue – how we make images about landscape using varied tools.

Selection and appropriation are more pressing concerns for generating ideas. Lawrence Halprin's assertion in the opening quote, of the conjoined nature of artistic production is key to enriching Digital Landscape. Thus the aim of the talk is to enrich digital landscape by looking at key computational inventions and artists movements. There are two ideas located within this enrichment. First that such computational invention led to specific hardware and software advances, and secondly that the artistic movements have brought interesting approaches and *USES* of such technology.

Openly admitting that such postmodern relativistic methods of combining and selecting parts of completely different design narratives and histories is questionable, the hope at least is to begin at a representational level a method of working that combines multiple approaches in the goal of creating and communicating information (thus the choice of a manifesto format) (NICOLESCU, 2002; NICOLESCU, 2008). If this is achieved then perhaps the method could be forgiven as moving towards a transdisplinarity as "goal is the understanding of the present world, of which one of the imperatives is the unity of knowledge" (NICOLESCU, 2002, p. 44). This is similar to what can be called courage of technology, a risk, as Ivan Sutherland would state (SUTHERLAND, 1996, p. 7).

Fig. 1: Aberdeen City Gardens, Satellite Image, Competition Resources, Malcolm Reading Consultants and Neale Smith 2011

For the recent design competition in the UK for Aberdeen Gardens (2011), ideas generated by firms at a representational level constructed landscape in a peculiar virtuality. Aberdeen receives on average 856 mm (33.7 in) of precipitation annually or 71 mm (2.8 in) each

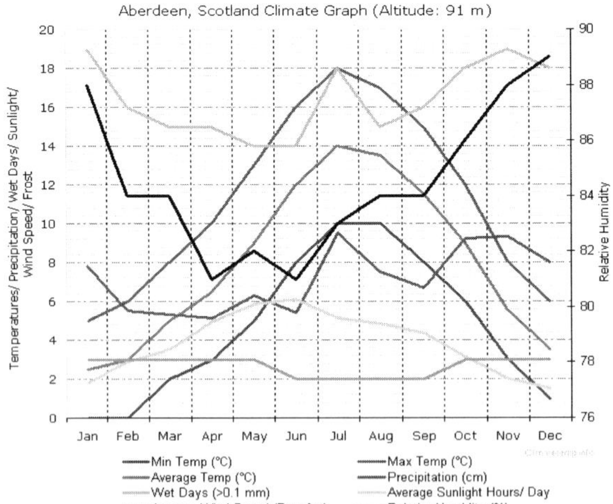

Fig. 2: Aberdeen Climate, Met Office UK (Office, n. d.)

month. However, the majority of visualisations created a climate akin to a Portuguese summer. At the same time, public reception drives such consumption for this productivity, undermining the credibility of such visualisations of considered representations of design intent.

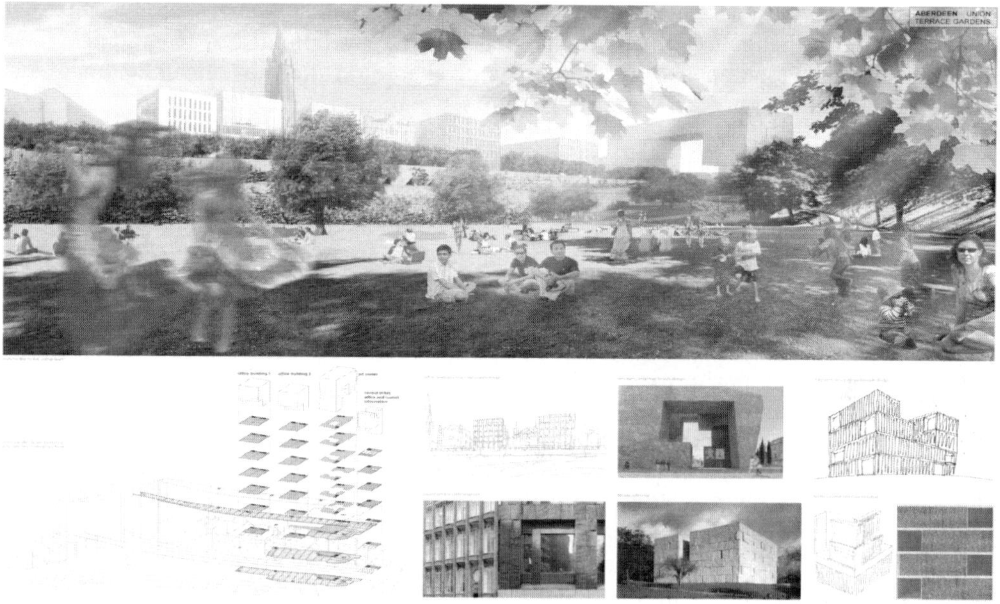

Fig. 3: West8 Competition Entry Page 3, Aberdeen City Gardens (visited 3/5/2012; http://www.malcolmreading.co.uk/architecturalcompetitions/citygarden)

Such work's effectiveness for communication of ideas limits its own possibility – a construct which shows freedom, however limited within its own prison of standardised representation. Paradoxically the GUI computational tools to allow such construction have marked a very different and more radical history of invention. Such representation is important to challenge as it links with wider landscape issues, as Christopher Tilley states,

> A centred and meaningful space involves specific sets of linkages between the physical space of the non-humanly created world, somatic states of the body, the mental space of cognition and representation and the space of movement, encounter and interaction between persons and between persons and the human and non-human environment (TILLEY, 1997, p. 10).

2 Lettrist Movement and Psycho-Geography

Lettrism sought a new form of communication which blended various arts; poetry, cinematography, print and painting. Essentially aiming and reacting to bring poetry to the everyday. Predominantly lettrism's outputs focused on notational styles. Maurice Lemaître (SEAMAN, 1983) also described HyperGraphics as "ensemble of signs capable of transmitting the reality served by the consciousness more exactly than all the former fragmentary and partial practices (phonetic alphabets, algebra, geometry, painting, music, and so forth)" (FORD, 2005, p.20). Lettrism continued its experimental activity in the form of Psycho – Geography through a break off group. As Guy Debord terms "the study of the precise laws and specific effects of the geographical environment, consciously organised or not, on the emotions and behaviour of individuals" (DEBORD, 1955). Here, Situationists opened new graphic communication and critic of capitalist society architecture and urbanism out of a distinction of a particular undertow or psychological – geographical relief of urban centres which apparently discourage or enable certain movements and activities (MCDONOUGH, 2002, pp. 55-87).

Exploring maps as constructions and perceptions of space, the work (notably in the Naked City 1957) conveyed maps figured as narratives rather than as tools of "universal knowledge" (MCDONOUGH, 2002, p. 61). As Debord states, "the production of psycho-geographical maps may help to clarify certain movements of a sort that, while surely not gratuitous, are wholly insubordinate to the usual directives" (KNABB, 2007, pp. 5-8). The most constructive example from the Situationist international was the notation of the 'derive' or drift, in which atmospheric points or locales connect with others at unrecognised scales forming a composite map of fragmentation states and drifts between and around obstacles. This work created at a representational level and abstract reduction of the Parisian space to social relations, but this in itself, was more 'truthful' than the presented fidelity of a 'whole' Parisian map seeking a totalising impossible eye or 'voyeur' (FOUCAULT, 1991; CERTEAU, 2002) in which the terrain is laid bare. The maps became the art of actions and narratives, of localised scales which are hard to interpret. Such work runs with Henri Lefebvre's description of the production of space in which capitalist conditions contributed to a condition of the segmentation of space. Lefebvre's complete lifelong project sought to unify the abstract, perceived and lived (LEFEBVRE, 1991). Such work in a digital landscape context is challenging, as remote sensing data constitutes the same 'totality' in landscape design, post Ian McHarg (1920-2001) the desire to gather and simulate 'wholesale conditions of working', transforming the complexity of landscape into a 'neutral' or simulacrum theatre of operations for new production. What can be extracted from these earlier movements is the use of more localised and refined individual notions and perceptions of space towards more collective in the urban realm (REYNOLDS & FITZ-PATRICK, 1999).

Such activity is useful in contemporary digital context, where such notational systems can be applied. For example, Lawrence Halprin (1916-2009) movement notation – 'motation' apropos Rudolf Laban, Noa Eshkol & Abraham Wachman, John Cage, Morton Subotnick and of course Anna Halprin has arguably proved invaluable in landscape design. Given its fusion I am currently working on digital Mobile Application of 'Motation' which has a form of public participation in landscape planning developing a device which attempts to consider time and space, combining people and environmental movements or what could be called the 'choreography of the city' (HALPRIN 1972). Lettrism and Psycho-Geography challenge the objectivity of mapmaking and provides possibilities for a radical cartography to be liveable and familiar. These are new challenges for urban design processes in the study of the tactics or stories of users and their particular drawings and mappings, lines and crossing of stories, re-appropriations, narratives, movements, metaphors and second geographies for the city and for the urban designer themselves – "what the map cuts up, the story cuts across" (CERTEAU, 2002, p. 129).

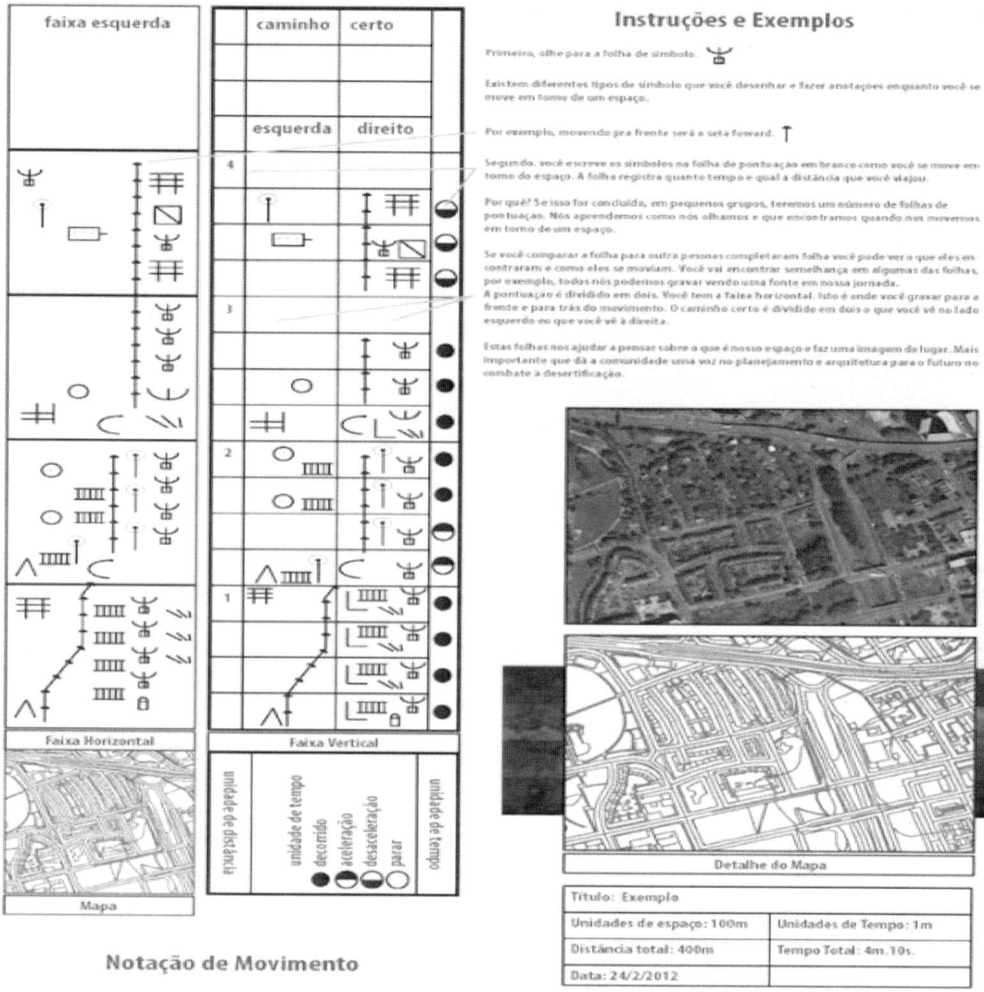

Fig. 4: Paul Cureton, Movement Notation After Lawrence Halprin, Digital Mobile & Tablet Development, Pre-Alpha Stage, 2013

3 Super Studio

Fig. 5: The Continuous Monument, An Architectural Model For Total Urbanisation, 1969

Superstudio was founded in Florence, Italy, by Adolfo Natalini and Cristiano Toraldo di Francia – recent graduates of the Florence School of Architecture – in 1966. Emerging from a critical cultural mielu,

> [Superstudio] saw that architects need to be involved in a different kind of thinking about what their profession is… Rather than just creating luxurious objects, or introducing people into the world of consumer objects, they should be worried about such political issues as, 'What is architecture?'
> (RINGEN, 2004).

Alongside similar groupings such as Archigram and Archizoom, Superstudio primarily conducted research outcomes in paper form. Utilising and layering photos they developed ideas about an all-encompassing architecture, the 'continuous monument' a black and white grid occupying all landscape thus rendering the world uniform (GARGIANI & LAMPARIELLO, 2010). This work marks an exemplary case for the development of critical ideational activity. The idea of the Superstudio was that an architectural practice could be conceptual and theoretical, concerned with cultural criticism rather than the production of buildings (BYVANCK, 2005). By extension the paper is focused on the development of a Landscape Architecture Superstudio interested in the theoretical, conception and criticism of built landscape. Superstudio's 'Anti Design' was a reaction against the failure of modernism and architecture in which the group blamed modernism for economic and environmental degradation (LANG & MENKING, 2003).

Corner identifies the need to address the estrangement of theory and practice from poetic value, in the constitution and embodiment of culture, stating that in landscape architecture "contemporary theory and practice have all but lost their metaphysical and mythopoetic dimensions, promoting a landscape architecture of primarily prosaic and technical construction" (CORNER, 1991, p. 116). Superstudio and Corner charts a method of working for digital landscape production, as further critical of its own outputs.

Further critical intention of digital landscape form is based on Bernard Tschumi's idea that "The paper representation of architecture will have the sole purpose of triggering desire for architecture" (WINES, 1988, p.186). Adopting such a motive as Tschumi states places Landscape architecture in new territories in which it can act.

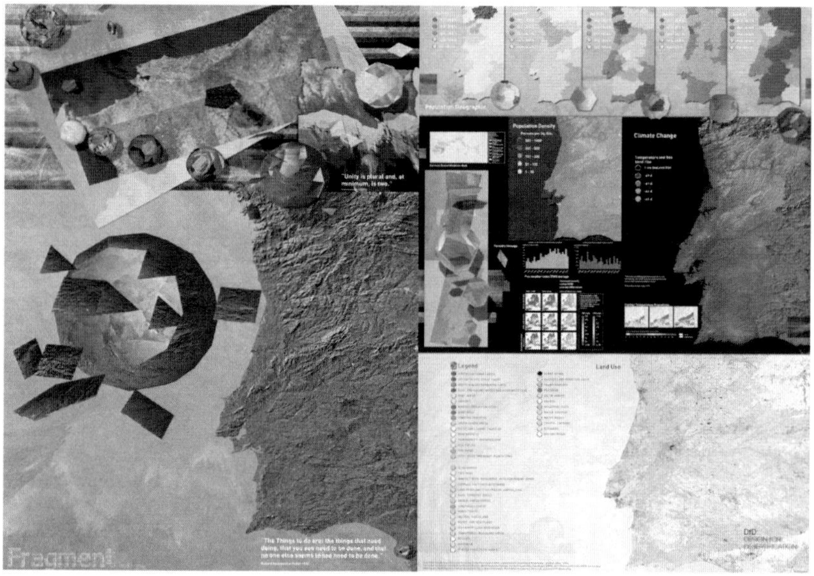

Fig. 6: Paul Cureton, Design for Desertification, '21st Century Rural Museum', Lisbon, Porto, Coimbra, November – July 2012- 2013, Group Exhibition

Such ambition formed part of experimental representation in this project. In Fig 5, Design for Desertification, a simple DEM model is combined with multiple data sets, unitary design philosophies of Buckminster-Fuller – an initial twist on hyper-graphic production but also bringing critic and attention to landscape practices and futurological land uses of segmented operations. Here, projections envisage the major centres to be flooded, vast areas victim of high temperatures, and limited liveable space.

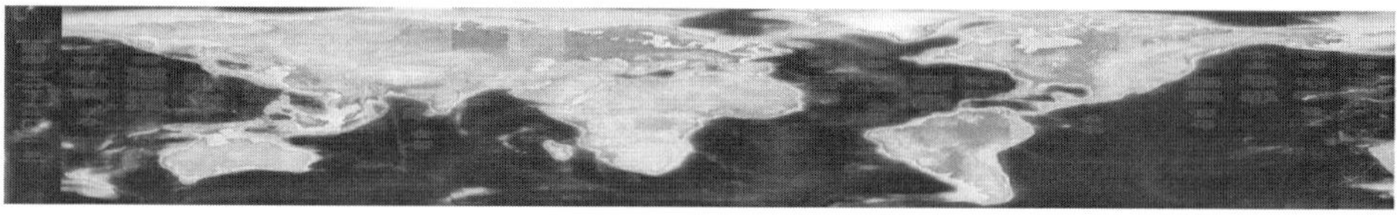

Fig. 7: Paul Cureton, Design for Desertification, '21st Century Rural Museum', Lisbon, Porto, Coimbra, November – July 2012- 2013, Group Exhibition

This is further explored in a distorted 17 metre by 2.5 metre world desertification map, titled 'Meantime ... Desert Poetics' overlaid with text by ecological artist David Haley and Portuguese poet Luis Quintais. The map was created from 283,000 tiled images, geo-referenced and reconstructed. Showing the world from within (mirrored map) the curved installation, centres visitors to the centre point of the earth. The suns position in the form of sequenced spot lights, light the panels as the same time the sun moves around the world. The impossibility of the view of the earth from within masks the real issue of an enveloping desert from which landscape is consumed.

It is worth making reference to three specific moments of computer invention that have had an impact of digital landscape production – Sketchpad, GML and Habitat. Each computational invention marked radical changes to the digital realm and evidences a continuing requirement for invention. This histories yield certain developments, like Hypergraphics and Superstudio that have a practical impact on digital landscape USE.

4 En Route to a Simple Line – Sketchpad (Ivan Sutherland 1963)

Fig. 8: Sketchpad developed beginning in 1961 by Ivan Sutherland at MIT is unveiled (Ref: Sutherland, Ivan E. Sketchpad: A Man-Machine Graphical Communication System. Proceedings of the AFIPS Spring Joint Computer Conference Washington, D.C., 1963, p. 329-346)

Sketchpad marked a specific computer development history. Sketchpad, based on the idea of napkin sketches, refined to more and more finished engineering drawings, formed the idea of the invention of the program. Working with a light pen, predecessor of the mouse, the user could point with the pen and interact with lines displayed on the screen. Ultimately, Sketchpad developed a graphical user interface that mediated between human and computer with real application to engineering solutions (SUTHERLAND, 2003, p. 3).

One of the earliest successes was to draw a cross which tracked the movements of the light pen (SUTHERLAND, 2003, p. 54). The cross formed the basis of the start of a line. This is what we can see still today in the use of AutoCad. The programming involved commands with buttons, for example pressing a button would fix the cross hair to a selected point to begin drawing a line. Further programming developed points, arcs, lines linked together to form objects (Blocks!). These could be stored in a library, scaled and rotated. Development of new displays and refresh buffers, as well as the might of large industrial conglomerates helped develop real world applications in aero and car manufacture.

Particular developments in patching and content awareness of imagery (moving and still) will push graphical interfaces further towards increasing virtuosity, most prevalent in Adobe CS7 and 8 future phases. Developments with new interfaces and more intuitive devices such as 3D immersion, Google glasses and Wacom's will increase the virtually in our delay lives. However, such sensory contact with landscape change and form is important to maintain. For example, in the Eye tracking drawings of Michel Paysant (CURETON, 2013). Landscape form is delimitated using eye trackers, primarily developed for people with limb paralysis, in a landscape context, the representation still maintains a connection with human construct. We view the lines of the eye as it moves around the page, unfocused, distracted, messy, un-glossed and unpatched. A truer representation of landscape experience than the uses of contemporary technology tools built with a propensity to stylise. Sketchpad was not such a product, it was more pragmatic.

5 Geographic Markup Language (Goldfarb, Mosher, Lorie 1969)

The Geography Markup Language (GML) is the XML grammar defined by the Open Geospatial Consortium (OGC) to express geographical features. The GML specification can encode dynamic features, spatial and temporal topology, complex geometric property types and coverage's (BRINK et al., 2012). GML has different goals compared to Hypergraphics or Superstudio. However, GML (GOLDFARB, 1973) serves as a modeling language for various geographic systems as well as an open interchange format for geographic transactions on the Internet. Users are also able to augment geo-spatial information supplied by a data producer with data of their own, creating new feature instances or modifying existing feature instances. The ability to integrate all forms of geographic information is key to the utility of GML. The ability of the democratisation of this language shows a successful vehicle and method in which new ideas can be shared and distributed.

Note that the concept of feature in GML[1] is a very general one and includes not only conventional 'vector' or discrete objects, but also coverages (see also GMLJP2) and sensor data. The geometries of those objects may describe, for example, roads, rivers, and bridges. InTIME (Marcoux & Sevigny, 1997) was the experimental version of the variant GML, and the core invention:

> The principle of separating document description from application function makes it possible to describe the attributes common to all documents of the same type. ... [The] availability of such 'type descriptions' could add new function to the text processing system. Programs could supply markup for an incomplete

[1] The key GML geometry object types in GML 1.0 and GML 2.0, are the following: Point, Line, String, PolygonGML 3.0 and higher also includes structures to describe "coverage" information, the "raster" model, such as gathered via remote sensors and images, including most satellite data.

> document, or interactively prompt a user in the entry of a document by displaying the markup. A generalized markup language then, would permit full information about a document to be preserved, regardless of the way the document is used or represented (GOLDFARB, 1996).

This coding and development was important as it brought a cultural significance to light. GML – has as a cultural effect – the ability to become 'the reconciliation of national differences' the portability of the file format between GIS allows trans-national working and opens up possibility in geographical knowledge sharing. Such ability to share and construct has not materialised as much as was intended. At a coding level, XML open source has the philosophy of Richard Buckminster- Fuller,

> … Because the meaning of design is that all the parts are interconsiderately arranged in respect to one another and because all the generalized principles are omni-interaccommodative which is to say that none ever contradict any others … (FULLER, 2008, p. 122).

Developments in OpenStreetMap which has its own XML, challenge and provide exciting connectivity, though such working remains constrained it the lack of 'purposeful' projects from the public at large. Software such as Mapbox is democratising what was a technical art form, though landscape architecture's role within this digital cartographic impulse needs to come to the fore, as in the GML case, founded on the basis of cross national dialogue and connectivity and sharing of data.

6 Habitat (Lucas Film 1986)

Fig. 9: A typical Habitat scene & Concept art for the box cover (© 1986 LucasArts Entertainment Company)

Quantum link was an online service for the commodore 64. Q-link enabled an interactive graphic resort island game called Habitat while in beta-testing. Later it was renamed Club Caribe. Lucasfilm's Habitat was created by Lucasfilm Games, a division of LucasArts Entertainment Company, in association with Quantum Computer Services, Inc.

> Habitat is ambitious in its scope. The system we developed can support a population of thousands of users in a single shared cyberspace. Habitat presents its users with a real-time animated view into an online simulated world in which users can communicate, play games, go on adventures, fall in love, get married, get divorced, start businesses, found religions, wage wars, protest against them, and experiment with self-government (BENEDIKT, 1991, p. 76).

Habitat as a software program made cyberspace a communications medium. It tested initial cyber development and the morality of its avatars and their interactions. As the developers Chip Morningstar and F. Randall Farmer stated (BENEDIKT, 1991), the introduction of weapons within the game led to a civilised area (gun free) and a 'wilderness' area (anything goes) in which players would be robbed and shot. In essence such development allowed networked operations and participation in virtual worlds. Such networkability will be key to future software operations and the democratisation of GIS systems and data.

The Habitat game trialled the first community within cyberspace. It is an example showing the results of interactivity, and such interactivity across practices will become more important in developing a greater forum of challenges and constraints of previous works. People define the parameters of their experience and such enabling of new digital realms is important to address the pressing ecological challenges what we make in the virtual and real.

7 Conclusion

Future motives for digital landscape architecture could be based on the following discussed principles. Lettrism sought new approaches to constructing images and increasing representational techniques could bring to the fore greater dialogues with the wider public. Psycho geography sought new mapping conditions and likewise greater interaction between fabrication and public participation which can yield a wider input into public space.

Superstudio sought a critical dialogue through representation and thus must be key to future work, creating, challenging, and developing ideas and applications working against standardisation. As Henri Lefebvre has warned in architectural production,

> ... we produce only the reproducible, and hence we only produce only by reproducing or imitating past production...because reproducibility is what ensures the renewal (or reproduction) of existing social relations (LEFEBVRE, 1991, p. 377).

Taking a moment out from the history of computer science we can see particular moments which mark unique inventions in the digital world, which would have profound impacts on future interfaces. Sketchpad allowed and marked the way to draw and interface with computers. The graphic interface would democratise our work to wider audiences and users. GML allowed the development and reference of geographic images to make real world visualisations at multi-scales, and most importantly in a standardised format to increase digital mapping systems availability and sharing of information systems. Habitat paved the way for increasing networking and tested the first virtual world. Its function and subversion by users gave first insights into future virtual worlds and the use of avatars in gaming environments. Such context can be appropriated for both software and hardware development, but also for the use of such tools for digital landscape.

The fate of digital landscape architecture can be located in the context of the computer instruction manual for Habitat which warned that the players and characters involved lost – "the spirit of adventure died away, and we became more and more content to do nothing" (LUCAS FILM LTD, 1987, p. 7).

References

Amoroso, N. & Hargreaves, G. (2012), *Digital Landscape Architecture Now*. London, Thames & Hudson.

Benedikt, M. (1991), *Cyberspace: first steps*. Cambridge, MA, MIT Press.

Brink, L. van den, Portele, C. & Vretanos, P. A. (2012), *Specification Profiles | OGC(R)*. http://www.opengeospatial.org/standards/profile (accessed 09/03/13).

Byvanck, V. (2005), *Superstudio – the Middleburg Lectures*. Middelburg, Zeeuwse Museumstichting.

Certeau, M. de (2002), *The Practice of Everyday Life* (2nd edition). University of California Press.

Corner, J. (1991), *A Discourse on Theory II: Three Tyrannies of Contemporary Theory and the Alternative of Hermeneutics*. In: Landscape Journal, 10 (2), 115-133.

Corner, J. (1992), *Representation and landscape: Drawing and making in the landscape medium.* In: Word & Image, 8 (3), 243-275.
Cureton, P. (2013), *Rhythm / Presence.* http://rhythm-presence.co.uk/ (accessed 09/03/13).
Debord, G. (1955), *Introduction to a Critique of Urban Geography.* Les levres Nues (6).
Deleuze, G. & Guattari, F. (1988), *A thousand plateaus: capitalism and schizophrenia.* London, Athlone Press.
Evans R (2000), *The Projective Cast: Architecture and Its Three Geometrics.* New Ed. Cambridge, MA, MIT Press.
Ford, S. (2005), *The Situationist International: A User's Guide.* London, Black Dog Publishing Ltd.
Foucault, M. (1991), *Discipline and Punish: The Birth of the Prison.* New Ed. London, Penguin.
Fuller, R, B, (2008), *And it Came to Pass – Not to Stay.* J. Snyder ed. Zürich, Lars Muller Publishers.
Gargiani, R. & Lampariello, B. (2010), *Superstudio.* Laterza, Laterza.
Goldfarb, C. (1973), *Design Considerations for Integrated Text Processing Systems.* IBM Cambridge Scientific Center. http://www.sgmlsource.com/history/G320-2094/G320-2094.htm (accessed 09/03/13).
Goldfarb, C. (1996), *The Roots of SGML – A Personal Recollection.* http://www.sgmlsource.com/history/roots.htm (accessed 09/03/13).
Harvey, D. (1969), *Explanation in geography.* London, Edward Arnold.
Knabb, K. (Ed.) (2007), *Situationist International Anthology.* Revised and expanded ed. Berkeley CA, Bureau of Public Secrets, U.S.
Kundera, M. (2005), *The Art of the Novel.* Rev. ed. London, Faber and Faber.
Lang, P. & Menking, W. (2003), *Superstudio: Life Without Objects: A Life Without Objects.* Milan, Skira Editore.
Lefebvre, H. (1991), *The Production of Space.* West Sussex, Wiley-Blackwell.
Lucas Film Ltd. (1987), *Habitat Instruction Manual.*
Marcoux, Y. & Sevigny, M. (1997), *Why SGML? Why Now?* In: JASIS, 48 (7), 584-918.
Mcdonough, T. (2002), *Guy Debord and the Situationist International: Texts and Documents.* Cambridge, MA, MIT Press.
Mitchell, W. J. (1989), *The Death of Drawing.* In: UCLA Architecture Journal, 2, 64-69.
Nicolescu, B. (2002), *Manifesto of Transdisciplinarity.* New York, State University of New York Press.
Nicolescu, B. (Ed.) (2008), *Transdisciplinarity: Theory and Practice.* New York, Hampton Press.
Office M, *Met Office: UK climate and weather statistics.* http://www.metoffice.gov.uk/weather/uk/climate.html (accessed 09/03/13).
Reynolds, B. & Fitzpatrick, J. (1999), *The Transversality of Michel de Certeau: Foucault's Panoptic Discourse and the Cartographic Impulse.* In: Diacritics, 29 (3), 63-80.
Ringen, J. (2004), *Superstudio: Pioneers of Conceptual Architecture.* http://www.metropolismag.com/story/20040106/superstudio-pioneers-of-conceptual-architecture (accessed 09/03/13).
Saffa, J. (1991), *Inner Landscapes.* San Fransico, KQED-TV.
Seaman, D. (1983), *Lettriste Pages – Isidore Isou Manifestos.* http://www.thing.net/~grist/l&d/lettrist/isou-m.htm (accessed 09/03/13).
Sutherland, I. (1996), *Sutherland, Technology & Courage, Perspectives 96-1.* In an Essay Series Published by SunLabs, April 1996 P7. Sun Microsystems Laboratories, 1-33.

Sutherland, I. (2003), *Sketchpad: A man-machine graphical communication system*. Cambridge, University of Cambridge. http://www.cl.cam.ac.uk/techreports/CAM-CL-TR-574.pdf (accessed 09/03/13).

Tilley, C. (1997), *A Phenomenology of Landscape: Places, Paths and Monuments*. London, Berg Publishers.

Wines, J. (1988), *De Architecture*. Illustrated edition. New York, Rizzoli International Publications.

Locative Media as a Tool for Landscape Interpretation

Elizabeth BRABEC

1 Introduction

New media in its various forms, including social networking, virtual reality, augmented or mixed reality, wikis, dynamic data, online communities and geo-tagging among others, have now been in use for more than a decade. While these various forms of new media hold tremendous opportunities for application to the interpretation of natural and cultural landscapes to the public, those with particular promise are applications that link informational content to specific physical locations. These falls into two types: locative media and augmented reality.

The term locative media was coined in the late 90's, as an application of geo-tagging focused on creating an understanding of the landscape of cities and communities (2012b). By connecting location-specific information about the history, meaning, characteristics and/or social culture via a mobile device (cell-phone and later GPS enabled devices), locative media was intended to make an urban environment more understandable. In its simplest form, locative media refers to the virtual portal, often a web site, and its geo-tagged content. Early applications were primarily visual arts projects (NOVA, 2004; Townsend, 2006) that engaged people in exploring and understanding their communities through the connection between artistic content and location. However, the potential for landscape and site interpretation, particularly in urban areas, was recognized by the early 2000's and projects and various software applications have proliferated since that time.

More recently, augmented reality, an alternate form of new media that also links content to location has also been promoted for its potential application to the visualization of landscapes (GRAHAM et al., 2012). Developed as a concept in the early 1990's, augmented reality differs from locative media in it's inclusion of the „real world" in the image displayed on the mobile device (STEUER, 1992). Augmented reality creates an image of a real site, often with Google street view, that is overlain with other images or floating tags – bubbles, arrows, etc., - that can be clicked to engage additional content about the feature. Although this may seem a minor distinction between augmented reality and locative media, the use and potential application of augmented reality for public information purposes has only recently become mainstream with the development of smart phone apps and augmented reality eye wear by Google and others (METZ, 2012), moving the application of the technology beyond bulky head and body wear that were the hallmarks of earlier gaming applications. However, both the technology and available software has developed very quickly in the past few years. Currently a number of software applications are available for smart phone applications (2013a), including open source software such as ARToolkit (2012a) to end-to-end branded apps such as Junaio (METAIO, 2013), WIKITUDE (2013b; Wikitude GmbH, 2013) and Layar (LAYAR, 2013).

Whether as locative media or augmented reality, by delivering content directly to a hand-held, GPS-enabled device, the downloaded content of these public information projects has the potential to create a user experience that is self-directed, has low infrastructure costs, and has minimal impact on the urban, cultural or natural landscape. Several key books, reports and journal articles have reviewed the use of multi-media projects in heritage (MARINOU, 2006), including interactive projects (AFFLECK & KVAN, 2008), games (CHAMPION, 2008) and digital visualization (JESSOP, 2008), and reviewed the opportunities and constraints of new media's applications to heritage resources (BRIZARD et al.; KALAY et al., 2008). However, the use and potential impacts of this technology, particularly with respect to diffuse resources such as heritage corridors and cultural landscapes, has yet to be fully explored. Landscape interpretation, particularly cultural landscapes which are difficult to make understandable to the public, show particular promise for this technology. Given the potential that this tool holds, a review of its applications is overdue. This paper reviews a series of twenty-one projects around the world, analysing the strengths, weaknesses, and hurdles to implementation of this developing technology.

2 Methods and Approach of the Study

2.1 Global Overview

A comprehensive web search was conducted to identify existing locative media and augmented reality projects implemented around the world. A total of twenty-one projects were found in a web search using the key words "locative media," "augmented reality" and "geo-tagging." Projects were limited to those that were specifically related to landscape heritage applications, linked content to a specific site location and conveyed heritage information to a physical or virtual visitor in a variety of formats. Information was gathered for each project on the number of physical sites included, date launched, technology used, and advertising of the site. Although an attempt was made to collect information from the developers of the sites about the number of website visitors and information downloads, data was only obtained for three of the websites.

The twenty-one locative media sites reviewed are detailed in Table 1. Eight countries are reflected in this review, covering the USA, Canada, Brazil, Ireland, Scotland, Norway, Australia, and the Czech Republic. Although these are the sites discovered during the web search, there are likely many more applications of this technology across the globe. While the "locative media" and augmented reality" webpages in Wikipedia provided a good location to start, and an organizational blog lists some additional locative media sites (primarily arts-focused) (Adam, n. d.) there is no central clearinghouse for these sites, nor an organization or conference that deals regularly with these applications of new media.

The identification of sites was also hampered by the fact that the technology has changed very quickly. One example of early technology was Mscape, developed by Hewlett Packard in 2002 under the name Mobile Bristol. Originally developed as a mobile media gaming platform, its use was expanded to include information delivery and was previewed at the 3GSM Conference in Barcelona in 2007 (HEWLETT PACKARD, 2007). The application was used at several sites including the Gulbenkian Gardens in Lisbon, a walk from the Royal Meridien Hotel to the Peoples Park in Shanghai, the Tower of London, and a walk through the Yosemite Valley (WIKIPEDIA, 2013). However the technology and software was discontinued sometime between 2010 and 2013 and all of the websites developed on this platform are currently unavailable. The Mscape technology has been replaced by other applications such as ARToolkit, Junaio, Layar, and Wikitude as the leading augmented reality applications. Although primarily developed for business applications, they are also currently supported for the delivery of general information. Specialized apps for these software programs have been developed for IPhone, Android, and Blackberry hardware, and are used on four of the sites reviewed. Barring the use of augmented reality, the state of technology has developed to the point that no specialized apps are necessary to deliver various types of information (video, audio, text) to cell phones and other GPS enabled hardware from a stand-alone website enabled by QR codes or GPS location.

2.2 Community Case Studies

The implementation of projects in the field was explored at four case study sites: Nove Hrady, Czech Republic (DOHAJE, 2012c); Chinatown (MURMUR, 2012) and Fraser Street in Vancouver, BC, (KLASSEN et al., 2012) and St. Helena Island, South Carolina (BRABEC, 2013). All four sites focused on providing historic and cultural information to visitors. Of these four projects, two used QR codes as the delivery system (Nove Hrady, Fraser Street, Vancouver) one relied on phone access to recorded stories (Vancouver Chinatown), and the fourth relied on direct download from a mobile website (St. Helena Island). Each case study site was evaluated based on ease of access/visibility of the signage, ease of access to the material, intrusiveness in the community and the publicity of the project.

3 Findings

While a total of 21 projects were discovered in the web search, six of those had been discontinued and were no longer functional leaving 15 projects to be fully reviewed in the study. These 15 projects had been developed between 2003 (Murmur Toronto) and the most recent five in 2012: Fraser Street Stories, Gullah Land and Community, Gloucester Harbor Walk, TXTile City, and Renaissance East Midlands.

3.1 Technology

As noted earlier, technology in this field has been changing rapidly over the past decade. The earliest projects were developed by the Murmur group, and relied on cell phone access to a phone number listed on signage in the city. The phone accessed audio recordings that the visitor could listen to while viewing the location where the story occurred. This application had, and continues to have, the virtue of being simple to implement, and inexpensive for both the user and the project developer. It does not require specialized smart phones, nor GPS or data access, a virtue felt particularly by visitors who face data roaming charges on their smartphones. The Murmur group has continued use this technology in its most recent project for the Textile Museum of Canada in Toronto, along with a specialized app for smart phones.

	Site Name	Location	Country	year started	# of sites	technology	format	content	associated sites/ social media	active
1	Murmur	Toronto	Canada	2003	173	Signage; phone number	audio	history	Website	yes
2	Murmur	Vancouver	Canada	2003	14	Signage; phone number	audio	history	none	no
3	Murmur City of Greater Sudbury	Sudbury	Canada	nd	7	Signage; phone number	audio	history	Website	yes
4	Murmur Kenora	Kenora	Canada	nd	16	Signage; phone numbe	audio	history	website; geocaching; map download	yes
5	Murmur Orange	Orange, New Jersey	USA	nd	20	Signage; phone number	audio	history	Website	yes
6	Murmur San Jose	San Jose	USA	nd	15	Signage; phone number	audio	history	Website	yes
7	Scape the hood	Mission District, San Francisco	USA	2005	10+	HP IPAQ Pocket PCs and GPS receivers	audio	history and culture	YouTube; Flickr	no
8	Murmur Sao Paulo	Sao Paulo	Brazil	2007	19	Signage; phone number	audio	history	none	no
9	Murmur Dublin	Dublin	Ireland	2007	36	Signage; phone number	audio	history	Website; brochure	no
10	Murmur Galway	Galway	Scotland	nd	14	Signage; phone number	audio	history	Website; brochure	yes
11	Murmur Edinburgh	Edinburgh	Scotland	2007	24	Signage; phone number	audio	history	Website	no
12	Mississippi Blues Trail	Mississippi; Chicago; Los Angeles; Memphis; Tallahassee; Rockland, ME; Helena, AK; Ferriday, LA; Notodden, Norway.	USA; Norway	2007	156	Website; smartphone apps; physical markers at sites	text and photos, 16 videos	Blues history	Website; Facebook; Wikipedia; Geocaching; Google Map	yes
13	dohaje.cz	throughout	Czech Republic	2007	2,500	website; signage with Tagg QR codes	text and photos	nature and history	Website; Facebook, Twitter	yes
14	Echo	Atlanta	USA	2008		unknown	audio	personal stories	none	no
15	Connecting Identities	Geelong	Australia	2008	65	website download; signage	video and audio	nature and history	Website	yes
16	Black History mobile	Washington, DC; Baltimore; Boston; Charleston; Chicago; Cleveland; New Orleans; Philadelphia; Phoenix; Richmond, VA.	USA	2008	10	Layar app ; Iphone app	text	history	Website, YouTube	yes
17	Fraser Street Stories	Vancouver	Canada	2012	10	website; signage; QR codes	text and photos/PDF	history	Website; dowloadable map; Flickr; Facebook; Twitter; Google Map	yes
18	Gullah Land and Community	St. Helena Island, South Carolina	USA	2012	9	Website download	videos, text, photos	history and culture	Website; Facebook; Vimeo; Google Map; YouTube	yes
19	Gloucester Harbor Walk	Gloucester, MA	USA	2012	41	website; phone app; signage; QR codes	videos and stories	history	Website; Facebook; Twitter; YouTube; downloadable map	yes
20	Txtile City	Toronto	Canada	2012	67	signage; phone number; smartphone app	audio and video	history	Website; YouTube	yes
21	Renaissance East Midlands	East Midlands	United Kingdom	2012	65+	smartphone app; QR codes	text; video	historic and cultural	Website; Google Maps	yes

Table 1: An overview of the 21 projects identified. Fifteen of the projects are currently active; six were either intended as limited-term installations, or have been discontinued and are indicated with grey shading in the table.

Table 1 – URLs:

1 – murmurtoronto.ca; 2 – murmurvancouver.ca; 3 – Sudbury.murmur.info; 4 – murmurkenora.ca; 5 – murmurorange.com; 6 – Sanjose.murmur.info; 7 – no URL available; 8 – Saopaulo.murmur.info; 9 – no URL available; 10 – galway.murmur.info; 11 – Edinburgh.murmur.info; 12 – msbluestrail.org; 13 – dohaje.cz; 14 – echoatlanta.com; 15 – geelongcity.vic.gov.au/connectingidentities/; 16 – hillretha.wix.com/blackhistorymobile; 17 – fraserstreetstories.com; 18 – gullahcommunity.org; 19 – ghwalk.org; 20 – txtilecity.ca; 21 – mpedia.org

Projects implemented since 2007 all utilize GPS and data access on smart phones, whether through the use of QRcodes, or a specialized smartphone app that links content with GPS phone location (figure 1). These configurations enable the use of both image and video streaming as part of the information delivery. Six of the projects – four of the most recent created in 2012 – use videos to deliver information. This has the advantage of being more engaging to an audience that grew up on YouTube, and also allows the delivery of information that cannot be delivered on a physical – and static – signboard.

Fig. 1: Screenshot of the website of the interactive map of sites in Nove Hrady, along with an image of site data on a cell phone screen. This is the most developed website of the group surveyed, allowing a visitor to create a customized tour from among the over 2500 locations listed.

3.2 Virtual Presence

Diversity in the web presence of each project was an important factor in their accessibility, and also their accessibility to potential visitors. Generally the projects that had the highest number of associated web (e.g. YouTube, Google Maps, Wikipedia, Geocaching) and social media presence (e.g. Facebook, Twitter, Flickr) were the easiest to identify for this study. The accessibility necessary to find these sites should also translate into higher visitor numbers, although the information collected was insufficient to complete this analysis. Although web count data for the sites is incomplete at this time, the most heavily networked sites also had the most linked articles, and showed high and escalating web counts where that information was available (e.g. the Mississippi Blues Trail). The [murmur] Toronto project, the longest site in existence in this study, showed an increase over time in online articles that discussed or linked to the site (figure 2). Whether it is the online presence that drives the publicity, or the online presence is merely a part of a robust advertising and outreach campaign is an important question for future research. However, the use of social networking seems to have potential: the Mississippi Blues Trail had the most developed Facebook page (set up in March, 2009) and has 11,735 likes (Mississippi Blues Trail, 2009). In comparison, [murmur] Toronto does not have a Facebook presence.

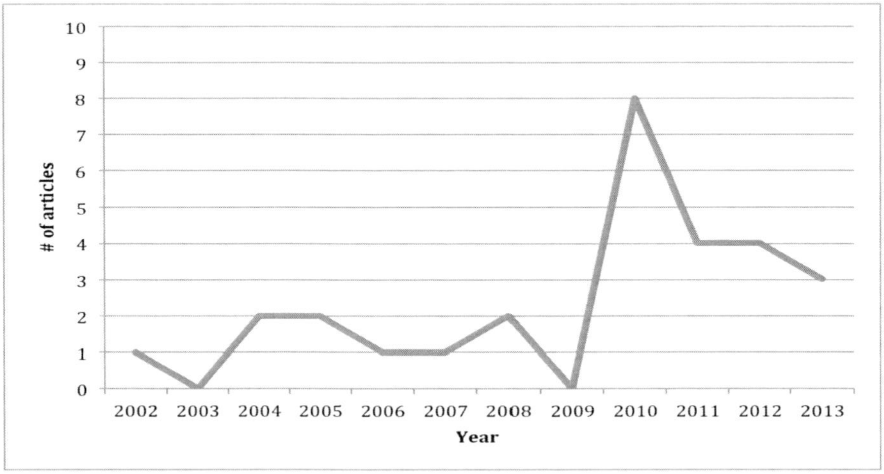

Fig. 2: The number of online articles per year referencing or linking to the [murmur] Toronto website between 2003 and 2013

3.3 Physical Presence in the Community Case Studies

The review of the four case study sites resulted in several significant findings for future project development. Key among the findings are an analysis of the potential to create a user experience that is self-directed, has low infrastructure costs, and has minimal impact on the urban, cultural or natural landscape.

The projects all offer the flexibility to the visitor, to access one, several or all of the information/story locations that are part of the project. The Czech project in particular provides the ability to search for and create a map of locations on their website, along a particular route or with a particular focus. Many of the recent projects, particularly those that use smartphone apps in conjunction with data that is pre-downloaded, provide the ability to access information based either on GPS location or through the use of QR codes. This is a strength in those areas where cell coverage is spotty or expensive.

All of the projects reviewed displayed a consistent logo and color format. However, even though the signage was well designed, placement in the field often resulted in a lack of visibility. Unless located at points of high pedestrian traffic, such as a bus stop, bus shelter or community signboard, project signage was uniformly difficult to locate on site. Particularly when signs were located with a proliferation of other municipal or directional signage, project signage tended to become invisible in the urban environment.

In many instances, the information conveyed by the mobile application was duplicative of the information on the physical sign. Merely repeating the text information located on the sign did not add value to the project, leaving project visitors with little incentive for downloading content. While text downloads can be easily read on a smart phone, it was clear that this was a less preferred method of information delivery: observed site visitors tired of reading very quickly, while their engagement with audio, photo or video content was more lengthy.

Few conflicts between user access to the project locations and private property owners were observed during site visits, underscoring the lack of intrusiveness of this type of landscape interpretation. However, the locations where these conflicts occurred were those where site visitors did not perceive private property boundaries. To avoid these conflicts, clear instructions should be given within the downloaded content, in addition to careful placement of signage to direct visitors away from private spaces.

4 Conclusions

While the technology holds promise for the interpretation of landscapes and their cultural, heritage and ecological resources, there are several significant issues in its implementation. Careful placement of site locations and signage, along with clear instructions for the boundaries of private property are essential in minimizing potential conflicts and privacy issues. The marketing and onsite visibility of projects must be addressed in multiple venues including extensive, on-going social and traditional media campaigns. Finally, the locations for signage and downloads must be carefully chosen to limit intrusion into private space, particularly in the urban environment. Signage that is set or directs the viewer to a private yard or garden runs the risk of unintentionally inviting visitors into private property. Finally, to engage the visitor, the downloads must have an added value for the viewer to what can be gained from onsite signage, and offer access to the community through video, voice or photographs.

References

2012a, *ARToolKit.* http://en.wikipedia.org/wiki/ARToolKit (Feb. 10, 2013).
2012b, *Locative Media.* http://en.wikipedia.org/wiki/Locative_media (May 5, 2012).

2012c, *Dohaje.cz ...napady na vylety!* http://www.dohaje.cz (May 5, 2012).
2013a, *List of Augmented reality Software.* http://en.wikipedia.org/wiki/List_of_augmented_reality_software (February 10, 2013).
2013b, *Wikitude.* http://en.wikipedia.org/wiki/Wikitude (Febraury 10, 2013).
Adam, F. (n. d.), *GPS Museum – a database about locative media and creativity.*
Affleck, J. & Kvan, T. (2008), *A Virtual Community as the Context for Discursive Interpretation: A Role in Cultural Heritage Engagement.* In: International Journal of Heritage Studies, 14 (3), 268-280.
Brabec, E. (2013), *Gullah St. Helena Island, South Carolina: Our Land, Our Community, Our Culture.* http://www.gullahcommunity.org (January 10, 2013).
Brizard, T., Derde, W. & Silberman, N., *Basic Guidelines for Cultural Heritage Professionals in the Use of Information Technologies: How can ICT support cultural heritage?* In: Know How Books: Heritage Sites (ed. by H. Gottlieb). Stockholm, Sweden, The Interactive Institute.
Champion, E. M. (2008), *Otherness of Place: Game-based Interaction and Learning in Virtual Heritage Projects.* In: International Journal of Heritage Studies, 14 (3), 210-228.
Graham, M., Zook, M. & Boulton, A. (2012), *Augmented reality in urban places: contested content and the duplicity of code.* Transactions of the Institute of British Geographers.
Hewlett Packard (2007), *HP mscape.* http://www.hpl.hp.com/mediascapes/ (May 30, 2012).
Jessop, M. (2008), *Digital visualization as a scholarly activity*, In: Literary & Linguistic Computing, 23 (3), 281-293.
Kalay, Y. E., Kvan, T. & Affleck, J. (2008), *New Heritage: New media and cultural heritage.* London/New York, Routledge.
Klassen, M., Wong, L., Arson, J. & Tylor, S. (2012), *Vancouver Street Stories: Find the Plaques to Make History.* http://www.vancouverstreetstories.com (June 4, 2012).
Layar (2013), What is Layar? *http://www.layar.com/what-is-layar/* (May 30, 2012).
Marinou, A. (2006), *The RI-SE Programme: Introducing an Innovative Approach to the Interpretation of Cultural Heritage Sites in Greece.* In: International Journal of Heritage Studies, 12 (3), 301-304.
metaio (2013), *Junaio.* http://www.junaio.com (February 10, 2013).
Metz, R. (2012), *Augmented Reality is Finally Getting Real.* http://www.technologyreview.com/news/428654/augmented-reality-is-finally-getting-real/ (August 2, 2012).
Mississippi Blues Trail (2009), *Mississippi Blues Trail* (May 5, 2012).
Murmur (2012), *Murmur Vancouver.* http://www.murmurvancouver.ca (June 4, 2012).
Nova, N. (2004), *Locative Media: A literature review.* In: CRAFT Research Report, Ecole Polytechnique Federale de Lausanne, Lausanne, 11.
Steuer, J. (1992), *Defining Virtual Reality: Dimensions Determining Telepresence.* In: Journal of Communication, 4 (2), 73-93.
Townsend, A. (2006), *Locative-Media Artists in the Contested-Aware City.* In: Leonardo, 39 (4), 345-347.
Wikipedia (2013), *mscape.* http://en.wikipedia.org/wiki/Mscape (May 30, 2012).
Wikitude GmbH (2013), *Wikitude.* http://www.wikitude.com/app (February 10, 2013).

Procedural Random Generation of Building Models Based Geobasis Data and of the Urban Development with the Software CityEngine

Christiane RADIES

1 Introduction

A two dimensional building plan contains much information about the classification of the building area into installation space, maximum permissible building height, building use and greening (Fig. 1). But only with a three dimensional view can the viewer obtain the most important information – the relation of the attributes to each other, as well as the spatial structures and the size relationships. Fig. 2 shows the two dimensional building plan as a three dimensional model of a possible simulation. This model allows area, sight or shadow analysis to be shown and cannot fail in modern urban planning.

There are different approaches for generation of 3D building models; laser scanning, terrestrial surveying or image analysis of satellite imagery. The procedural approach is a parameter-controlled system. This method is the basis for the implementation of 3D building models in this work.

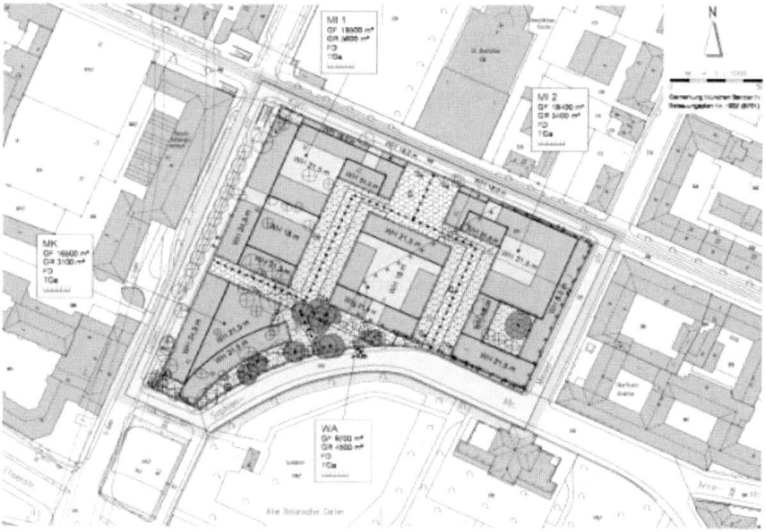

Fig. 1: 2D building plan

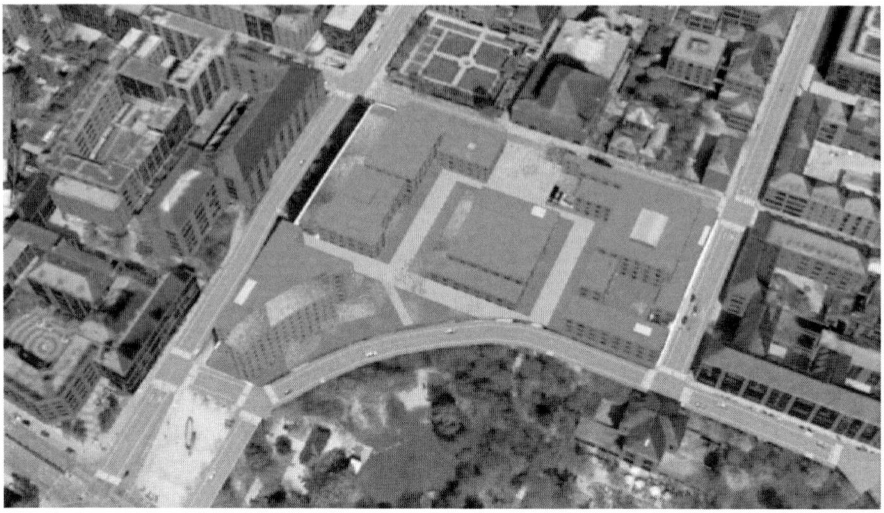

Fig. 2: 3D building model in CityEngine

2 Theoretical Basics

2.1 Procedural Modeling

The term ''procedural modeling'' is a concept of computer graphics which produces information objects (entities) which have let obtain access to their creation as a sequence of instructions. There are many approaches for the generation of procedural objects (HAVEMAN, 2005). The best known method is L-Systems. Especially in the fields of computer graphic and simulation, L-Systems allow an artificial generation of natural looking objects. Procedural modeling means rule-based modeling and it applies production rules to geometries, such as points, lines and polygons. This production rule follows the method of shape grammar. Principle of the procedural modeling is the iterative addition of more rules to geometry. The objects are, as a sum of rules, the result (PRUSINKIEWICZ et al., 1991). Based on existing or randomly produced data, the 3D objects can be generated extensively and automatically via production rule. Instead of saving the whole 3D model, storing the production rule as a ASCII-File with conditions and commands is sufficient.

L-Systems

L-Systems stand for Lindenmayer-Systems and were developed by the Hungarian biologist and botanist Aristid Linden Mayer as a formalism for simulation of bacterial and plant growth. Objects are describable as a successive sequence of production rules which have led to their creation. This formalism for generation of natural looking objects indicates an iterative refinement algorithm. This algorithm starts with a simple initial geometry and increases the level of detail by repeatedly overwriting parts of the object (rewriting systems) (PRUSINKIEWICZ et al., 1991). A L-System contains an initial geometry, the axiom ω, production rules P, constants S und an alphabet of existing variables.

The single elements are described with sign chains (string). On the basis of the initial geometry, the instructions are performed (LINDENMAYER, 1968).

2.2 CityEngine

CityEngine is a stand-alone software for efficient, extensive and fast generation of 3D building models based on data similar building outlines.

2.2.1 Operation

CityEngine works based on the formalism of the L-Systems. The initial geometry and the axiom ω are attributed geometric rule sets which change the intial state via parameter or contants. Axioms containing polygons or points are designated as initial shapes and are the start elements for all production rules (P). The production rules are described as CGA rule files and are stored as ASCII-format. CGA stands for Computer Generated Architecture and generates as a shape grammar with different production rules detailed 3D objects out of simple geometric outlines (Fig. 3).

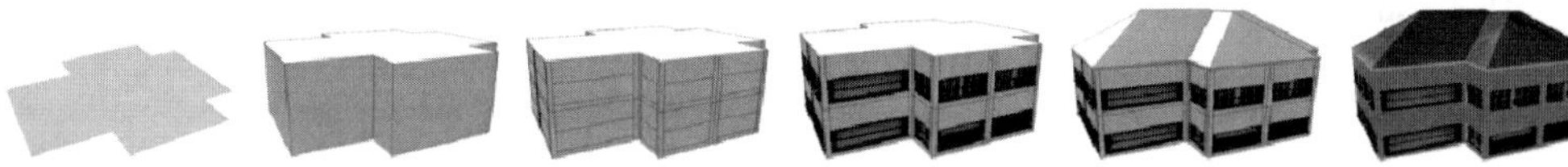

Fig. 3: Principle of procedural modeling

The application of geometric production rules is iterative. At first the building outline as the initial shape is extruded to a block model. This block model is divided into facades and a roof. The new areas are also called shapes and reserved for further operations, such as splitting into elements and adding information, i.e. textures or colors, to these new elements. (PARISH & MÜLLER, 2001).

2.2.2 Shapes

A shape is a finite amount of geometric elements, like points, lines and polygons (COHN, BLUNSOM, 2010). The geometry of the shapes is within a minimal bounding box and is defined as scope in CityEngine. This scope is combined with the intial shape over the pivot vector (pivot). The shape receives geo-reference through this. The pivot vector contains the axis of the local coordinate system in CityEngine. Fig. 4 shows the application of several production rules to one building outline. The initial shape was rectangular and was extruded to a box along the y axis via production rule. The pivot sets the x axis to red, y axis to green and the z axis to blue.

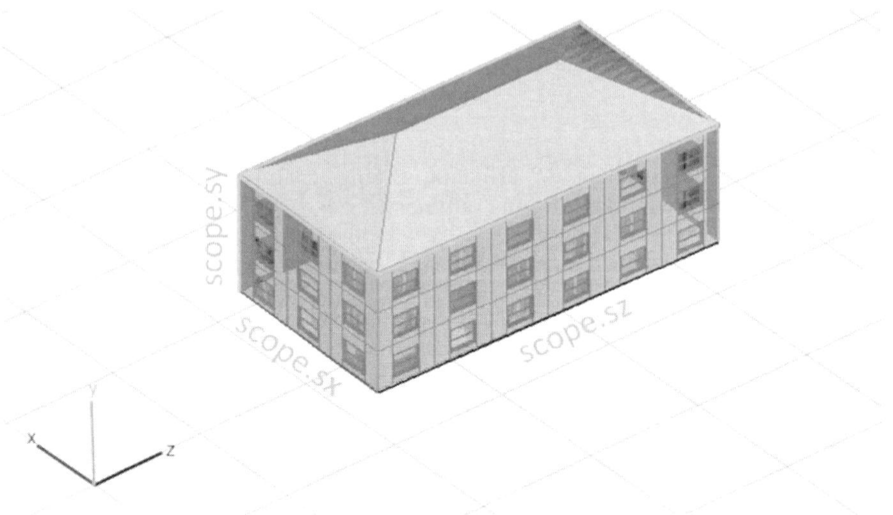

Fig. 4: Shape with Scope in CityEngine

The sides of the box are divided into shapes, and rules can be applied again. Through repeated generation of following shapes (successor) out of previous shapes (predecessor) a tree structure grows. The visualized rule shows this structure in Fig. 5. The initial shape is *Lot* within the orange box. Successors and predecessors are marked in the blue boxes.

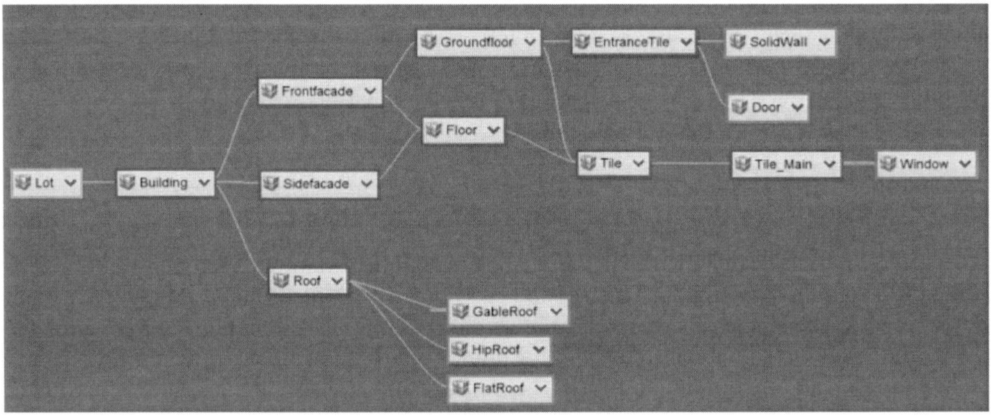

Fig. 5: Tree structure of the shapes within procedural process

Door and *Window* in the green boxes of Fig. 5 define the Leaf-Shapes. Those can be set with textures or colors and are the visual result of the object.

CGA-Rule-File

The rule set bases on the descriptive programming language computer generated architecture *CGA shape grammar.* Shape grammars are applied to two dimensional spacial shapes (GIPS, 1999). The CGA shape grammar is a shape grammar especially for modeling architectural models (MÜLLER et al., 2006). This grammar contains several predefined operations. For example, operations for generation of roof types with transfer of the corresponding roof angle. Via repeated application of these rules, sets to shapes complex object models can be generated. Through the addition of details, the level of complexity grows. Fig. 6 shows the principle of CGA shape grammar.

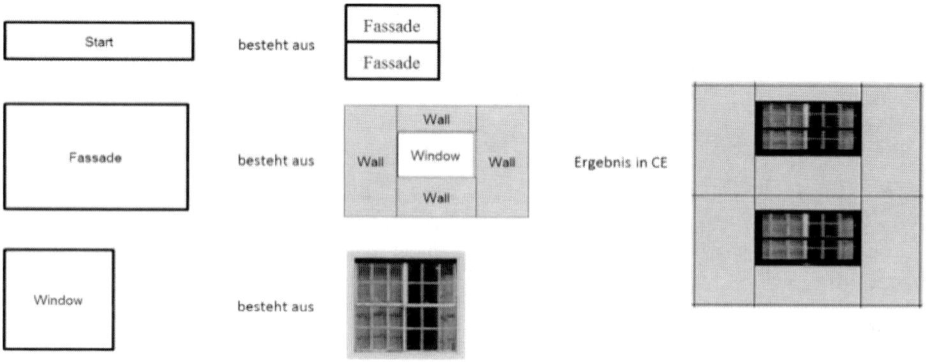

Fig. 6: Principle of adding details iteratively

A start rule is applied to the initial shape (parent) and divides it into two facade components. The facade itself is split into elements such as wall or window (child). These elements are complemented with attributes such as color or texture.

3 Material and Methods

3.1 Data foundation

The geo-basis data comes from the Munich city surveying office. The data of urban development comes from the planning unit Munich and contains the following data with the coordinate system DHDN/3-Degree Gauß-Krüger Zone 4:

- A digital elevation model (DEM) as a triangulated irregular network (TIN) with a mesh size of 20 metres. The geographical extent includes 25 x 20 kilometres of the city border of Munich.
- An aerial view as a true-orthophoto mosaik with a longitudinal and transversal overlap of 80 percent. The orthogonal projection provided a digital surface model. It contains a data volume of 2 GB a spacial extent of 30000 x 20000 pixel. Ground resolution of one pixel is about 10 cm.
- A zoning plan no. 1937 for the region called „Lenbachgärten“ near Munich central station with the format Computer Aided Design (CAD). Including a legend as statutory text.

3.2 Modelling approach

This chapter explains the approach of the procedural random generation of a three dimensional building scenario and the necessary data preprocessing of the urban planning data. Each polygon has to contain the corresponding attributes and geometry information. This information is used in the CityEngine to visualize as close to reality as possible. The main issue of this work is to provide maximum space and floor use for each polygon according to the defined values for building use. Furthermore, the result should be a dynamic model within the regulations of the legal framework of the Bavarian building regulations. This work gives evidence if it is possible to generate a procedural random building simulation within the given framework with the software CityEngine.

3.2.1 Clarification of the building plan's framework

Fig. 7 shows the legal framework of the zoning plan like floor and space ratio, maximum permitted height, type of building use, possible civil engineering for basement and building borders in blue boxes. Maximum of permitted height is marked by wall height (WH) within the corresponding polygons. The following architektonisch work is shown by red polygons.

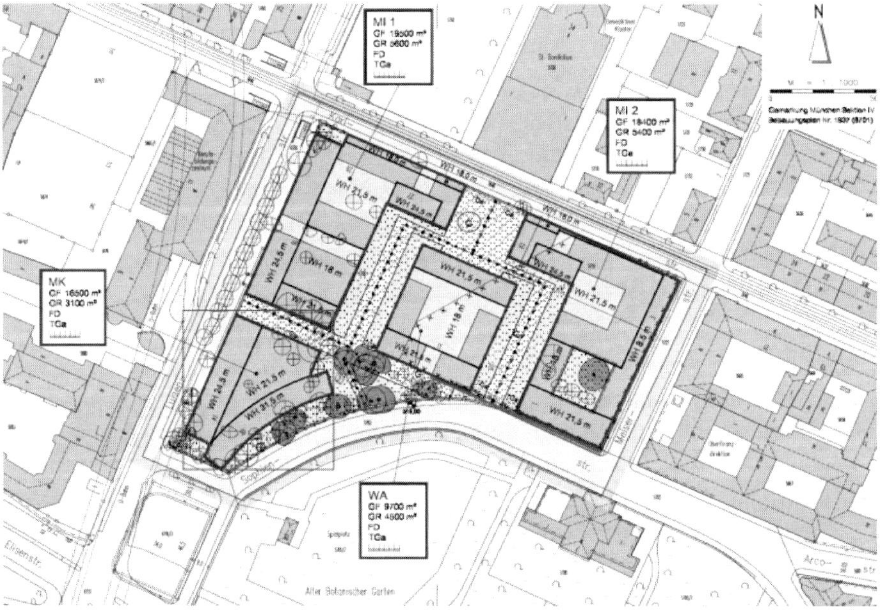

Fig. 7: Zoning plan no 1937

A simulation of development makes sense after defining the legal framework but prior to the following architectural draft. It is not the aim to model the already planned draft (red). Instead, the use of the random simulation achieves new building drafts within the blue building borders based on the legal framework. In the following work, the possibilities of construction are explained which are regulated in the statutory text and contain development of buildings, greening and basement garages.

3.2.2 Data pre-processing in ArcGIS

CityEngine supports an import of CAD files but without access to the attributes. Because of the necessary access to all information, a transformation from CAD to shape in ArcGIS is used, followed by a semantic classification between geometry and attributes. The geometries are stored as polygons, the attributes (heights, floor and ground area) as points. Result is a join of geometry and attribute, as well as the creation of separate object classes like development, greening and basement garage. Later, those object classes can be switched on and off in the CityEngine.

3.2.3 Random procedural generation

For a rule-based random generation of a three dimensional development scenario, the attribute and geometry information from the preprocessed data are used as rules for visualizing. The maximum ground and floor area are implemented in CityEngine. Furthermore, the location of the model is variable and the volume model changes dynamically. Tab 1 explains the implementation of the zoning plan's regulations with corresponding CGA rule.

Table 1: Implementation of the zoning plan's regulations in CityEngine

Regulations of the zoning plan	Explanation of the CGA rule
Type of building use	• Request to building type of the polygons • Allocation of facade textures and wall colour
Roof type	• Procedural generation of roof type • Request of roof type, allocation of roof angle
Ground area	• Determination of maximum permitted area of each polygon • Iterative split-function reduces the polygon area to the maximum buildable building area • Recursive implementation checks spare area after split
Floor area	• Iterative filling with floors based on reduced ground area for each polygon
Height	• Recursive implementation checks spare floor area and max. permitted height after floor building

Ground area recursion and floor area recursion

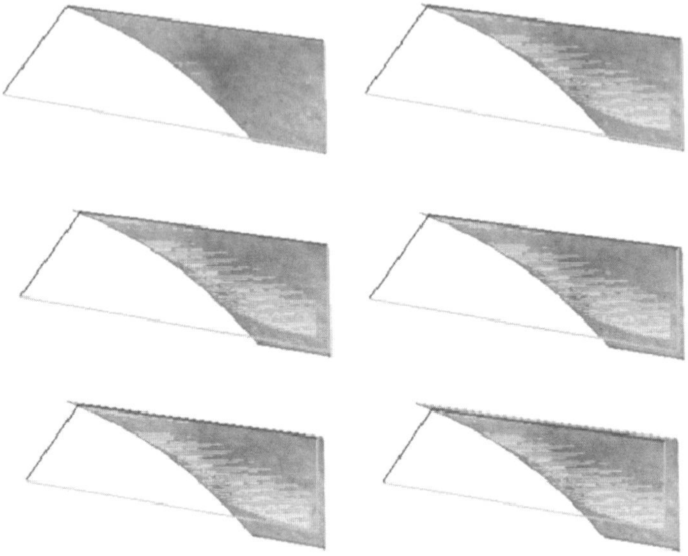

Fig. 8: Ground area recursion

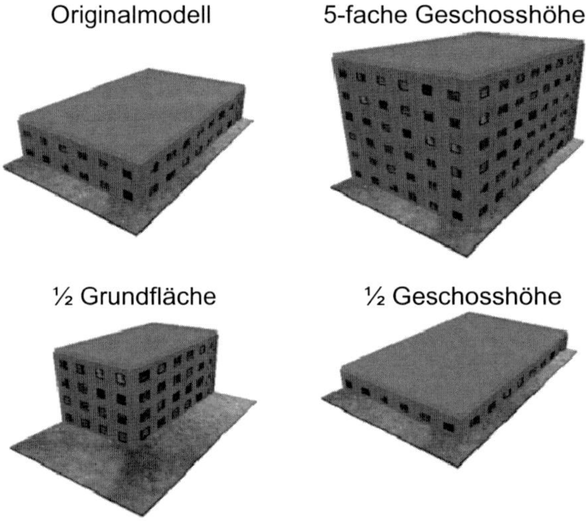

Fig. 9: Floor area recursion

4 Results

The result is a procedural random generation of development scenarios based on the regulations of a zoning plan with implementation of CGA-rules in CityEngine. Conditions are two recursive implemented functions for the allocation of ground and floor area. Both work in dependencey and allow a maximum of permitted use of ground and floor area of each polygon with the same type of use. Fig. 10 and Fig. 11 show the random simulation of the zoning plan number 1937 with five different object class buildings, street network, greening, planting and basement garage (Fig. 11 obscured by Terrain).

Fig. 10: Result showing the procedural generated simulation of a 3D zoning plan

Fig. 11: Basement Garage

5 Conclusions and Outlook

This work gives evidence that a 3D simulation of a zoning plan as a dynamic model within the regulations (height, areas, type of building use) of the zoning plan is implementable with the rules in CityEngine. Recursive functions generate CGA volume models with maximum allocation of ground and floor area and variable position. The shape grammar does not allow a transfer between the child element and the parent element. Respect to spatial relations of neighborhood is not possible. An iterative filling until the maximum of possible ground area allocation is achieved is only possible for each polygon and not practicable for global determinations of more polygons within a type of use. Regarding this step, the simulation of reality is questionable.

The data of urban planning consists of a zoning plan in CAD and statutory text. After data preprocessing every geometry consists of regulation of the zoning plan in the form of attributes in the attribute table.

For the first time, data in urban planning found application in the procedural modeling with CGA rules.

The result is a dynamic model where regulations can be changed in manual ways in CityEngine. Furthermore, additional elements like basement garage can be modeled.

CityEngine stands in contrast to established approaches and does not fit into well-known workflows of modeling tools. With respect to automatic fast and efficient generation of 3D building models, CityEngine cannot fail in modern urban planning.

The manual work of data preprocessing can be minimized by the standard format for urban planning data XPlanung. E-Government Projekt XPlanung follows this approach with the development of an object oriented data exchange format XPlanGML. This enables an exchange of planning data without losses as a standard for Germany (IAI, 2013). Standard attribut tables with standard column names would make a generation of procedural models easier.

More attribute information, such as building year, type of facade, situation, or information about renovation of each of the building geometries, could generate more detailed models. The depth of detail is dependent upon extra information in the form of attributes to the building polygons.

References

Cohn, T. & Blunsom, P. (2010), *A Bayesian Model of Syntax-Directed Tree to String.* Grammar Introduction. School of Informatics. University of Edinburgh EH8 9AB. Scotland, United Kingdom.

Gips, J. (1999), *Computer Implementation of Shape Grammars.* Boston College, Computer Science Department.

Haveman, S. (2005), *Generative Mesh Modeling.* PhD thesis. Technical University Braunschweig.

Lindenmayer, A. (1968), *Mathematical Models for Cellular interactions.* Parts I and II. In: Journal of Theoretical Biology, 18, 280-315.

Mueller, P., Wonka, P., Haegler, S., Ulmer, A. & Van Gool, L. (2006), *Modeling of Buildings. A Shape Grammar for CGA Architecture.*

Parish, Y. I. H. & Müller, P. (2001), *Procedural Modeling of Cities.* In: ACM SIGGRAPH 2001, ACM Press, E. Fiume (Ed.), 301-308.

Prusinkiewicz, P. & Lindenmayer, A. (1996), *The algorithmic beauty of plants.* New York, USA, Springer Verlag New York, Inc.

Wonka, P., Wimmer, M., Sillion, F., & Ribarsky, W. (2003), *Instant Architecture.* In: ACM Transactions of Graphics, 22 (3), 669-677.

Digital Communication in Landscape Architecture

Application of Social Media in a Regional Design Competition – A Case Study in the Netherlands

Renée DE WAAL, Annet KEMPENAAR, Ron VAN LAMMEREN and Sven STREMKE

1 Introduction

Many scholars argue for significant stakeholder involvement in landscape planning and design[1] (for instance STEINITZ, 2012; BOOHER & INNES, 2002; CRAIG, 1998). Facilitating public engagement can be a challenging task, which involves the sharing of information, enabling citizens to form opinions, the exchange of opinions, and community building to create trust and acceptance (MACINTOSH, 2008). When applying traditional methods for participation – such as participation meetings and workshops – accessibility and inclusiveness of the process can be problematic due to boundaries in time and space. Web-based methods offer participants the opportunity to engage without being physically present, at a time and place that suits them, and anonymously if necessary (BRABHAM, 2009; MACINTOSH, 2008; VAN LAMMEREN et al., 2007). Moreover, new technologies allow interactions with the public at relatively low cost and high levels of efficiency (MACINTOSH, 2008; KAPLAN & HAENLEIN, 2009; BRABHAM, 2009, KINGSTON et al., 2000).

Several publications explore the potential use of web based technologies to enhance public involvement in spatial planning and design, for instance in the form of online visualization tools (VERVOORT et al, 2010; SHEN, 2009), serious games (POPLIN, 2012), so-called argumentation maps (RINNER et al., 2008) and crowdsourcing (JEANSSON et al., 2012, LIGTENBERG & VAN LAMMEREN, 2012; HAMMON & HIPPNER, 2012; BRABHAM, 2009); all advocating further development of such technologies to enable online participation and collaboration in planning and design. The extent to which existing platforms, such as social networking sites of MySpace and Facebook, can support participation and collaboration is rather underestimated in planning and design literature and practice. As we are witnessing increased use of social media in society – and the impact of social media on society – their use as a means to enable citizen participation and collaboration needs to be explored. In this paper, we analyse the role of social media in the Eo Wijers regional design competition in the Netherlands, which provides examples of the use of social media from the perspective of integrated design/ planning teams. We studied A) which social media the teams proposed and applied in the competition entries and B) the level of interaction with the public that the teams aimed for. By analysing competition entries, we aim to give insight in the knowledge and attitude of the professional designers and planners concerning

---

[1] Although landscape planning and design vary in their approaches, to many of us a combination of both seems most fruitful for finding solutions to wicked problems such as planning and design of sustainable landscapes (see STOKMAN & VON HAAREN, 2012; STREMKE et al., 2012; DE JONGE, 2009). The desire to create solutions for existing or projected spatial problems, in collaboration with the people involved, is what both disciplines have in common and what we focus on if we talk about 'planning and design' in this paper.

the use of social media in interaction with the public, and the possible value of social media as a tool for collaborative creation in landscape planning and design.

Social media are characterized by Kaplan and Haenlein as *"a group of Internet based applications that build on the ideological and technological foundations of Web 2.0, and that allow the creation and exchange of User Generated Content"* (2009, p. 61). To specify 'user generated content', they distinguish between:

- Social networking sites (e.g. Facebook, Twitter)
- Collaborative projects (e.g. Wikipedia)
- Blogs (e.g. Blogger, WordPress.com)
- Content communities (e.g. Flickr, YouTube)
- Virtual game worlds (e.g. SimCity)
- Virtual social worlds (e.g. Second Life)

Social networking sites are platforms that people use to present themselves on the web and sustain relationships with others. These sites often allow many forms of content, such as photo's, video's and text. In the context of social media, collaborative projects are platforms which enable joint and simultaneous creation of content, which should lead to better results than each participant could achieve on his own. A well-known example of an online collaborative project is Wikipedia, where members can add, change and remove content. Blogs, or weblogs, are websites where date-stamped content is placed in reverse chronological order. Content communities, such as YouTube and Flickr, offer platforms to share content between users. To conclude, virtual game worlds and virtual social worlds offer three-dimensional environments in which people can respectively game or simulate themselves.

A classic work on public involvement is Sherry Arnstein's 'Ladder of Citizen Participation' (1969), which represents levels of power that citizens can have. Arnstein placed 'manipulation' and 'therapy' on the lowest rungs and summarized these as non-participation. On rungs three to five we find 'informing', 'consulting' and 'placation', summarized as 'tokenism'. She concludes with 'partnership', 'delegated power' and 'citizen control' on rungs six to eight, which enable 'citizen power'. For studying nowadays policy practice in the Netherlands, Pröpper (2009) translated and updated Arnstein's classification, laying more emphasis on interaction with citizens than citizen power. He distinguished between the levels 'no interacting', 'informing', 'consulting', 'advising', 'co-creating' and 'co-decision making', which we visualised in a scaffold-like figure (see fig. 1).

When adjusting Pröpper's distinction for the purpose of our analysis, we come to five possible levels of interaction: 'no interaction', 'one-way communication', 'two-way communication', 'collaborative decision making' and collaborative creation' (see fig. 1 again). If interaction is defined as "reciprocal action or influence" (Oxford Dictionaries Online), the levels no interaction and one-way communication can technically not be called interaction, but for analysing the usefulness of social media in the competition it is valuable to distinguish them. Two-way communication is communication that takes place in two or multiple directions, for instance on the exchange of local knowledge, narratives, ideas and so on. Collaborative decision making means that the public can inform and influence decisions to be taken in the planning and design process, for instance on the programme or allocation of projects. Finally we distinguished collaborative creation, whereby creation is defined as the act of "putting elements together to form a novel, coherent whole or make an original product" (KRATWOHL, 2002, p. 215). Acts of collaborative creation can range from the level of ideas for activities or interventions to their actual implementation.

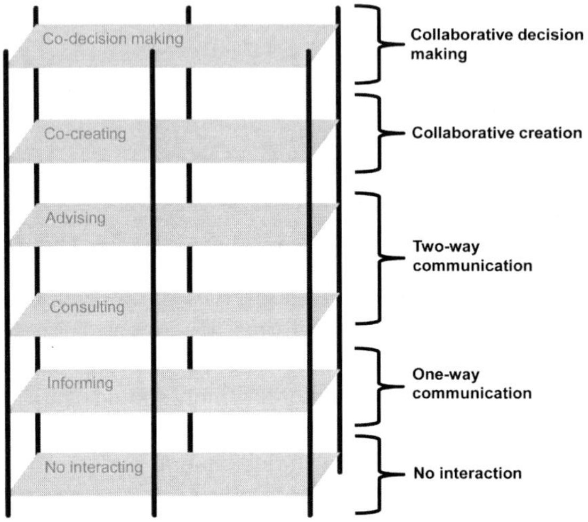

Fig. 1: Levels of interaction with citizens after PRÖPPER (2009), placed in a scaffold-like figure and regrouped into five levels of interaction

In section 2 we further describe the Eo Wijers competition and its assignment, followed by our methodology in section 3. The results are presented in section 4, beginning with an overview of the different social media platforms that were proposed and applied in the design competition (4.1), followed by an analysis of the levels of interaction to which the proposals should lead (4.2). One of the two entries that actually applied social media is then studied closer, to evaluate implications and perspectives of the use of social media for interaction with the public in planning and design (4.3). In section 5 we describe our conclusions and an outlook for further research and practice.

2 The Eo Wijers Competition for Regional Design

This paper presents a case study on the 9th EO Wijers regional design competition which took place in 2011/2012. The Eo Wijers competition is the most prestigious competition for regional design in the Netherlands and has significant influence on both the development of the profession as well as the development of particular regions (DE JONGE, 2008). This 9th edition focused on finding innovative solutions for energy transition, population shrinkage, agriculture development and water management in the Veenkoloniën, a region in the North of the Netherlands. For specific information on the competition and its organisation, see table 1.

Table 1: Specific information on the 9th edition of the Eo Wijers competition (Sources: EO WIJERS-STICHTING, 2012; KEMPENAAR et al., 2012)

Competition title	Nieuwe Energie voor de Veenkoloniën, op zoek naar regionale comfortzones [New energy for the Peat Colonies, in search for regional comfort zones].
Associated competition principals	Province of Drenthe, Agenda voor de Veenkoloniën [a cooperative of local governments in the region].
Time period	June 2011-March 2012.
Region	De Veenkoloniën [Peat Colonies], in the northeast of the Netherlands.
Number of entries	36
Average team size per entry	Six team members.
Team composition	Multidisciplinary. Most participants had a background in urban planning, landscape architecture, spatial planning, and/or architecture. Also disciplines like history, management consultancy, economy, industrial engineering, energy consultancy, communication and social service were represented.
Requirements per entry	1) Three A0 posters to represent the physical designs on local, regional and supra-regional scale level. 2) An essay of max. 1500 words to explain and amplify the physical and process design, and a proposal to communicate this with citizens and business in the region. 3) A medium free to choose to facilitate communication with the region.

Participants were asked to devise a method for regional development, capable to include local initiatives and involve the inhabitants of the region (Part A of the competition assignment, see text box below). Further, participants were explicitly asked to take the stories and experiences of the inhabitants and users of the region into the account while creating their entry (Part B of the assignment, see text box below). The emphasis on public involvement was so strong due to the little lasting success of earlier development plans and investments for this region (EO WIJERS-STICHTING, 2011).

Competition assignment (translated from Dutch; see Eo Wijers-stichting, 2011):

A. *Develop a method for sustainable value creation to come to 'regional comfort zones' in the Veenkoloniën. Utilize with that the identity and uniqueness of the region, and the opportunities in the energy, agricultural and water systems, in order to enhance the independence of the region and empower its inhabitants.*

B. *Depart from the narratives and experiences of the inhabitants and users. Take these into account in developing the method.*

C. *Investigate and visualise your ideas on the possible outcome of the method on three scale levels:*
   - *Radius approx. 5 km: The local scale of projects and local initiatives.*
   - *Radius approx. 20 km: The scale of the 'regional comfort zone'.*
   - *Radius approx. 100 km: The regional scale of the Veenkoloniën and beyond.*

D. *Clarify how you want to proceed when you win the competition, and give a matching planning for 2012 and further. Specify the implementation strategy you advocate, as well as an accompanying strategy for in- and external communication.*

Although it was not specified in the assignment that social media should be considered by the teams, twenty out of thirty-six entries *proposed* application of social media as tool for interaction with the public in the competition area (see fig. 2). Going further than proposing social media, two of these teams *actually* tried out social media for planning and design, to base their entry on. Sixteen entries did not apply social media at all; these entries were not further analysed for this paper.

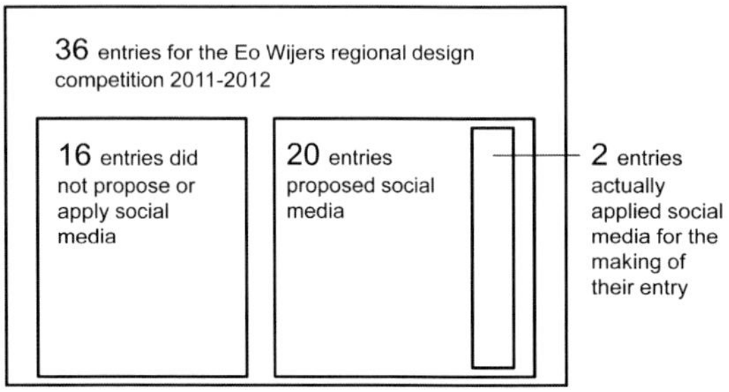

Fig. 2:
Overview of social media proposals and application in the Eo Wijers design competition entries

3 Methods

In this case study, we analysed A) which social media the teams proposed and applied in the competition entries and B) the level of interaction with the public that the teams aimed for by proposing social media in their competition entry, or even realised in making their competition entry.

For analysing the different types of social media that were proposed in the Eo Wijers competition, Kaplan and Haenlein's categories have been used (see introduction). We adopted an interpretive approach to study the competition entries, making use of a textual coding technique (see for instance SCHWARTZ-SHEA & YANOW, 2012; CROTTY, 2009). First, we added Kaplan and Haenlein's categories (i.e. collaborative projects, blogs, content communities, social networking sites, virtual game worlds and virtual social worlds) to a codebook, plus their synonyms and additional terms (for instance 'Facebook' for social networking sites and 'YouTube' for content community). Next, we used this collection of terms to scan the texts in the entries on words and phrases which could indicate use of the different social media platforms. This has been done manually, and checked by using the search function in Adobe Acrobat X Pro to determine whether the analysis was complete. Finally, we added up occurrences of the different social media platforms and summarized the findings.

For each of the twenty competition entries in which social media were proposed, we analysed the level of interaction with the public that was aimed for, based on the description of the levels as given in the introduction (no interaction, one-way communication, two-way communication, collaborative decision making and collaborative creation).

Finally, we illustrate how application of social media actually led to collaborative creation in the process of making a competition entry. Based on a study of the entry itself, and the interactions on the social media platforms that were created for this entry, we describe:

- The target groups for interaction.
- The purpose of interaction.
- The social media platforms applied.
- Other ways to interact with the public proposed and used.
- Purpose of social media application (such as crowdsourcing for narratives, ideas).
- Aimed levels of interaction and to what extent this was realised.
- How social media application is embedded in the wider planning/design process.

4 Results

4.1 Different social media categories

In the twenty entries that proposed social media to facilitate interaction with the public (see fig. 2), a total of thirty-one platforms is proposed (see table 2). This means that several teams proposed a combination of two or more different platforms. Next, interaction via social media is often proposed in combination with more traditional ways for public involvement such as personal meetings and communications.

Table 2: Proposals for social media in the competition entries, divided into the 6 categories of social media as described by Kaplan and Haenlein (2009)

Social media categories	Occurrence in entries
Social networking sites	10
Collaborative projects	9
Blogs	8
Content communities	4
Virtual game worlds	0
Virtual social worlds	0

The category of social networking sites occurred most frequently. The platforms that were proposed (i.e. Facebook, Twitter, LinkedIn and Hyves) are the most popular of their kind in the Netherlands. Social networking sites were proposed to share information with the public, but most of all, to let the public form and share their opinion online. Also collaborative projects have been proposed often. In this category, the teams tended to devise platforms themselves rather than using an existing platform, such as Wikipedia or uploading photos to Google Maps. An example of a self-devised collaborative project can be found in the entry 'Boeren, Burgers & Buitenlui' ['farmers, citizens and country people'] of TU Delft students Wouter Keizer and Ule Koopmans. Keizer and Koopmans present 'Index'; a 'knowledge and communication centre' which consists among other things of a digital interactive map[2]. 'Index' works and looks a bit similar to uploading photo's to Google Maps.

[2] For the purpose of the competition, this application and its working is imagined and not (yet) developed.

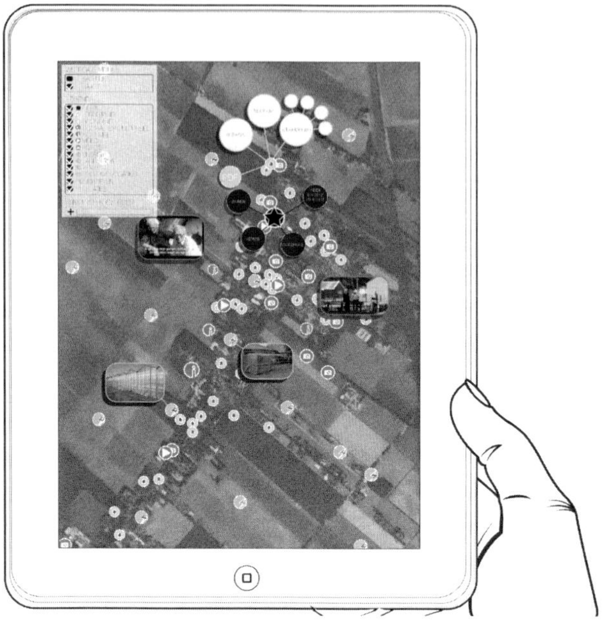

Fig. 3:
Image of 'Index', the interactive digital map from the 'Boeren, Burgers & Buitenlui' entry. Source: Wouter Keizer and Ule Koopmans.

Blogs were also frequently proposed and mainly used to share information with the public about the competition, or the entry in progress. At itself, a blog is not interactive, but many blogs in the competition were combined with possibilities for giving feedback and/ or opinion polling, which lend themself for interaction. Content communities were less frequently proposed; only four entries propose YouTube to share video content with the public. Since this was a regional design competition, in which pictures and design images play an important role, one could have expected more proposals for photo sharing sites such as Flickr. To conclude the list of different social media categories, no virtual social worlds were proposed and only one entry made reference to the possibility of creating virtual games.

4.2 Levels of interaction

The proposals for facilitating interaction with the public via social media, as categorised in section 4.1, aim for different levels of interaction. For each of the proposed platforms we analysed whether they could lead to one or more of the five levels of interaction that we distinguished in the introduction (no interaction, one-way communication, two-way communication, collaborative decision making and collaborative creation). It became clear that facilitating one- and/ or two-way exchange of information is much more aimed for than collaborative decision making and collaborative creation (see table 3).

These outcomes support that, for the purpose of this competition, the majority of the teams deemed the use of existing social media effective to provide and/ or exchange information. In several entries, it is mentioned that social media have been applied to understand 'what the people want'. In that sense, the entries responded well to the assignment which asked to 'depart from the narratives and experiences of the inhabitants and users'. Two entries

Table 3: Levels of interaction aimed for by social media proposals in the Eo Wijers competition

	No interaction	One-way: sending of information	Two-way: exchange of information	Collaborative decision making	Collaborative creation
Proposals for application of social media in future planning and design in the region	–	20 entries	18 entries	2 entries	11 entries

propose collaborative decision making but only via online polls, concerning small and unambiguous issues. In these entries, it is not made clear how this process should relate to formal decision making. Eleven teams tried to use the potential of social media to organise collaborative creation. Among these were the nine entries that proposed collaborative creation by means of a collaborative project (see table 2). Next, two of the ten entries that proposed social networking sites did it in such a way that not only provision and exchange of information was facilitated, but that the public could also form and share their opinions, bring forward ideas and start up activities and initiatives with others.

The digital interactive map from the entry 'Boeren, Burgers & Buitenlui' (see also section 4.1), is one of the collaborative projects that aims to facilitate collaborative creation. The self-devised 'Index' offers social, economic and sustainable energy stakeholders the possibility to upload ideas and initiatives to the interactive map, which would enable them to connect, exchange and collaborate. The platform aims for one- and two-way communication and collaborative creation by connecting citizens, providing them with information and knowledge (among others about sustainable energy) and a place where they can share their own knowledge and ideas, to eventually implement (some of) these. In other words, a form of crowdsourcing is organised here for finding optimal and supported solutions for the transition to sustainable energy, one of the core themes of the competition. This entry, like many others, is not specific about how their proposal for collaborative creation is to be embedded in current planning and design procedures. It is only indicated that ''Index' is a government initiative which should lead to bottom-up initiatives'.

4.3 Implications and perspectives

'Ondermekoar' ['by ourselves'] is one of the two entries that not only proposed social media, but *actually* used social media for the making of their competition entry. The public could learn about, inform, influence and collectively create parts of the entry via a blog, Facebook, Twitter and YouTube (for an overview of their activities, see table 4).

The team illustrated that in a relatively short time, and with almost no budget, local (and non-local) people can be reached by making use of existing social media platforms. The blog (http://ondermekoar.wordpress.com/over-ons/) provided information about the team and the competition, links to Facebook and Twitter to crowdsource ideas for the competition, and offered the possibility for giving feedback. On Twitter the team initiated interaction between professionals and inhabitants among other things about wind energy and the local public transport (ONDERMEKOAR, 2012). The account had, just before the submission deadline, 150 followers. Facebook polls were used to survey opinions, about the local landscape, social securities, politics and so on. This resulted in 45 likes and over 50 responses on polls (ONDERMEKOAR, 2013a). Note that between the moment of starting up Facebook (1 December 2011) and Twitter (6 December 2011) and the competition deadline (6 January 2012) was only five weeks. Further, the team concentrated on current regional affairs in a short movie, as a trailer for a new regional soap, and shared this via Facebook and YouTube. By now, January 2013, this movie is viewed 604 times (ONDERMEKOAR, 2013b).

Table 4: Actual use of social media in 'Ondermekoar'

Design team	*HKB stedenbouwkundigen*: Jeroen Leemans, Henk Bouwman, Dominic Tegelbeckers, Shingrong Wu, Sacha Schram. *De Mannen van Schuim*: Rico Zweers, Niels de Vries Humel. *Territoria*: Karin Peeters. *Fundament All Media*: Boris Geheniau.
Target groups for interaction	Citizens of the Veenkoloniën who are active on internet in general, Facebook and/ or Twitter.
Purpose of interaction	To develop ideas for their competition entry.
Social media platforms applied	Blog, Facebook (community page and polls), Twitter, trailer of regional soap on YouTube.
Other interaction	Interviews with inhabitants.
Purpose of social media	Harvesting local knowledge and narratives; let the public form and share opinions; harvesting and testing of ideas.
Aimed levels of interaction	Two-way communication happens on all platforms. Collaborative decision making and creation is aimed for by the design-team, but the number and content of reactions on the various platforms do not support this.
How is social media application embedded in the wider planning/design process	The team sees their entry as a new way for planning and design, but does not explicitly relate, or fit in, social media application to existing planning and design procedures. However, all platforms are kept alive after the competition deadline on 6 January 2012. In this way, the outcomes of interactions taking place on the several social media platforms can still inform planning and design processes going on in the Veenkoloniën the same as they informed the making of the 'Ondermekoar' entry.

The history of the interaction and communication through social media is still visible online, although not all interactions that the team refers to in their entry are traceable. For instance, the 'Tank en proat plek' ['place to refuel and chat'] - is a central idea in the 'Ondermekoar' entry (see fig. 4) and the team claims in their entry that the idea originated from interactions with the public via social media. The 'Tank en proat plek' is presented in the short movie on YouTube, and the blog invited the public to react on the ideas shown in the movie. However, reactions on the movie are hard to find on the blog, Facebook or Twitter. Also it cannot be traced that the idea originated from earlier social media interactions.[3] Further, from the entry we know that the team of 'Ondermekoar' did face-to-face interviews with local inhabitants. The social media platforms in 'Ondermekoar' make no reference to this, nor is it clear how the outcomes of the interviews are related to the application of social media. Did the interviews provide input for, or were they conducted to verify or specify output of the social media interactions?

Fig. 4: Image of the 'Tank en Proat plek' ['place to refuel and chat'] from the 'Ondermekoar' entry. Source: HKB stedenbouwkundigen, De Mannen van Schuim, Territoria and Fundament all media.

Although the process lacked openness on several aspects, 'Ondermekoar' shows how existing social media platforms can be applied to crowdsource local knowledge, narratives and ideas. Some of these have been localised and concretized in the process of making a design competition entry; a process that models real planning and design. By actually applying social media and not only proposing it, 'Ondermekoar' responded very well to the competition assignment which asked teams to depart from the narratives and experiences of the inhabitants and users. This was acknowledged by the competition jury who awarded this entry with an honourable mentioning for its innovative communication strategy.

5 Conclusions and Outlook

In this paper we studied the role of existing social media platforms in the Eo Wijers regional design competition, which provides examples of the use of social media from the perspective of integrated design/planning teams. For the twenty competition entries that proposed social media to facilitate interaction with the public, we A) categorized the types

3 This invisibility might be related to the character of a design competition. Since other teams were in the same process of developing their entry, the authors might have kept central ideas to themselves.

of social media platforms and B) indicated the levels of interaction the teams aimed to facilitate via social media. The entry 'Ondermekoar' is further described to illustrate how, for the making of the competition entry, social media have actually been applied.

Our analysis was limited to the study of the proposals for and application of social media in the competition entries. Evaluating the value of social media compared to traditional ways of facilitating interaction, and with that public participation and collaboration, was beyond the scope of this paper. Next, the analysis was quite directed by the fact that we studied (the making of) entries for a design competition. Further study of the use of social media in practice is needed, to critically evaluate its additional value to public participation and collaboration in real planning and design processes. When practice would be under study, the framework for analysis should be able to address more criteria and evaluate higher levels of complexity. The framework could be added by criteria concerning the method(s) for interaction, such as flexibility and feasibility, clarity of its communication and to which levels of engagement the methods lead (see VERVOORT et al., 2009).

Nevertheless, in answer to the competition assignment that required that the entries should include local initiatives, involve the local inhabitants and users of the region and depart from stories and experiences of the inhabitants and users of the region, it appeared that more than half of the teams proposed social media for interaction with the public (twenty out of thirty-six). Next, the teams showed that they are aware of a variety of existing platforms in different social media categories and that they can imagine new platforms that would meet the demands for interaction with the public in this specific case (see section 4.1). When looked at the levels of interaction that the teams aimed to facilitate via social media platforms, it appeared that one and two-way communication was aimed for frequently, whereas collaborative decision making and collaborative creation were proposed much less (see section 4.2). Notwithstanding, two teams experimented with applying social media to collaboratively create their competition entry itself. The team 'Ondermekoar' initiated interactions and polls on their blog, Facebook and Twitter and shared a short movie via YouTube. From the accounts of these interactions we concluded that one- and two-way communication certainly took place. The processes of (informal) collaborative decision-making is kept vague, which makes it hard to judge the actual input from the public in the process of collaboratively create the competition entry. When we return to Kratwohl's definition of creation – "putting elements together to form a novel, coherent whole or make an original product" – we should conclude that none of the entries proposed or realised this synthesizing act via existing social media platforms. It may be due to the fact that existing social media platforms are not suited for this; they are directed to exchanging text and ready-made images but do not enable, for instance, sketching a design together. Other reasons can be that the teams were not aware of the possibility to outsource also this part of the process to the 'crowd', or that they did not wish to do so, which would be an interesting question for further research.

We would still argue however, that existing social media platforms can be very valuable for designers and planners to crowdsource for local knowledge and narratives in a relatively quick, easy and cost-effective way. Although collaborative creation of plans and designs as described above may go further, narratives brought forward by inhabitants and users of an area can form a rich base of knowledge and inspiration for planning and design (see also VAN HULST 2012; SANDERCOCK, 2003; POTTEIGER & PURINTON, 1998). We think that the competition entries, including the endeavour of 'Ondermekoar', illustrate this potential of social media, and as long as sufficient attention is being paid to the openness and transparency of the process (for the importance of this for citizen engagement see also HEALY, 1997 and ARNSTEIN, 1969) social media have the potential to enable collaborative decision making and collaborative creation as well. This supports the conclusion that the potential for social media application in planning and design (see for instance MACINTOSH, 2008; KAPLAN & HAENLEIN, 2009; BRABHAM, 2009; KINGSTON et al., 2000) is recognized by many competition participants, but still not fully exploited by all. We suggest that options to support collaborative design intentions by social media embedded webservices should be extensively explored and communicated by landscape designers and planners.

6 Acknowledgements

We should like to thank the Eo Wijers-stichting for putting the competition entries to our disposal and the cooperation in our plan analysis process. Further, we are grateful to the competition participants whose entries we used and displayed. Last but not least, we should like to thank the anonymous referees for their useful comments.

References

Arnstein, S. R. (1969), *A Ladder of Citizen Participation*. In: JAIP, 35 (4), 216-224.

Booher, D. & Innes, J. (2002), *Network Power in Collaborative Planning*. In: Journal of Planning Education and Research, 21, 221-236.

Booher, D. & Innes, J. (2008), *Contemporary movements in planning theory*. In: Hillier, J. & Healy, P. (Eds.), Critical essays in planning theory, volume III. Aldershot, Ashgate Publishing Company.

Brabham, C. (2009), *Crowdsourcing the Public Participation Process for Planning Projects*. In: Planning Theory, 8 (3), 242-262.

Craig, W. J. (1998), *The internet aids community participation in the planning process*. In: Computers, Environment and Urban Systems, 22 (4), 393-404.

Crotty, M. (2009), *The foundations of social research*. London, Sage.

De Jonge, J. (2008), *Een kwart eeuw Eo Wijers-stichting, ontwerpprijsvraag als katalysator voor gebiedsontwikkeling*. Gouda, Habiforum.

De Jonge, J. (2009), *Landscape Architecture between Politics and Science*. Wageningen, Blauwdruk, PhD thesis Wageningen University.

Eo Wijers-stichting (2011), *Eo wijers-prijsvraag 2011-2012. Nieuwe energie voor de Veenkoloniën, op zoek naar regionale comfortzones. Brochure voor de ideeënfase over krimp, energietransitie en ruimtelijke kwaliteit*. Deventer, Eo Wijers-stichting.

Eo Wijers-stichting (2012), *Eo Wijers-prijsvraag 2011-2012. Nieuwe energie voor de Veenkoloniën, op zoek naar regionale comfortzones. Juryrapport*. Deventer, Eo Wijers-stichting.

Hammon, L. & Hippner, H. (2012), *Crowdsourcing*. In: Business & Information Systems Engineering, 3, 163-166.

Healy, P. (1997), *Collaborative Planning. Shaping places in Fragmented Societies*. Houndmills, Macmillan Press.

Jeansson, E., Telldén, E. & Farmanbar, K. (2012), *3D Online: Crowdsourcing for City Planning*. In: GIM International, 12 (26), 28-31.

Kaplan, A. & Haenlein, M. (2009), *Users of the world, unite! The challenges and opportunities of Social Media*. In: Business Horizons, 53, 59-68.

Kempenaar, A., De Waal, R. & Van den Brink, A. (2012), *9e Eo Wijersprijsvraag 2011-2012. Nieuwe energie voor de Veenkoloniën, op zoek naar regionale comfortzones. Plananalyse*. Deventer, Eo Wijers-stichting.

Kingston, R., Carver, S., Evans, A. & Turton, I. (2000), *Web-based public participation geographical information systems: an aid to local environmental decision-making*. In: Computers, Environment and Urban Systems, 24, 109-125.

Kratwohl, D. (2010), *A Revision of Bloom's Taxonomy: An Overview*. In: Theory Into Practice, 41 (4), 212-218.

Ligtenberg, A. & Van Lammeren, R. (2012), *Sociale media als bron voor ruimtelijke informatie*. In: Geo-Info, 9, 16-20.

Macintosh, A. (2008), *The emergence of digital governance*. In: Significance, 12/2008, 176-178.

Ondermekoar (2012), [online] available at: https://twitter.com/ondermekoar (accessed 31 December 2012).

Ondermekoar (2013a), [online] available at: https://www.facebook.com/veenkolonien (accessed 14 January 2013).

Ondermekoar (2013b), [online] available at: http://www.youtube.com/user/regiosoap/videos?view=0&flow=grid (accessed 14 January 2013).

Poplin, A. (2012), *Playful public participation in urban planning: A case study for online serious games*. In: Computers, Environment and Urban Systems, 36, 195-206.

Potteiger M. & Purinton, J. (1998), *Landscape Narratives: Design practices for telling stories*. New York, John Wiley & Sons.

Pröpper, I. (2009), *De aanpak van Interactief beleid: elke situatie is anders*. Bussum, Coutinho.

Rinner, C., Kessler, C. & Andrulis, S. (2008), *The use of Web 2.0 concepts to support deliberation in spatial decision-making*. In: Computers, Environment and Urban Systems, 32, 386-396.

Sandercock, L. (2003), *Out of the Closet: The Importance of Stories and Storytelling in Planning Practice*. In: Planning Theory & Practice, 4 (1), 11-28.

Schwartz-Shea, P. & Yanow, D. (2012), *Interpretive research design. Concepts and processes*. New York, Routledge, Taylor & Francis Group.

Shen, Z. & Kawakami, M. (2010), *An online visualization tool for Internet-based local townscape design*. In: Computers, Environment and Urban Systems, 34, 104-116.

Steinitz, C. (2012), *A Framework for Geodesign, Changing Geography by Design*. Redlands, Esri Press.

Stokman, A. & Von Haaren, C. (2012), *Integrating science and creativity for landscape planning and design of urban areas*. In: Richter M. & Weiland, U. (Eds.), Applied Urban Ecology. Chichester, Wiley-Blackwell.

Stremke, S., Van Kann, F. & Koh, J. (2012), *Integrated visions (part I), Methodological framework for long-term regional design*. In: European Planning Studies, 20, 305-320.

Van Hulst, M. (2012), *Storytelling, a model of and a model for planning.* In: Planning Theory, 11 (3), 299-318.

Van Lammeren, R., Ligtenberg, A., Serpa, J., Abreu, J. & Plezier, I. (2007), *Geo-visualisation: the e-interaction factor in spatial planning.* In: Van den Brink, A. et al. (2007), Imaging the future: geo-visualisation for participatory planning in Europe. Mansholt publication series, 3, 89-105.

Vervoort, J., Kok, K., Van Lammeren, R. & Veldkamp, A. (2010), *Stepping into Futures: exploring the potential of interactive media for participatory scenarios on social-ecological systems.* In: Futures, 42 (6), 604-616.

State-of-the-Art Online Communication for Landscape Architecture

Petra SCHOELKOPF and André EIßER

1 Landscape Architecture within the Context of the Evolving Use of the Internet

Only a small number of people can identify what is involved in landscape architecture beyond the laying out of private gardens, let alone know the variety of services that belong to the profession of landscape architecture, or what many positive effects landscape architecture generates in both rural and urban contexts. An innovative supplying of information cannot only increase the social and political relevance of landscape architecture but also improve appreciation of the built environment and the professions which play a role in it.

Connectivity and collaboration are current approaches in product marketing which can also be applied to culture and services marketing. "*The possibility for dialog with and the relationship to the consumer count more than the price of the product*" and "*The creation of value is a by-product of fulfilling the customer's wishes*" (WIPPERMANN, 2012) are central to the transfer of information. State-of-the-art online communication for Landscape Architecture must acknowledge the evolving means of communication, thereby taking the changes of internet use into account. A new kind of internet access (mobile) and a transformed focus (local) (s. RINGEL & RADNÜZ, 2011) are greatly having an effect on communication in landscape architecture, and improving it.

Today the internet is seen by many experts as the greatest change in the transfer of information since the invention of the printing press. Without doubt, it has an enormous impact on many areas of our daily life. "*The internet is changing from a medium of supplying information to a medium of generating information; that is to say that first through the interaction, the commentary, the evaluation, and with that, an overview of the demand, does a 'message' obtain its value.*" (EISSER & SCHRÖDER, 2012). The internet itself is beginning to change in the direction of communication – the "calf and cow configuration" (SEARLS, 2011) as classic server-client communication is starting to dissolve. Web 2.0 social media (s. SCHÜRING, 2011) is the catchphrase for a series of interactive and collaborative elements of the internet. Here, the user not only consumes the contents, rather he also creates through user generated content. Today, it is no longer about using centrally edited – mostly linear and static – websites as information media, rather it is increasingly about the development of uses of the web in which the demands and expectations of the provider of the information (landscape architects and planners) as well as those searching for information are both taken into consideration.

2 bdla Concept – Online Communication 2012

The German Association of Landscape Architects (bdla) uses both traditional and new media to present itself. Until now, its web presentation has mainly concentrated on communicating association related news. The presentation of products and services of its members – current landscape architecture projects – is seen only in certain information modules such as on the website for the bdla landscape architecture award. At an online communication workshop, the conclusion was reached that landscape architecture must, including reasons already mentioned, be made more accessible, and that this needs to be done on emotional, personal, human, concrete, and tangible levels. With its concept, Online Communication 2012, the bdla has laid the groundwork for adapting its public relations work and supplying of information to new user behavior and current technological developments.

Important strategic points include:

- generating additional values for users through the means of individual communication components (i.e. culture, tourism)
- establishing an integrated and flexible technical and editorial system with a central project databank for the multiple use of data in a variety of communication components which allows for the involvement of decentralized authors
- platform independent publishing with portability of the software to multiple platforms e.g. in mobile internet devices which meets current standards
- making available position sensitive services
- allowing the development of social relationships and creating opportunities for dialog

Under the motto "Communicating landscape architecture spatially related, emotionally appealing and professionally grounded", the bdla is striving to develop access to landscape architecture that is open, inclusive of (building) cultures, and possibly offers a low threshold of entry to a variety of target groups.

3 Components of Online Communication 2012

3.1 Component 1: ParcView Portal

The first component of the new online communication was the launching of the principally (culture) tourist-oriented portal ParcView.de (SCHOELKOPF, 2012) for BERLIN[er] KULTUR[-]LANDSCHAFT[en].

The ParcView portal particularly addresses the cultural meaning of landscape architecture in Berlin. Berlin art, cultural assets, and cultural institutions have always been closely tied to urban open space. They define and inspire each other and mirror their intentions of varied forms of expression and the language of form. The coinciding historic and modern open spaces with their integrative character take on the function of a common experiential space which opens the multitude of cultural facets and attractions to the urban society.

Together they make up a kind of urban cultural landscape, which in Berlin is particularly evident and exciting. The cultural potential that arises from this interaction was formerly not sufficiently made available in Berlin. The goal of ParcView is therefore to develop a new type of access that is highly informative for the interested layman, whether inhabitant or tourist. Within the context of the trend of garden tourism, presenting the tourist's angle of view is a fitting addition in so far that open spaces are seen to have an important potential for the tourism strategy of the State of Berlin (SENATOR FÜR WIRTSCHAFT, TECHNOLOGIE UND FRAUEN, 2011). Because of its culturally interlinking nature, as well as for other reasons, the portal ParcView is being financially supported through funds from the State of Berlin as well as the European Social Fund for Regional Development from the Program to strengthen and exploit innovative potential in culture (INP).

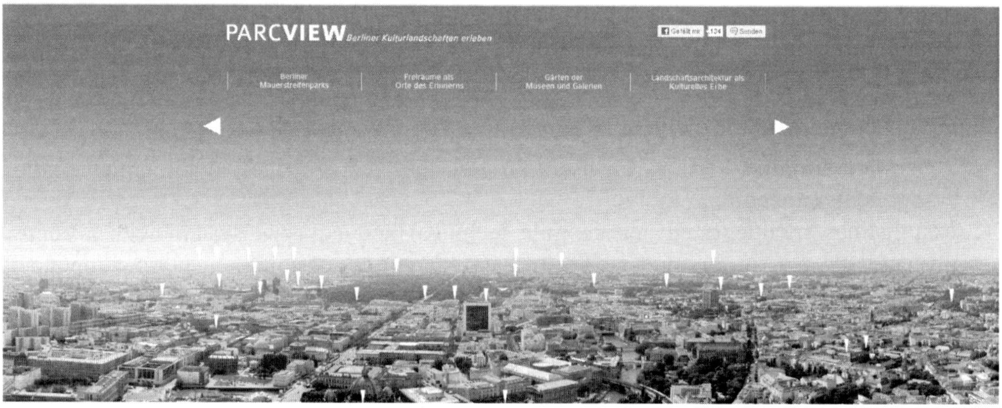

Fig. 1: The 360° pivoting panorama of the city of Berlin on the start page of ParcView for intuitive viewing (screenshot www.parcview.de)

3.2 Component 2: The Portal Landscape Architecture Today

In contrast to ParcView, the portal www.landschaftsarchitektur-heute.de *(landscape architecture-today)* focuses on the direct presentation of the services of the profession and is oriented primarily to professionally interested target groups. Landscape architecture is seen here in the sense of professional concerns as the linking of all the services offered by landscape architects spanning the spectrum from object planning to landscape planning. The information presented covers the sphere of activity of the association nation-wide and, because of the body of work of its members, projects all over the world. The portal concentrates on three main points: offices, projects and topics. An existing module on the bdla.de webpage for searching for planning offices has been further developed and extended and built into Landschaftarchitektur-heute as the menu item *offices*. Here the user finds an interactive map to be used by means of keywords or by filtering the focus of main activities of the planning office. To present the offices, short portraits, project examples, and contact data are available. Every office can present its projects and through texts supplementing construction data and service information, as well as various media formats, offer a more in-depth view of its planning ideas and the concepts and drafts that belong to them. The projects are shown independently in the category projects. They are also linked to the office profiles as references in the database. Additionally the editors from LA-heute can assign the projects to main categories or online exhibition themes in the database and can publish this in the menu item "Themen" (topics). Regardless from which focal point the user approaches the portal, he will always be referred to other content-related focuses.

Fig. 2: The start page of Landschaftsarchitektur-heute.de offers a stage for current topics of landscape architecture (screenshot www.landschaftsarchitektur-heute.de)

3.3 Component 3: Information Hub for the Association and Offices

As part of the new online communication, the bdla's website with its 13 subdomains of state chapters is currently being restructured. The new technical structure and editing concept make it possible that all contents can be accessed in all of the different portals Parcview, LA Heute and bdla.de, as well as in the newsletters, using modules for editing. In addition to association specific communication, widgets can be integrated giving access to the central databank for use in various online projects at the regional or state levels. This widget can additionally be used by individual authors and increases the motivation for using the bdla's project databank. The information and photos of the projects will be only included once, and along with the publication on the bdla portals can be shown on the internet site of the individual offices, for example in the form of a lightbox.

4 Added Values, Multiplicators and Networking

4.1 Online Exhibitions and Special Modes of Narration

Recognizing that the medium alone does not bring attention, intuitive-playful operability, visual presentations and special modes of narration are incorporated into both portals to waken the interest of the public. "*Only then, do you get clicks*" (MUDRA, 2007).

Topic clusters form the basis of the communication and increase the understanding of landscape architecture through thematic links. They are interpreted in the form of spatially concrete, real and virtually experiential online exhibits. Individual topics are presented and augmented with thematically appropriate projects from the database. To date, four online exhibits have been published on ParcView. The online exhibit Mauerstreifenparks *(wall-strip parks)*, for example, shows very different public spaces along the former Berlin Wall which act as a kind of memory for the city. In the portal www.landschaftsarchitektur-heute.de there are initially seven topics and clusters such as urban spaces, developing landscape, and the residential environment. The communication in both portals is supported through special narratives used to enhance the experience of landscape architecture: "gallery", "story", and the "tour", as well as "time transitions". The "project presentation with gallery" is targeted to the specific behavior of the internet reader. Pages are skimmed over and only apparently important anchor points merit a brief look. Longer texts are of less interest than short precise teasers which encourage one to read more. Following this kind of reading behavior, a concept has been developed which allows for the individual and playful discovering of a place through powerful images and short texts. The use of videos speaks to the emotions of the user and allows one to make one's own picture of the intentions and working method of the landscape architect. The "story" provides more in-depth information about specific aspects of a place, for example plant use. The spatial context of the information can be presented in the "story" by simple means. The idea of making the invisible seeable in the sense of augmented reality was the inspiration for this kind of narrative. The experience of landscape architecture is the experience of space and spatial relations. The narrative of the "tour" acknowledges this in that the user is given the possibility to virtually move through space. The new kind of presentation of spatial series of pictures demands a special technical handling of the task and requires the borrowing of online game techniques. The "time transitions" places the focus on the development of a place. "Time transitions" bring the user closer to the historic levels, individual construction phases, or the before and after of a planning.

4.2 Mobile Story-Telling and Location-Based Service

Both portals have been combined with a mobile website (mobil.parcview.de; mobil. landschaftsarchitektur-heute.de) which supports all aforementioned information and multi-media formats of the portals. This allows for a flexible and spontaneous access to information on landscape architecture and its spatial and cultural context. "What current examples of landscape architecture are nearby that I can look at?", "How do I get there?", "Can I find out more when I'm there?" are typical user needs that are supported by the mobile websites. Using the browser of a smart phone with internet access, one can set up a tour of an area without having had to previously install an app.

For its development strategy, the BDLA has chosen to use the concept of web app. When considering marketing however, it's certainly advisable that the lifestyle image and the distribution channels of a native app are used. The possibility of a greatly simplified native app will be explored with an "icon install app".

Fig. 3: Web app mobil.landschaftsarchitektur-heute.de (Perspektivmedien UG)

4.3 Cultural and Tourism Partnerships and Social Media

To increase awareness of the new information services offered, supporting measures for the web and for print have been conceived that come into effect shortly or medium dated after activating. Decisive is that in addition to expert relevant channels of information, other users from other contexts are made aware of the unique value of the portals. The ParcView project is, for example, a partner of many renowned cultural institutes in Berlin such as the Topography of Terror Foundation and the Berlin Wall Foundation. Through linking and data exchange with visitBerlin.de, the official tourism portal of Germany's capital, Berlin's open spaces can presented for the first time in international marketing as the cultural places of interest that they are. What's more is that the bdla, considering the changing perception of the internet, has asked the question to what ends is using the marketing strategies of social media relevant to providing information. At a workshop for developing the new online communication concept, the consensus was reached that to date landscape architecture-related target groups don't make much use of the possibilities of Web 2.0. This can lie in the age structure, or the affinity for the medium of the potential target group, but also in the lack of specific offerings. Exactly for this reason, the use of state-of-the-art technological services needs to be explored for increasing the attractiveness of the built environment for young people. Therefore the bdla has decided to gather first experiences of effectiveness of the social media in the internet and for this to initially use external infrastructures. In both portals the visitor can recommend contents to others via Facebook, and on youtube videos about landscape architecture projects can be commented on.

5 Current Technical Developments and User-Friendliness

The portability of the software to multiple platforms and modular design are the most important challenges in the incremental development process so that both portals offer flexible, media-continuous tools which fulfill the entire future online communication of the bdla.

The basis for implementing the portals and the mobile web pages is RegionCreator, originally developed for the Open Source Content Management System (CMS) Joomla!, an extension developed for the integration of the Joomla platform 11.x. framework using the concept *model, view, controller (MVC).* This extension also works without the CMS and therefore can also be used for other applications. The focus of the development was to make patterns available for an object-oriented entry and output of data by the user. Whereas each pattern is based on a data-structure generalized catalog with classes and methods with which a variety of individual objects can be flexibly generated and made available. The basic data structure of the patterns is a content builder, developed similarly to content construction kits of the CMS Frameworks Drupal type. With this, new content types can be defined and immediately:

1) decentral contents from different authors in differentiated access levels can be added and
2) these contents can be not only retrieved through the above named application programming interface (API) but also combined and linked to (new) objects.

With the last phase of implementing a query syntax (parameter), an object-oriented API was created that allows for the retrieval and output of regional and/or thematic information, similar to the Graph API on Facebook. All the add-ons (components and modules) within the CMS for the central editors also work on the basis of this API. They not only deliver the necessary objects on the data level but also vary the templates, depict the results of project queries in a grid/raster, simultaneously deliver the data to the map API and if desired transfer all data to a JavaScript Object Notation (JSON) interface. This way a single CMS can also administrate and operate external applications, such as mobile web apps do.

JQuery in JavaScript framework mobile builds the basis for using the data in mobile web pages so that it is already possible today to implement many HTML 5 functions, even if this is not yet fully supported by the browser manufacturer. Additionally, jQuery mobile offers a series of multiplatform functions and display options (for Andoid, Windows, iPhone, Blackberry, etc.) which comply with user behavior of the internet and smartphones (buttons, list views, page transitions, map views, etc.) The mobile applications can be structured and administrated directly in the CMS of the portal. The JSON interface of the above mentioned extensions produces the data which then can be processed and displayed within the mobile application.

6 Summary and Outlook

It was the idea that bdla members could present their own activities that brought about the development of the new online communication of the bdla. Through the interplay of searching for planning offices and the possibility that presenting individual projects could serve as the nucleus for the communication in the most varied of contexts, two platforms came into being which fill the gap (cf. SCHOELKOPF & RINGKAMP, 2008) that had previously existed in the professional and cultural arenas. In addition, the platform Landschaftsarchitektur-heute has the potential of becoming a renowned online salon for the current works of landscape architecture.

Indeed, the proof of the visual-emotional approach of development seems to be already there. A first analysis of the online visits to ParcView shows a much longer length of stay on the website than the average. The great acceptance of the site www.landschafts architektur-heute.de by bdla members already during the development stage has led to around 800 project entries and 160 office profiles. This is extremely unusual for a web project that is based on the participation of decentralized authors. A great number of page views of ParcView have come through external sources of the social media (Facebook) and we are eager to see whether this will also be the case for Landschaftsarchitektur-heute. The analysis will provide more insight into a further development in the direction of social media functionality. Within the first two months of operation, Landschaftsarchitektur-heute had over 10,000 visits and around 141,000 page views. This initial feedback allows for a positive view of future visitor numbers, especially as extensive communication measures on the web and in print will follow in the coming months. Whether the information services are accepted not just by the providers but also by the user, and if they are to become a viable pillar for future online communication remains to be seen. In any case, the initial positive effects encourage the bdla to consider developing further features that support the sustainable providing of information.

References

Eisser, A. & Schröder, T. (2012), In: Rein, H. & Schuler, A. (Hrsg.), *Tourismus im ländlichen Raum*. Wiesbaden, Springer Gabler.

Kahl, M. (2009), *Online-Kommunikation 2012*. In: Landschaftsarchitekten, 2/2009, 22.

Mudra, T. (2007), *Öffentlichkeitsarbeit 2012. Verträgt sich ein modernes Marketingkonzept mit dem Selbstverständnis des bdla?* In: Landschaftsarchitekten, 1/2007, 24.

Ringel, T. & Radünz, A.-L. (2011), *SoLoMo – Die Social Local Mobile Bewegung*. In: Schwarz, T. (Hrsg.) (2011), Leitfaden Online Marketing. Band 2.

Senator für Wirtschaft, Technologie und Frauen (2011), *Tourismuskonzept Berlin – Handlungsrahmen 2011+*.

Schoelkopf, P. (2012), *Landschaftsarchitektur emotional präsentiert*. In: Stadt und Grün, 9/2012, 49.

Schoelkopf, P. & Ringkamp, C. (2008), *Deutsche Datenbank der Gärten und Parks – Vision oder notwendige Herausforderung?* In: Buhmann, E., Pietsch, M. & Heins, M. (Eds.), Digital Design in Landscape Architecture 2008. Heidelberg, Wichmann.

Searls, D. (2011), *Habt Geduld*. In: brand eins, Wirtschaftsmagazin, 2/2012, 43.

Websites

http://www.henningschuerig.de/blog/2010/social-media-statt-web-20/ (20.12.2012): Schuering, H.: *Social Media statt Web 2.0*.

http://www.trendbuero.de/index.php?f_articleId=3414 (29.09.2012): Wippermann, P.: *Der Quellcode der Netzwerkökonomie: Connectivity + Collaboration + Coopetition + Co-Creation*.

Zeitgemäße Online-Kommunikation für Landschaftsarchitektur

Petra SCHOELKOPF und André EIßER

1 Landschaftsarchitektur im Kontext der veränderten Wahrnehmung des Internets

Nur ein kleiner Teil der Bevölkerung kann mit Landschaftsarchitektur Inhalte verbinden, die über das Anlegen von Hausgärten hinausgehen, geschweige denn, dass er weiß, welche Leistungen zum Berufsbild von Landschaftsarchitekten gehören oder welche vielschichtigen, positiven Effekte Landschaftsarchitektur im ländlichen und urbanen Kontext auslöst. Eine zeitgemäße Informationsvermittlung kann die gesellschaftliche und politische Relevanz von Landschaftsarchitektur stärken und letztlich zu einer verbesserten Wertschätzung gebauter Außenräume und des zuständigen Berufsstandes beitragen. Connectivity and Collaboration sind aktuelle Ansätze, die sich aus dem Produktmarketing auf das Kultur- und Dienstleistungsmarketing übertragen lassen. *„Die Inszenierung von Gesprächsanlässen und die Beziehungen zu den Konsumenten zählt mehr als die Anpreisung des Produkts"* und *„Die Wertschöpfung wird ein Nebenprodukt bei der Erfüllung von Kundenwünschen"* (WIPPERMANN, 2012) sind dabei zentrale Einschätzungen, die die Informationsvermittlung bestimmen. Eine zeitgemäße Online-Kommunikation zur Landschaftsarchitektur muss sich dieser geänderten Kommunikation stellen und dabei grundlegende Veränderungen der Internetnutzung berücksichtigen. Eine neue Art des Internetzugangs (mobil) und ein gewandelter Fokus (lokal) (vgl. RINGEL & RADÜNZ, 2011) wirken sich auch im hohen Maße auf die Kommunikation von Landschaftsarchitektur aus und begünstigen diese zudem.

Heute wird das Internet von vielen Experten als die größte Veränderung des Informationswesens seit der Erfindung des Buchdrucks gesehen, zweifelsohne mit enormen Auswirkungen auf viele Bereiche unseres täglichen Lebens. „Es wandelt sich das Internet vom Medium der Informationsvermittlung zum Medium der Informationsgenerierung, *d. h. erst durch die Interaktion, den Kommentar, die Wertung und damit durch eine Abbildung der Nachfrage, wird die 'Botschaft' werthaltig"* (EISSER & SCHRÖDER, 2012). Dabei beginnt sich das Internet selbst in der Richtung der Kommunikation zu wandeln – die „Kalb- und Kuh-Konfiguration" (SEARLS, 2011) als klassische *One-Way* Server-Client-Kommunikation beginnt sich mehr und mehr aufzulösen. Web 2.0 „Social Media" (vgl. SCHÜRING, 2011) sind dabei die Schlagwörter, für eine Reihe interaktiver und kollaborativer Elemente des Internets. Hierbei konsumiert der Nutzer nicht nur den Inhalt, sondern er stellt als Prosument („User Generated Content") selbst Inhalte zur Verfügung. So geht es heute nicht mehr um die Umsetzung von redaktionell-zentral – meist linear und statisch – aufbereiteten Webseiten als Informationsmedium, sondern es geht nunmehr verstärkt um die Entwicklung von Webanwendungen, in denen sich sowohl die Ansprüche und Erwartungen der Absender von Informationen (Landschaftsarchitekten und Planer) als auch der Informationssuchenden zeitgemäß wiederfinden.

2 bdla-Konzept „Online-Kommunikation 2012"

Der Bund Deutscher Landschaftsarchitekten (bdla) betreibt zwar Öffentlichkeitsarbeit in den klassischen und Neuen Medien. Das Webangebot konzentriert sich allerdings auf die Kommunikation verbandspezifischer Inhalte. Die Darstellung der Produkte und Dienstleistungen seiner Mitglieder – realisierte Projekte aktueller Landschaftsarchitektur - erfolgt nur in einigen ausgewählten Informationsmodulen wie z. B. in der Website zum Landschaftsarchitekturpreis. Ein Workshop zur Online-Kommunikation ergab, dass die Auseinandersetzung mit Landschaftsarchitektur, auch aus den eingangs skizzierten Gründen, emotionaler, persönlicher, menschlicher, konkreter und greifbarer gestaltet werden soll (KAHL, 2009) Mit seinem Konzept zur „Online-Kommunikation 2012" hat der bdla die Weichen gestellt, seine Öffentlichkeitsarbeit und Informationsvermittlung im Internet grundlegend auf das geänderte Nutzerverhalten auszurichten und an die aktuelle technische Entwicklung anzupassen. Wichtige strategische Eckpunkte sind:

- Generierung von Mehrwerten für den Nutzer und Überführung in unterschiedliche, eigenständige Kommunikationsbausteine (z. B. Kultur, Tourismus).
- Etablierung eines integrierten, flexiblen technischen und redaktionellen Systems mit zentraler Projektdatenbank für die Mehrfachdatennutzung in unterschiedlichen Kommunikationsbausteinen, das die Einbindung dezentraler Autoren zulässt.
- Plattformunabhängiges Publizieren im mobilen Internet mittels neuer Standards.
- Bereitstellung positionssensitiver Dienste.
- Ermöglichung sozialer Beziehungen und Schaffung von Gesprächsanlässen.

Unter dem Leitbild „Landschaftsarchitektur räumlich verortet, emotional ansprechend, fachlich fundiert vermitteln" strebt der bdla an, einen (bau-)kulturübergreifenden, offenen und vielleicht auch niedrig-schwelligen Zugang zu Landschaftsarchitektur für die unterschiedlichen Zielgruppen zu ermöglichen.

2 Bausteine der Online-Kommunikation 2012

2.1 Baustein 1: Portal „ParcView"

Als erster Baustein der neuen Online-Kommunikation wurde das eher (kultur-)touristisch orientierte Portal ParcView.de (SCHOELKOPF, 2012) zum Erleben der BERLIN[er] KULTUR[-]LANDSCHAFT[en] im Internet gelauncht.

Das Portal „ParcView" thematisiert insbesondere die kulturelle Bedeutung von Landschaftsarchitektur in Berlin. Berliner Kunst, Kulturgüter und Kultureinrichtungen sind seit jeher eng verwoben mit dem urbanen Freiraum. Sie bedingen und inspirieren sich gegenseitig und spiegeln ihre Intentionen in unterschiedlichen Ausdrucksformen und Formensprachen. Der gleichwohl historische wie zeitgenössische Freiraum mit seinem integrativen Charakter übernimmt dabei die Funktion des gemeinsamen Erfahrungsraumes, aus dem heraus sich die städtische Gesellschaft die unterschiedlichen kulturellen Facetten und Angebote erschließen kann. Zusammen bilden sie eine Art urbane KulturLandschaft, die in Berlin besonders deutlich ausgeprägt und sehr spannungsreich ist. Das kulturelle Potenzial, das sich aus diesem Zusammenspiel ergibt, war bislang nur unzureichend in Berlin erschlossen. Ziel des Vorhabens „ParcView“ ist es deshalb, einen neuartigen und erkenntnisreichen Zugang zu entwickeln, der im hohen Maße für interessierte Laien z. B. kulturinteressierte Bürger und Touristen von Bedeutung ist. Vor dem Hintergrund des steigenden Trends zum Gartentourismus stellt der touristische Betrachtungswinkel eine sinnvolle Ergänzung dar, zumal Freiräume auch in den touristischen Strategien des Landes Berlin als wichtiges Potenzial gesehen werden (SENATOR FÜR WIRTSCHAFT, TECHNOLOGIE UND FRAUEN, 2011). Das Portal „ParcView“ wird u. a. aufgrund seines kulturell vernetzenden Charakters aus Mitteln des Landes Berlin sowie aus Mitteln des Europäischen Fonds für Regionale Entwicklung des Programms zur Stärkung und Nutzung des Innovationspotenzials in der Kultur (INP) gefördert.

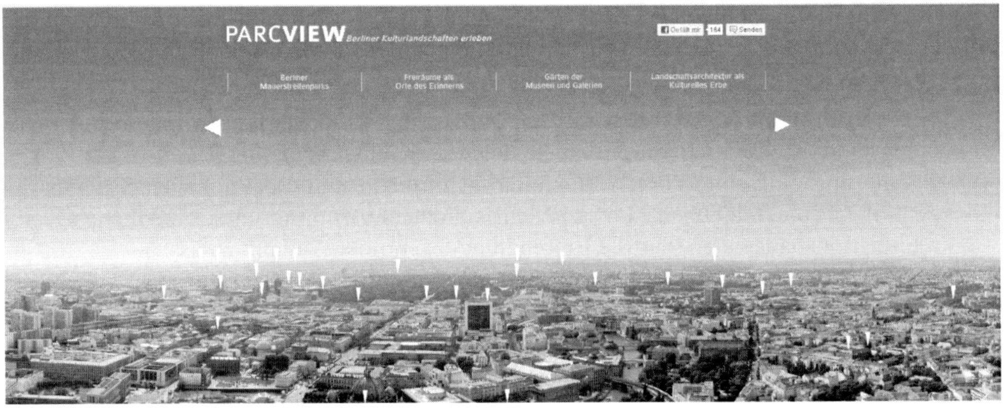

Abb. 1: Das 360° schwenkbare Panorama der Stadt Berlin auf der Startseite des Portals ParcView zur intuitiven Annäherung (screenshot www.parcview.de)

2.2 Baustein 2: Portal „Landschaftsarchitektur-heute“

Im Gegensatz zu „ParcView“ fokussiert sich das Portal www.landschaftsarchitektur-heute.de auf die unmittelbare Darstellung der Leistungen der Profession und richtet sich vorrangig an fachlich interessierte Zielgruppen. Landschaftsarchitektur wird dabei im Sinne der berufsständischen Auffassung als Klammer für alle Leistungsfelder der Landschaftsarchitekten gesehen, die beispielsweise von der Objektplanung bis Landschaftsplanung reichen. Das Informationsangebot ist entsprechend dem Wirkungskreis des Verbandes bundesweit ausgerichtet, umfasst aber aufgrund der Tätigkeit seiner Mitglieder Projekte in aller Welt. Inhaltlich konzentriert sich das Portal auf die drei Schwerpunkte „Büros“, „Projekte“ und „Themen“. Aus dem Ansatz der Planungsbürosuche auf der Verbandsseite entstand der Schwerpunkt „Büros“. Hier findet der Nutzer über eine interaktive Karte, per Stichwortsuche oder Filterung nach Tätigkeitsschwerpunkten das Planungsbüro, das er sucht. Zur Präsentation der Büros stehen u. a. Kurzporträts, Projektbeispiele und Kontaktdaten zur Verfügung. Jedes Büro kann seine Projekte vorstellen und über Textbeschreibungen in Ergänzung mit Baudaten und Serviceinformationen sowie unterschiedlichen Medienformaten vertiefte Einblicke in die Planungsidee und dazugehörigen Konzepte und Entwürfe eröffnen. Im Sinne vernetzter Inhalte werden die Projekte den Büros zugewiesen. Darüber hinaus können sie übergeordneten thematischen Bezügen zugeordnet und daraus aktuelle Themen kreiert werden (siehe 3.1). Egal, aus welchem Interessenschwerpunkt sich der Nutzer dem Portal nähert, er wird immer wieder auf die anderen inhaltlichen Schwerpunkte verwiesen.

2.3 Baustein 3: Büro- und verbandsbezogene Informationsvermittlung

Im Zuge der neuen Online-Kommunikation befindet sich auch die Verbandsseite des bdla mit seinen 13 Subdomains der Landesgruppen in der Umstrukturierung. Es findet eine technisch-inhaltliche Verknüpfung aller Kommunikationsinhalte über alle Trägermedien statt. Neben der verbandsspezifischen Kommunikation können hier Widgets eingebunden werden, die Landschaftsarchitektur im Sinne des neues Kommunikationsleitbildes in unterschiedlichen Online-Projekten auch auf regionaler oder landesbezogener Ebene aus der zentralen Projektdatenbank vermittelt. Dieses Widget kann darüber hinaus als Service für die einzelnen Autoren eingesetzt werden und erhöht die Motivation zur Nutzung der bdla-Projektdatenbank. Dabei werden Daten und Fotos der Projekte nur einmal erfasst und können neben der Publikation in den bdla-Portalen auf der Internetseite des jeweiligen Planungsbüros z. B. in Form einer sogenannten Lightbox angezeigt werden.

3 Mehrwerte, Multiplikatoren und Vernetzungen

3.1 Online-Ausstellungen und besondere Erzählweisen

Wohl wissend, dass das Medium allein keine Aufmerksamkeit schafft, werden mit intuitiv-spielerischer Bedienbarkeit, visueller Präsentation und besonderen Erzählweisen Angebote in beiden Portalen geschaffen, die das Interesse der Menschen wecken sollen. „Nur dann wird angeklickt." (MUDRA, 2007). Themencluster bilden die Grundlage der Informationsvermittlung und schärfen das Verständnis für Landschaftsarchitektur durch das Aufzeigen thematischer Bezüge. Sie werden in Form von räumlich konkreten, real und virtuell erfahrbaren Online-Ausstellungen übersetzt. Dazu wird das jeweilige Thema redaktionell vorgestellt und mit thematisch passenden Projekten aus der Datenbank untersetzt. Bislang wurden für „ParcView" vier Online-Ausstellungen publiziert. Die Online-Ausstellung „Mauerstreifenparks" beispielsweise zeigt ganz unterschiedliche Orte, die den öffentlichen Raum im Verlauf der ehemaligen Mauer als Gedächtnis der Stadt inszenieren. Im Portal Landschaftsarchitektur heute werden vorerst sieben Themen und Cluster wie z. B. Urbane Plätze, Landschaft entwickeln oder Wohnumfeld präsentiert.

Die Informationsvermittlung wird in beiden Portalen unterstützt durch besondere, auf das Erleben von Landschaftsarchitektur zugeschnittene Erzählweisen: die Galerie, die Story, der Rundgang sowie die Zeitblende. Die „Projektdarstellung mit Galerie" ist auf das spezifische Leseverhalten im Internet ausgerichtet. Seiten werden überflogen und nur an scheinbar wichtigen Ankerpunkten wird ein kurzer Halt eingelegt. Lange Texte werden weniger wahrgenommen als kurze prägnante „Teaser", die Interesse zum „mehr lesen" auslösen. Ausgehend von diesem Verhalten wurde ein Konzept entwickelt, bei dem die spielerische und individuelle Entdeckung des Ortes über aussagekräftige und ansprechende Bilder und kurze Texte im Mittelpunkt der Darstellung steht. Die Einbindung von Videos unterstützt die emotionale Ansprache des Nutzers und ermöglicht es, sich ein Bild von den Intentionen und Arbeitsweisen der Landschaftsarchitekten zu machen. Die „Story" vermittelt tiefer gehende Informationen zu besonderen Aspekten eines Raumes z. B. zur Pflanzenverwendung. Der räumliche Kontext zur Information kann bei der „Story" auf einfache Weise hergestellt werden, da bei der Entwicklung dieser Erzählweise die Idee, das Unsichtbare sichtbar im Sinne sogenannte Augmented Reality zu machen, die Inspiration bot. Das Erleben von Landschaftsarchitektur ist eine Erfahrung von Raum und räumlichen Zusammenhängen. Die Erzählweise des „Rundganges" berücksichtigt dies, in dem der Nutzer die Möglichkeit erhält, sich virtuell durch Räume zu bewegen. Die neuartige Darstellung von raumbezogenen Bildfolgen erforderte auch einen besonderen technischen Umgang mit der Aufgabenstellung und machte Anleihen bei der Programmierung von Online-Games erforderlich. Die „Zeitblende" stellt die Entwicklung des Ortes in einzelnen Abschnitten in den Mittelpunkt der Darstellung. Mit der Zeitblende werden dem Nutzer historische Schichten, einzelne Bauabschnitte oder das Vorher und Nachher einer Planung näher gebracht.

3.2 Mobile Story telling und Location-based Services

Die beiden Portale wurden jeweils mit einer mobilen Website (mobil.parcview.de; mobil.landschaftsarchitektur-heute.de) kombiniert, die alle eingangs beschriebenen Informationen und Mulitmediaformate der Portale unterstützt. Damit wird ein flexibler und spontaner Zugriff auf Informationen zur Landschaftsarchitektur und auf den jeweiligen räumlichen und kulturellen Kontext ermöglicht. „Welche Orte aktueller Landschaftsarchitektur kann ich in der Nähe anschauen?", „Wie komme ich am besten dorthin?" und „Kann ich vor Ort mehr erfahren?" sind typische Nutzerbedürfnisse, die die mobilen Websites unterstützen. Sie können über den Browser eines Smartphones mit Internetzugang ohne vorherige Installation für Erkundungstouren vor Ort eingesetzt werden.

Abb. 2: WebApp mobil.landschaftsarchitektur-heute.de (Perspektivmedien UG)

Der bdla hat sich bei der Entwicklungsstrategie für das Konzept der Web-App entschieden. Aus Überlegungen des Marketings will man allerdings das Lifestyle-Image und die Vertriebskanäle einer nativen App dennoch nutzen. Mit einer Art Icon-Install-App sollen die Möglichkeiten einer stark vereinfachten nativen App ausgelotet werden.

3.3 Kulturelle und touristische Kooperationen und Social Media

Um die Bekanntheit der neuen Informationsangebote wirksam zu steigern, wurden flankierende Kommunikationsmaßnahmen für Web und Print konzipiert, die kurz- bis mittelfristig nach der Freischaltung umgesetzt werden sollen. Entscheidend dabei ist, dass neben den fachrelevanten Informationskanälen auch Nutzer aus anderen Kontexten über bestimmte Mehrwerte auf die Portale aufmerksam werden. Das Vorhaben ParcView wird beispielsweise von vielen renommierten Berliner Kultureinrichtungen wie der Stiftung Topographie des Terrors sowie der Stiftung Berliner Mauer als Kooperationspartner begleitet. Durch Vernetzung und Datenaustausch mit visitBerlin.de, dem offiziellen Tourismusportal der deutschen Hauptstadt, können sich die Berliner Freiräume nun erstmals als kulturtouristische Sehenswürdigkeiten in der internationalen Vermarktung beweisen.

Dem bdla stellte sich darüber hinaus vor dem Hintergrund der veränderten Wahrnehmung des Internets die Frage, inwieweit der Einsatz von Strategien des Social Media Marketings für die Informationsvermittlung erforderlich ist. Bei einem Expertenworkshop zur Entwicklung der neuen Online-Kommunikation bestand Konsens in der Einschätzung, dass Landschaftsarchitektur-affine Zielgruppen bisher wenig Gebrauch von den Möglichkeiten des Web 2.0 machen. Das kann u. a. an der Altersstruktur, der Medienaffinität der potenziellen Zielgruppen aber auch an dem Fehlen spezifischer Angebote liegen. Gerade vor diesem Hintergrund sollte modellhaft geprüft werden, ob eine zeitgemäße Technologieversorgung die Attraktivität baukultureller Arbeitswelten für junge Menschen steigern kann. Der bdla hat sich deshalb entschlossen, erste Erfahrungen zur Wirksamkeit der sozialen Netzwerke im Internet zu sammeln und dafür vorerst externe Infrastrukturen zu nutzen. In beiden Portalen kann der Besucher Inhalte, die ihm gefallen z. B. eine bestimmte Online-Ausstellung anderen Interessierten bei Facebook empfehlen und Videos zu Landschaftsarchitekturprojekten können auch bei YouTube kommentiert werden.

4 Aktuelle technische Entwicklungen und Benutzerfreundlichkeit

Plattformunabhängiges Publizieren und modularer Aufbau begleiteten als wichtigste Anforderung den inkrementellen Entwicklungsprozess, um mit der Freischaltung der Portale möglichst flexible, medienbrucharme Werkzeuge zur Verfügung zu stellen, die der gesamten zukünftigen Online-Kommunikation des bdla gerecht werden.

Die Basis für die Umsetzung der Portale und der mobilen Webseiten bildet der Region-Creator, eine ursprünglich für das Open-Source-Content-Management-System (CMS) Joomla! Nach dem Konzept *Modell – Präsentation – Steuerung* (MVC) entwickelte Extension, welche jedoch auch über die Einbindung der Joomla Plattform 11.x. (Framework) ohne das CMS funktioniert und auf diese Weise bei Bedarf auch für weitere Applikationen genutzt werden kann. Im Fokus der Entwicklungsarbeit stand die Bereitstellung von Lösungsschablonen („Patterns“) für eine objektorientierte Ein- und Ausgabe der Informationen. Wobei sich die einzelnen Pattern eines auf die Datenstruktur generalisierten Kataloges an Klassen und Methoden bedienen mit denen eine Vielzahl an individuellen Objekten flexibel er- und bereitgestellt werden können. Die den Pattern zu Grunde liegende Datenstruktur ist ein an den „Content-Construction-Kit" des CMS Frameworks Drupal angelehnter sogenannter Content-Builder, mit dem sich beliebige neue Inhaltstypen definieren und damit bereits unmittelbar

(1) Inhalte dezentral durch verschiedene Autoren in differenzierten Zugriffsebenen hinzufügen und diese
(2) Inhalte über die o. g. API nicht nur abfragen, sondern auch zu (neuen) Objekten zusammenfassen und verknüpfen

lassen.

Über die Implementierung einer eigenen Abfrage-Syntax (Parameter) entstand im letzten Schritt eine objektorientierte Programmierschnittstelle (API) mit der sich Informationen, ähnlich der Graph API von Facebook, etwa in ihren regionalen und/oder thematischen Bezügen, abfragen und ausgeben lassen. Auf Basis dieser API arbeiten in der Konsequenz auch alle für die zentrale Redaktion innerhalb des CMS bereitgestellten Erweiterungen (Komponenten und Module). Jedoch liefern diese hier nicht nur die benötigten Objekte auf der Datenebene, sondern variieren in der Darstellungsebene (Template), stellen also die Ergebnisse einer Projektsuchanfrage in einem Grid/Raster dar, liefern gleichzeitig die Daten an die Karten-API und übermitteln bei Bedarf zudem auch alle Daten an eine JSON-Schnittstelle, so dass über ein und dasselbe CMS auch externe Applikationen, wie etwa die mobile Web-App, administriert und gesteuert werden können.

Zur Nutzung der Daten in der mobilen Webseite bildet das in Javascript-Framework „JQuery mobile" die Basis, mit dem es bereits heute möglich ist viele HTML-5-Funktionalitäten umzusetzen, auch wenn diese durch die Browserhersteller noch nicht vollständig unterstützt werden. Zudem bietet „JQuery mobile" eine Reihe an plattformübergreifenden Funktionen und Darstellungsvarianten (also für Android, Windows, iPhone, Blackberry, etc.), welche dem Nutzerverhalten des Internets auf Smartphones entsprechen (Buttons, Listenansichten, Seitenübergänge, Kartendarstellungen, etc.). Direkt im CMS des Portals kann die mobile Applikation strukturiert und verwaltet werden. Die JSON-Schnittstellen der oben beschriebenen CMS-Erweiterungen liefern die Daten die dann innerhalb der mobilen Applikation entsprechend verarbeitet und aufbereitet (dargestellt) werden.

5 Fazit und Ausblick

Die neue Online-Kommunikation entstand aus der Idee, dass sich bdla-Mitglieder mit ihrer Tätigkeit präsentieren können. Aus dem Zusammenspiel der Planungsbürosuche und der Möglichkeit, dass das einzelne Projekt als Nukleus der Informationsvermittlung in unterschiedlichen Kontexten publiziert werden kann, entstanden zwei Plattformen, die eine bislang vorhandene Lücke in der Online-Informationsvermittlung zur Landschaftsarchitektur (vgl. SCHOELKOPF & RINGKAMP, 2008) auf fachbezogener und kultureller Ebene schließen. Die Plattform Landschaftsarchitektur-heute hat darüber hinaus das Potenzial, ein renommierter Online-Salon mit einer Werkschau für aktuelle Landschaftsarchitektur im Netz zu werden. Bereits kurzfristig scheint sich der visuell-emotionale Entwicklungsansatz zu bestätigen. Eine erste Analyse der Zugriffszahlen von „ParcView" zeigt eine deutliche höhere Verweildauer auf der Website als durchschnittlich üblich ist. Die starke Akzeptanz der bdla-Mitgliedsbüros auch schon während der Entwicklungsphase führte bei Landschaftsarchitektur-heute zu der hohen Anzahl von rund 800 Projekteinträgen und 160 Büroprofilen, die äußerst ungewöhnlich für Webprojekte ist, die auf die Mitwirkung dezentraler Autoren ausgerichtet sind. Besonders viele Seitenzugriffe auf „ParcView" wurden über die externen Informationsquellen der sozialen Medien (Facebook) erreicht und lassen gespannt sein, ob sich dieser Ansatz auch für Landschaftsarchitektur-heute bestätigen wird. Die Analyse wird wichtige Hinweise für eine Weiterentwicklung in Richtung Social-Media-Funktionalitäten liefern. Für das Portal Landschaftsarchitektur-heute konnten in den ersten zwei Monaten nach Freischaltung über 10.000 Besuche mit rund 141.000 Seitenansichten ermittelt werden. Diese erste Zwischenbilanz lässt eine positive Entwicklung der Besucherzahlen erwarten, zumal umfassende Kommunikationsmaßnahmen in Web und Print erst in den folgenden Monaten umgesetzt werden und Wirkung zeigen sollen. Letztlich muss sich mittelfristig zeigen, ob das Informationsangebot nicht nur angebotsseitig, sondern auch nutzerseitig akzeptiert und zu einer tragfähigen Säule der zukünftigen Online-Kommunikation werden wird. Die ersten positiven Effekte bestärken den bdla in jedem Fall, die Entwicklung weiterer Features in Erwägung zu ziehen, die zudem das nachhaltige, redaktionelle Betreiben der Informationsangebote auch in wirtschaftlicher Hinsicht befördern sollen.

Quellen

Eisser, A. & Schröder, T (2012), In: Rein, H. & Schuler, A. (Hrsg.), *Tourismus im ländlichen Raum*. Springer.

Kahl, M. (2009), *Online-Kommunikation 2012*. In: Landschaftsarchitekten, 2/2009, 22.

Mudra, T. (2007), *Öffentlichkeitsarbeit 2012. Verträgt sich ein modernes Marketingkonzept mit dem Selbstverständnis des bdla?* In: Landschaftsarchitekten, 1/2007, 24.

Ringel, T. & Radünz, A.-L. (2011), *SoLoMo – Die Social Local Mobile Bewegung*. In: Schwarz, T. (Hrsg.), Leitfaden Online Marketing, Band 2.

Senator für Wirtschaft, Technologie und Frauen (2011). *Tourismuskonzept Berlin – Handlungsrahmen 2011+* , 17.

Schoelkopf, P. (2012), *Landschaftsarchitektur emotional präsentiert.* In: Stadt und Grün, 9/2012, 49.

Schoelkopf, P. & Ringkamp, C. (2008), *Deutsche Datenbank der Gärten und Parks – Vision oder notwendige Herausforderung ?* In: Buhmann, E., Pietsch, M. & Heins, M. (Eds.), Digital Design in Landscape Architecture 2008. Heidelberg, Wichmann Verlag.

Searls, D. (2011), *Habt Geduld.* In: brand eins, Wirtschaftsmagazin, 2/2012, 43.

Webseiten

http://www.henningschuerig.de/blog/2010/social-media-statt-web-20/ (20.12.2012): Schuering, H.: *Social Media statt Web 2.0.*

http://www.trendbuero.de/index.php?f_articleId=3414 (29.09.2012): Wippermann, P.: *Der Quellcode der Netzwerkökonomie: Connectivity + Collaboration + Coopetition + Co-Creation.*

Konzeption und Aufbau eines Crowdsourcing-Portals zur Visualisierung geocodierter Meldungen eines sozialen Nachrichtendienstes am Beispiel von Twitter

Concept and Construction of a Crowdsourcing Portal to Visualize Geocoded Messages of a Social Network Based on the Example of Twitter (English summary)

Stefanie MÄNNL, Mareike WENDTLAND und Martin BECKER

Summary

Maps are changing our world and the view on complex issues in ways we could hardly imagine just a few years ago. Nowadays, the combination of communication channels of Web 2.0 with cloud services and crowdsourcing enables completely new ways of building interactive maps and information visualization e.g. real-time generated information maps. In connection with the spreading number of GPS sensors in mobile devices, also untrained people get the chance to publish GPS supported images, tracks and messages on the internet. Challenges in the enormous potential of usage are shown by the way of structuring, harmonizing and quality checks of the huge amount of collected data. Beyond every crowdsourcing application the basic concept of filtering the amount of information consists of three general steps. The typical process is divided into the elevation of information by the community, as well as the filtering and weighting as the second and third step of collecting information.

The following paper offers the concept, construction, technical design and essential workflow of a prototype, which handles and visualizes geospatial messages of the news broadcast Twitter in a web application. Because of the various scenarios of usage, the development of geo mashups, which is how applications like this are called, experienced an upswing on the internet in the recent years. In comparison to existing examples in the web, the potential of the introduced application lies in the persistent storage of crowdsourcing information in a spatial database and the resulting opportunities of structuring and functional harmonization of the collected data. New and various evaluated potentials could be emerged by linking these information with other spatial data. Another advantage would be the possibility of temporal spatial analyses by using such stored data. The project has shown that the development of a crowdsourcing application could completely be built out of OpenSource software and that the potential of usage is given by various amounts of scenarios e.g. in sustainable land management or landscape architecture.

1 Einleitung

Karten verändern die Welt in einer Art und Weise, wie es vor wenigen Jahren nur schwer vorstellbar war. Informationskanäle des Web 2.0 in Kombination mit Cloud-Diensten und Crowdsourcing ermöglichen völlig neue Formen der Kartenerstellung bspw. aus Echtzeitdaten generierte Informationskarten. In Verbindung mit der sich ständig ausbreitenden Zahl von mobilen, GPS-gestützten Erfassungssensoren können auch wenig ausgebildete Nutzer der Technik raumgestützte Informationen wie z. B. GPS-gestützte Fotografien, GPS-Tracks oder auf GPS- basierende Textmeldungen erstellen und über das Internet verfügbar machen. Die Herausforderung ist dabei, das riesige Potenzial dieser Daten unterschiedlicher Art und Qualität durch Strukturierung und Verknüpfung in nutzbare Information zu wandeln. Positive Anwendungsbeispiele, u. a. im Crisis Mapping mit der Bündelung und Visualisierung von Informationen aus Krisengebieten in Karten (FRICKEL, 2010), verdeutlichen die Eignung und das Potenzial die Daten aus solchen Erfassungsmethoden auszuwerten. Der vorliegende Artikel stellt das Konzept und den Aufbau einer Anwendung dar, welche Meldungen des Nachrichtendienstes Twitter geocodiert und in einer prototypischen Webapplikation visualisiert. Das Projekt wurde im Rahmen einer Bachelorarbeit an der Hochschule Anhalt am Institut für Geoinformation und Vermessung im Jahr 2011 realisiert. Die Ziele waren:

- Beschreibung des wissenschaftlichen und technischen Standes in Bezug auf Geoapplikationen im Web 2.0.
- Erstellung von Konzepten für den Aufbau einer Internetplattform (Client-Server Architektur).
- Prototypische Umsetzung des Crowdsourcing-Portals mit der Visualisierung geocodierter Twitter-Meldungen.

2 Stand der Wissenschaft

Der Begriff Crowdsourcing wurde erstmals im Jahr 2006 geprägt und reflektiert sich aus den Wörtern „Crowd", einer unbestimmten Masse von Menschen und „Outsourcing", der Auslagerung von Prozessen und Aufgaben ins Internet (HOWE, 2008). Crowdsourcing als „Baustein einer virtuellen Wertschöpfungskette" (ROSKOS, 2010) ist ein Prozess, welcher durch das Nutzen des Internets, orts- und zeitunabhängige Inputs seitens der Nutzer zulässt. Die Einbindung der „Kraft, der Kreativität der Vielfalt der Menschen im Internet mit deren Ideen, kulturellen Backgrounds, beruflichen Qualifikationen und Erfahrungen" führt zu neuen Ergebnissen und Mehrwerten, welche bisher nicht möglich waren (ROSKOS, 2010). Eine grundsätzliche Voraussetzung für das Funktionieren stellen die Informationskanäle des Web 2.0 dar (ROSKOS, 2009). Speziell für die freiwillige, kollektive Erfassung von Geodaten hat sich der Begriff „Volunteered Geographic Information" (VGI) etabliert, welcher Menschen als „human sensors" beschreibt (GOODCHILD, 2007). Aufgegriffen und erweitert wurde der Begriff durch (RAMM, 2009), der den Crowdsourcing-Begriff strukturiert und Beispielprojekte zuordnet (vgl. Abb. 1).

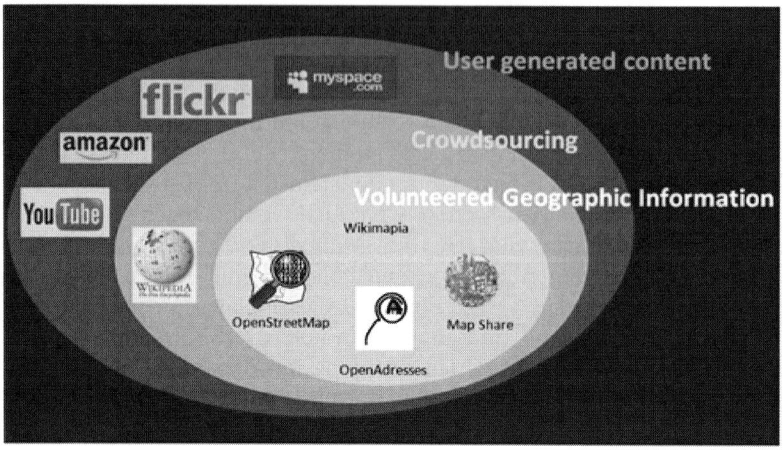

Abb. 1: Strukturierung des Crowdsourcing Begriffs (RAMM, 2009)

Als Standards für Übertragungskanäle von Geodaten haben sich GeoSMS[1], GeoRSS[2] und Twitter[3] etabliert, welche sich u. a. in Projekten wie ushahidi[4] und dem USGS Twitter Earthquake Detector[5] bewährt haben. Für die prototypische Umsetzung wurde sich auf den Nachrichtendienst Twitter beschränkt. Beim 2006 gegründeten Nachrichtendienst Twitter steht bei den Anwendern die zentrale Frage „Was machst du gerade?" im Mittelpunkt. Diese Echtzeitempfindungen oder -meldungen, welche sowohl von Privatpersonen, als auch Institutionen und Unternehmen zur Kommunikation von aktuellen Ereignissen genutzt werden, bilden mit der maximal 140 Zeichen umfassenden Nachricht – auch bezeichnet als Microblogging – eine prägnante und fest strukturierte Statusmeldung für Nutzer oder auch Anwendungsentwickler.

Der Umstand, dass die geposteten Feststellungen, Fakten, Erfahrungen und Ereignisse der Nutzer mit dem gesamten weltumspannenden, sozialen Netzwerk geteilt werden und anschließend durch die offengelegte Entwicklerschnittstelle analysiert und ausgewertet werden können, definiert das Netzwerk bereits als Crowdsourcing-Portal – einer Plattform zur Integration, Auswertung und benutzerspezifischen Anpassung von Inhalten, Diensten und Funktionen zu bereitgestellten Informationen im Web (vgl. KIRCHHOF, 2004). In Verbindung mit GPS-Informationen führte diese Technologie bereits zu einer Reihe von Applikationen im Internet – sogenannten GeoMashups – welche die Visualisierung von Ereignissen mit Raumbezug implizieren. Beispiele dafür bilden die Anwendungen My TWEET Map[6], Middle East Protest – Tweets Mapped[7] oder twittervision[8].

1 Siehe http://www.opengeospatial.org/standards/opengeosms
2 Siehe http://georss.org/Main_Page
3 Siehe https://dev.twitter.com
4 Siehe http://www.ushahidi.com
5 Siehe http://recovery.doi.gov/press/us-geological-survey-twitter-earthquake-detector-ted
6 Siehe http://www.mytweetmap.com
7 Siehe http://www.mibazaar.com/meprotests.html
8 Siehe http://twittervision.com

Eine umfangreiche Reflexion über den Nutzen von Crowdsourcing-Technologien im Umfeld der Landschaftsarchitektur zeigt (HEWITT, 2012). Auf Grundlage von weltweiten Analysen von Webseiten wurden Schwerpunktthemen und Technologien aus den Bereichen Architektur, Landschaftsarchitektur und Städtebau identifiziert um daraus soziale und geographische Hotspots und Betrachtungen zu den Themengebieten definieren zu können.

3 Konzeption der Webanwendung

Für die Konzeption der Webapplikation namens „geo.tweet“ wurden unterschiedliche Bearbeitungsbereiche identifiziert, welche in einen Arbeitsworkflow (siehe Abb. 2) gemündet

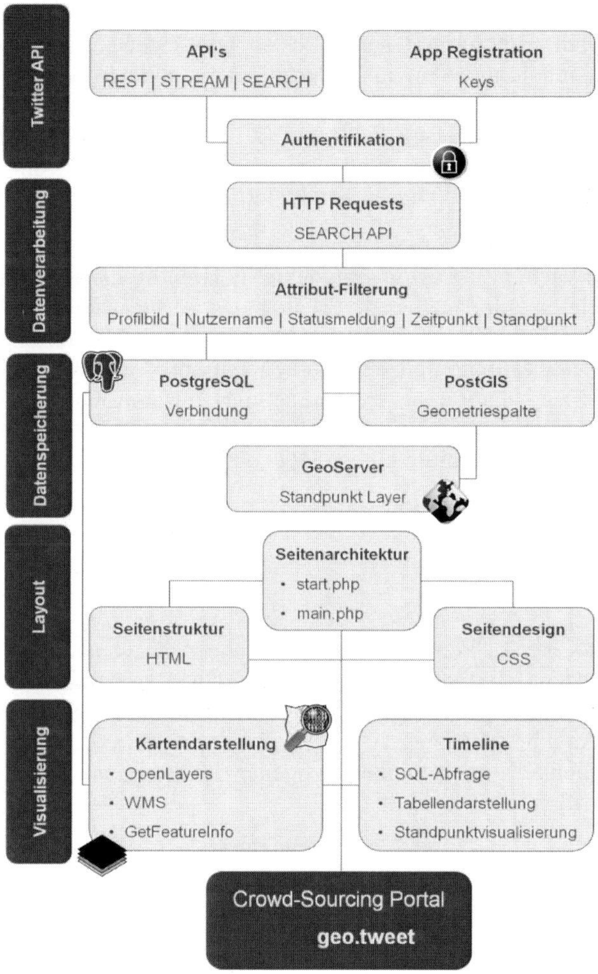

Abb. 2: Workflow zur prototypischen Umsetzung der Internetplattform (eigene Darstellung)

sind. Nachfolgend werden die Herangehensweisen in der Entwicklung für die durchgeführten Arbeitsbereiche erläutert. Hauptansatzpunkte waren dabei die Twitter API sowie Aufbau und Design, welches sich wiederum in die Bereiche Datenverarbeitung, Datenspeicherung, Layout und Visualisierung untergliedert.

3.1 Twitter API

Das „Twitter Application Programming Interface" ist eine von Twitter offengelegte Programmierschnittstelle mit der Möglichkeit, auf Twitter-Daten zugreifen zu können.

Der Zugriff auf die Twitter API ist mit allen gängigen Programmiersprachen und den dafür existierenden Bibliotheken möglich. Für die Umsetzung des Projekts wurde die Programmiersprache PHP und die dazugehörige Bibliothek von Abraham Williams[9] genutzt. Der Twitter Web Service ermöglicht die Nutzung von drei unterschiedlichen API's. Die REST (Representational-State-Transfer) API stellt alle Basisfunktionen bereit, die auch unter http://twitter.com zur Verfügung stehen. Hierzu zählen das Posten von Tweets, das Ansehen einer Timeline (Liste von Statusmeldungen) sowie die Möglichkeit, anderen Nutzern zu folgen oder sie der eigenen Freundesliste hinzuzufügen. Um die Twitter-Suche zu ermöglichen, wird die SEARCH API verwendet. Nach WEIGERT (2010) „erscheinen neue Tweets nicht unmittelbar (...), sondern werden in Intervallen von einigen Sekunden oder manchmal auch Minuten in 'Paketen' aktualisiert." Diesen Effekt soll die im Sommer 2010 erschienene STREAM API verhindern und Tweets in Echtzeit liefern.

Aus Gründen der Sicherheit und Überwachung ist es für Entwickler notwendig die eigenen Applikationen bei Twitter zu registrieren, um nach dem Erhalt eines Authentifikationsschlüssels den Zugriff auf die Daten zu gewährleisten. 2010 gab es dazu eine Umstellung von der einfachen Basic Authentication auf die komplexere Authentication. Vormals reichten Benutzername und Passwort zur Identifikation aus, welche jedoch mit dem neuen, sicheren Einloggen über den erwähnten Schlüssel angepasst wurde. Demnach werden die Zugangsdaten des Nutzers der Applikation nicht mitgeteilt, sodass keine Speicherung dieser Daten erfolgen kann.

3.2 Aufbau und Layout

Datenverarbeitung

Mittels HTTP-Abfragen in der SEARCH API können über spezifische Abfrageparameter z. B. Gebietsabhängigkeit gefiltert werden. Hierfür wurde die SEARCH API verwendet mit der unter anderem gebietsabhängige Filterungen möglich sind. Der Befehl zur Abfrage lautet „geocode" und beinhaltet die Angabe von Längen- und Breitengrad sowie dem Radius. Des Weiteren besteht die Möglichkeit, nach themenbezogenen Hashtags zu filtern. Eine Beispielabfrage sieht folgendermaßen aus:

```
http://search.twitter.com/search.json?q=&tag=DLA&geocode=52%2C12%2C100km
```

Das Resultat eines Requests sind 15 Nutzer-Profile, dessen Informationen separat abgefragt werden können. Für das Projekt relevante Angaben sind das Profilbild, der Name des Nutzers, der Zeitpunkt, der jeweilige Tweet und der Standort. Letzteres bildet das Hauptkriterium zur Filterung, falls der Standort in Koordinatenform vorliegt.

[9] Siehe https://github.com/abraham/twitteroauth

Datenspeicherung

Nach der Filterung der Daten ist eine Speicherung in einer Datenbank vorgesehen. Dazu wurde die Open-Source Software PostgreSQL in Verbindung mit PostGIS genutzt. Um Standpunkte in einer Karte visualisieren zu können, wird ein MapServer benötigt. Allein durch die standardmäßige PostGIS-Anbindung erwies sich der GeoServer als vorteilhaft und ermöglichte eine unkomplizierte Verbindung zwischen Datenbank und MapServer. Des Weiteren ist das Erstellen eigener Styles möglich, das für die Erzeugung eines Symbols für die Standpunkte benötigt wurde.

Layout

Die grundlegende Struktur des Portals beschränkt sich auf zwei Seiten. Nachdem sich der Nutzer auf der Startseite mit seinem Twitter-Log-in angemeldet hat, wird er auf die Hauptseite zur Datenvisualisierung geleitet. Neben der Funktionalität wurde auf eine ansprechende Gestaltung des Layouts geachtet. Hierfür ist die Trennung von Struktur und Design ausschlaggebend. Während die logische Struktur der Seite über HTML realisiert wird, bedient man sich beim Design der Cascading Style Sheets.

Visualisierung

Die Informationsvisualisierung bezieht sich einerseits auf die Darstellung einer Karte mit den geocodierten Meldungen und andererseits auf die einer Liste von Statusmeldungen, in einer sogenannten Timeline. Für die Kartendarstellung wird die JavaScript-Bibliothek OpenLayers und der freie Kartendienst OpenStreetMap genutzt. Die im GeoServer vorgehaltenen Standorte der Statusmeldungen werden mittels eines Web Map Service (WMS) in die Karte integriert. Für die Anzeige von Informationen über einen bestimmten Punkt in der Karte ist die getFeatureInfo-Abfrage des WMS zuständig. Hierbei wird eine Abfrage-URL generiert und an den GeoServer gesendet. Über einen Proxy-Server lassen sich die Rückgabewerte empfangen, die in einem Popup-Fenster visualisiert werden. Für die Timeline-Darstellung werden die Daten (Profilbild, Koordinaten, Nutzername, Status-Meldung und Zeitstempel) aus der Datenbank abgefragt und neben der Karte in Tabellenform, zeitlich absteigend dargestellt.

4 Visualisierungsergebnis

Das Endprodukt des erstellten Crowdsourcing-Portals geo.tweet ist in Abbildung 3 ersichtlich. Im Header-Bereich befindet sich ein Info-Symbol, welches bei Klick Auskunft über das Projekt gibt. Im Footer-Bereich wird über Icons auf verwendete Technologien verwiesen, die zusätzlich mit einem Hyperlink zur entsprechenden Homepage versehen sind. Der linke Content-Bereich beinhaltet die Kartendarstellung mit der Standpunktvisualisierung der geocodierten Twitter-Meldungen. Ergänzend dazu ist im rechten Contentbereich die Timeline als interaktive Tabelle dargestellt. Um Tabelleneinträge lokalisieren zu können, wird bei Klick auf eine Zelle der dazugehörige Standpunkt in der Karte rot markiert. In Analogie dazu besteht ebenso die Möglichkeit, die Standpunkte in der Karte zu identifizieren. Wählt man einen Standpunkt in der Karte, so erscheint ein Popup-Fenster mit Nutzername und Statusmeldung.

Abb. 3: Darstellung des Crowdsourcing-Portal „geo.tweet“ im Browser (eigene Darstellung)

5 Zusammenfassung und Ausblick

Crowdsourcing ermöglicht durch die Vielzahl an Nutzern, in kurzer Zeit mit vergleichbar geringen Kosten eine enorme Fülle an Informationen zu sammeln. Der Arbeitsaufwand für den einzelnen innerhalb der Community ist gering, da ohne große Hilfsmittel mitgewirkt werden kann. Nachteilig ist der Umstand, dass nicht jede Information als sinnvoll und brauchbar angesehen wird oder Fehler entstehen – ob mutwillig oder ungewollt. Hierfür sind bei einer projektbezogenen Umsetzung Vorgehensweisen zur Qualitätssicherung zu schaffen. Ansätze hierbei sind das Erkennen von Dubletten, das Filtern über Hashtags sowie das Ignorieren von Tweets ohne Angabe exakter Koordinaten.

Eine schematische Darstellung der grundlegenden Funktionsweise der Anwendung ist in Abbildung 4 ersichtlich. Die Community gibt über Twitter Statusmeldungen ab. Die Funktionen des Portals ermöglichen es diese Meldungen zu filtern und in einer Datenbank zu speichern. Ruft der Nutzer geo.tweet auf, erfolgt eine Anmeldung über die Log-in-Daten von Twitter. Anschließend werden die Datensätze aus der Datenbank mithilfe einer Web-Mapping-Anwendung im Portal dargestellt.

Verbesserungen des Portals können beispielsweise durch eine Optimierung der realisierten WMS-getFeatureInfo Abfrage geschaffen werden. Anstatt eines Popup-Fensters würde bei Klick auf einen Standpunkt in der Karte, der betreffende Timeline-Eintrag markiert und darauf gescrollt werden. Eine weitere denkbare Zusatzfunktion wäre außerdem eine Zeitskala – so könnten Tweets manuell in Abhängigkeit von der Zeit ein- und ausgeblendet werden. Vorteilhaft wäre außerdem eine Realisierung weitergehenden Auswertemöglichkeiten in Form von Grafiken oder Diagrammen.

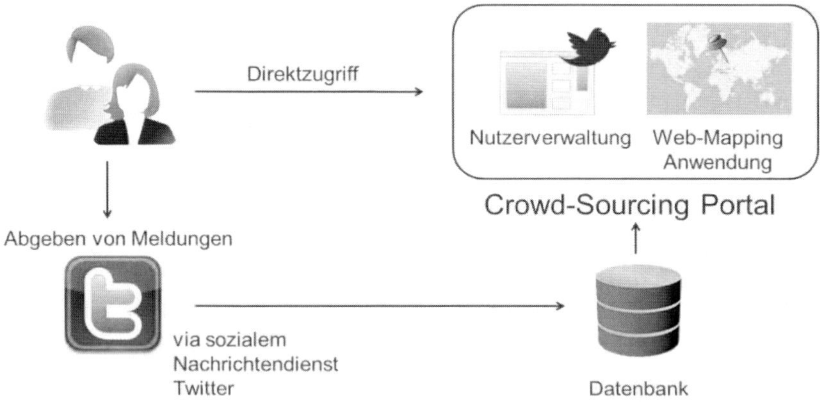

Abb. 4: Darstellung der grundlegenden Funktionsweise des Crowdsourcing-Portals (eigene Darstellung)

Da es sich hierbei um einen Prototyp handelt, ist das Portal nicht online erreichbar. Das Potenzial zur Nutzung der Plattform zeigt sich einerseits in der rasanten Entwicklung der Bereitstellung von Informationskarten mit Echtzeitdaten im Internet, und andererseits durch die Vielzahl an GPS-Sensoren in heutigen, mobilen Endgeräten als Grundvoraussetzung der Datenerfassung. Daraus ergeben sich vielfältige Anwendungsszenarien, sodass die Plattform auch in weiteren Hochschulprojekten ausgebaut und genutzt werden wird. Der innovative Faktor und große Vorteil gegenüber vergleichbaren Anwendungen besteht in der persistenten Speicherung der Crowdsourcing-Informationen in einer räumlichen Datenbank mit den sich daraus resultierenden Möglichkeiten der Strukturierung und fachlichen Harmonisierung der Daten sowie der Ansatz der raumzeitlichen Abbildung der Meldungen über einen individuell auswählbaren Zeitraum. Ähnliche Applikationen verfolgen nur das Konzept der Realtime Darstellung von Informationen. Jedoch können sich mit der Verknüpfung zu weiteren Geodaten neue, vielfältige und noch nicht absehbare Auswertungspotenziale für verschiedene Nutzergruppen herauskristallisieren.

Insbesondere für die Landschaftsarchitektur ergibt sich mit dem Portal eine Vielzahl von denkbaren Anwendungsfällen. So könnten landschaftsbezogene Twitter-Meldungen über Hashtags gefiltert und anschließend räumlich visualisiert werden, sodass möglicherweise regional unterschiedliche Diskussionsthemen und Trends rund um die Landschaftsarchitektur erkennbar wären. Auch in der Phase des Nachgangs der Fertigstellung eines Projektes, bspw. in der Umgestaltung einer Parkanlage, ließen sich neben der ökologischen Neugestaltung auch die sozialen Aspekte sammeln und untersuchen. Nachrichten mit einem bestimmten Stichwort oder der räumlichen Eingrenzung rund um die Parkanlage können extrahiert werden und in die direkte Bewertung des Projektes eingehen. Umgekehrt können im Vorfeld einer geplanten Gestaltungsmaßnahme die Tweets der Leute räumlich eingegrenzt gefiltert und die Informationen in einem Monitoringverfahren erworben werden. Diese Art von Nutzungs- und Stimmungsanalyse kann wiederum in landschaftsarchitektonische Planungen einfließen und die Zustimmung zu einem Projekt steigern.

Literatur

Frickel, C. (2010), *Krisen-Plattform Ushahidi – Datensammler für den guten Zweck.* http://www.focus.de/digital/internet/tid-18595/krisen-plattform-ushahidi-datensammler-fuer-den-guten-zweck_aid_518248.html (15.02.2013).

Goodchild, M. F. (2007), *Citizens as sensors: The world of volunteered geography.* http://www.ncgia.ucsb.edu/projects/vgi/docs/position/Goodchild_VGI2007.pdf (14.02.2013).

Hewitt, R. (2012), *Finding Multi-Centers: Using Crowdsourcing Technologies to Define Communities of Landscape Architecture.* Council of Educators in Landscape Architecture.

Howe, J. (2008), *Crowdsourcing: Why the power of the crowd is driving the future of business.* Crown Publishing Group.

Kirchhof, A. et al. (2004), *Was ist ein Portal?* http://pub-379.bi.fraunhofer.de/Images/Was_ist_ein_Portal_tcm379-129040.pdf (15.02.2013).

Ramm, F. (2009), *Crowdsourcing Geodata.* http://www.geofabrik.de/media/2009-09-08-crowdsourcing-geodata.pdf (17.06.2011).

Roskos, M. (2009), *Was ist Crowdsourcing? 12 Eckpunkte, die den Rahmen abstecken.* http://www.socialnetworkstrategien.de/2009/08/was-ist-crowdsourcing-12-eckpunkte-die-den-rahmen-abstecken (17.06.2011).

Roskos, M. (2010), *Crowdsourcing – Mehrwerte, Chancen, Definition. http://www.socialnetworkstrategien.de/2010/07/crowdsourcing-mehrwerte-chancen-definition* (15.02.2013).

Weigert, M. (2010), *Anschnallen für den Echtzeitstream.* http://netzwertig.com/2010/07/30/twitters-neue-user-streams-api-anschnallen-fuer-den-echtzeitstream/ (15.08.2011).

Collaboration in Landscape Architecture

The Role of Values Assigned to Water Shaped Landscapes in Collaborative Landscape Planning – Karasu River (Upper-Euphrates) Case

Emel BAYLAN and Nilgül KARADENİZ

1 Introduction

Overlooking how the public relates to the landscape during a landscape planning process, results in conflict among stakeholders which in turn hinders planning implementation. There is an increasing consensus that the planning process only focuses on expert judgments thereby limiting the participation of the community and other related stakeholders in the decision making process (LUZ, 2000; OLIVEIRA & DINEBOSKA, 2004). It is therefore paramount to alter the planning process to allow for stakeholder participation in all the decision making phases of planning. In this framework, collaborative approaches have proven promising and vital in bringing together various stakeholders in a joint decision making process, accommodating their different views and conflict interests (INNES & BOOHER, 2010). In a collaborative platform, problem solution and conflict resolutions are addressed by stakeholder negotiations and consensus instead of an authoritarian top-down process. In this context, collaboration is taken as a process of joint thinking, joint problem solution, joint planning and joint participation among stakeholders having equal rights during the process (OVERALL, 2005).

Collaboration is defined as a "working relationship based on trust, among two or more equal participants in the processes of joint thinking, joint problem solution, joint planning and joint creation" (OVERALL, 2005). By "*…equal participants…*" it is meant that the every participant is equal and has the equal right in the decision making process. According to GRIMBLE & WELLARD (1997) and MARGERUM (2002), collaborative planning is an interactive process which uses active stakeholder and community participation in decision making and consensus building, to work out the defined common problem(s). Common decisions are one of the main outputs of this process.

Common problems and common decisions related with landscapes are connected with the factors such as expectations, needs, and interests of the stakeholders, and land use types in the landscape. These factors result in value assignment to the landscape among stakeholders or among the community in general. Therefore, collaborative landscape planning process calls for determining the values assigned to the landscape by different stakeholders and by the local communities, as well as addressing conflicts among these values. Landscape values are the meanings attributed by people to the landscape features related to areas where they live, work, visit and use for other different purposes (ALESSA et al., 2008). Depending on the landscape feature, these values vary from a symbolic context to functionality. The values may as well differ in respect to relationship of individuals and communities with the landscapes. Therefore, different stakeholders may have different values for the same given landscape or landscape unit.

Values attached to landscapes have varying dimensions. They are effective in communication among stakeholders during planning, when geographically referenced or visualized. Geographically referenced data and values of stakeholders in large numbers and in various interests are currently making Geographical Information System (GIS) an essential tool for planning tasks. Landscape values that are geographically referenced and visualized through GIS output maps are the main communication tools for planning (SIMÃO et al., 2009)

In this regard, this paper examines the potential role of landscape values assigned to landscape by using landscape values mapping by workshop process and GIS in the collaborative landscape planning process. By determining these roles, it is aimed to guide the process of active participation of locals in the landscape planning on the case of Karasu River and near surroundings. In this context, paper attempts to propose the landscape value mapping as a procedure for making easier the inclusion of different stakeholders' valuations and judgments on the landscape they are in relation with to the landscape planning process in the context of existing legislations.

In this scope, according to the methodology and the findings, paper also attempts to contribute to the long term objective defined as *"to contribute in the development of approaches and applications of participatory landscape planning"* in the project titled as "**Water Resources Management and Definition of Landscape Quality Objectives Within the Scope of Collaborative Landscape Planning: Karasu River (Upper-Euphrates Basin-Erzincan) Case**". The specific objective of the project which is desired to be achieved in the end of the project is *"to determine the framework of collaborative landscape planning for water resources management."* Within the scope of this objective one of the sub-objectives that are planned to be reached is the "determination of landscape values of different interest groups in the case area".

River systems serve the human society in various ways and have various values for the communities in different scales. River systems function as one of the main dominant feature of the landscape by shaping bio-physical, cultural and social-symbolic values of the landscape, and therefore shaping life in the landscape also. For both Turkey and Middle East Countries, River Euphrates is one of the two major rivers playing a significant role in this unique landscape composed by diverse natural and cultural features. River Euphrates is formed by the confluence the Karasu River and the Murat River. Karasu River rises in Erzurum Province, and then flows west through Erzincan Province, and receiving various tributaries before flowing into the Keban Dam Lake in the southeast of the Euphrates. Karasu lies in the west of the two major sources of the Euphrates in North-Eastern Anatolia, and shapes the Upper Euphrates Basin and the Erzincan Plain, while dividing the plain into two. Karasu River, also known as "Euphrates" by locals, is the key driver component of life and landscape in Erzincan Plain for its irrigation, energy and water sports purposes. At the same time, regarding its life-sustenance and life-threating properties, Karasu also holds a cultural and social-symbolic significance in the area. This cultural significance and influence of Karasu River can be seen in village names, in proverbs related with floods or flow style of Karasu, in local legends and mystical/religious beliefs among local people.

2 Material and Methods

2.1 Mapping of landscape values

The methodology used in mapping stakeholders landscape values for this study were based on concepts described in BROWN (2007), ALESSA et al. (2008) and ZHU et al. (2010). In this study however, the difference is in the implementation phase of the mapping process, where workshop was used for interaction of the represented stakeholders. The workshop was held in the case study area. Among the sectors and institution stakeholders who participated include; central government's local administrative units of forestry, water, agriculture, environment, city planning, culture and tourism, local municipalities, irrigation unions, agricultural cooperatives, and local university.

According to the 2010 census data for the area, the determined minimum sample size for the mapping was 43 at 95% confidence level. The confidence interval was set at 15% to correct uncertainty associated with a sample estimate of a population parameter.

50 representatives from different sectors participated in the workshop but only 40 of them participated in the mapping process. After the workshop, 5 more individuals from different villages participated in the mapping study with the help of project assistants. In total 45 local people participated in the landscape value mapping survey. The participants were asked to point and locate on the maps the places and features that they thought represented one or more of the given 12 landscape values, along the Karasu River and its surrounding. These 12 landscape values in this study were same as the ones originally given by other studies referenced above. The surveyed landscape values were described thus;

- **Aesthetic value:** I value these places for their attractive views, smells, sounds or pleasing natural scenes.
- **Economic value:** I value these places for their role in economic benefits e.g, income generating activities such as agriculture, tourism, industry or commerce.
- **Recreation value:** I value these places for supporting outdoor activities and leisure.
- **Life sustaining value:** I value these places because they help safeguard and protect human life and/or they are important for renewing air, water, and soil.
- **Learning (knowledge) value:** I value these places because of their role in nature and environmental education.
- **Biological diversity value:** I value these places because their role in supporting and provision of biodiversity e.g, plant variety, animals, aquatic organisms and other living organisms.
- **Spiritual value:** I value these places because they have a spiritual significance or are sacred places.
- **Intrinsic value:** I value these places because they are valuable at the observers point of view, independent from thoughts about them or whether they are actually used; valuable just for being.
- **Heritage value:** I value these places because they have features and/or elements related with natural and human history.
- **Future value:** I value these places because they allow future generations to know and experience them as they are now.

- **Therapeutic value:** I value these places because they helped people feel physically and/or physcologically better.
- **Wilderness value:** I value these places because of their wild nature (BROWN & RAYMOND, 2007; ALESSA et al., 2008; ZHU et al., 2010).

Steps of landscape value mapping

a. Workshop phase

a.1. At the beginning of the workshop, a short presentation was done on landscape values and on the steps to be followed during the mapping process to acquaint the workshop participants with the topic.

a.2. To obtain value judgments of locals for each landscape value in spatial format, 1/150.000 scaled 12 topographic maps of the area were prepared and hang on the workshop room. Statements and descriptions for each landscape value were written on each of the topographic map separately.

a.3. Each workshop participant was asked to select at least one location or a maximum of five locations and to point these selected locations. Scoring of 5, 10, 25 or 50 points for each landscape value was used. Each participant would score a maximum of 100 points for one landscape value. Each of the different scores were marked in four dots with different colors stickers (Fig. 1.)

b. Office phase (data transfer to GIS and analyses)

b.1. The x; y coordinates of the locations which the workshop participants selected on the 12 topographic maps were imported to GIS using ArcGIS 10 (ArcMap) software. This process was done for each landscape value. A database consisting of the name of locations and total points of the selected location for each landscape value was created.

b.2. The established database was used for analyses and later, maps for the 12 landscape values were generated. In this step, *symbology* and *geostatistic* tools were used. Different symbols were assigned to each spatial data by using *Graduated symbols* feature in *Quantities* section. By using natural breaks feature, different colors were assigned to the different symbols and score groups for each landscape value.

b.3. Finally, 12 maps, representing the 12 landscape values and showing the locations valued in different score intervals were prepared.

2.2 Analysis for prior landscape values and hot spots in the rivers-wetlands landscape character area

The rivers-wetlands landscape character area (LCA) in the research area covers the Karasu River, its banks, main streams connecting to the River and wetlands in the near surroundings. First, total general average for 12 landscape values were calculated from total points for all values given by the mapping participants. Then, general average for each of the 12 landscape values were calculated from related points. The landscape values having the general average score above the total general average score for landscape values in all over the case area were considered as prior landscape values for local stakeholders. The locations having scores above the general average score of the related landscape value were considered as the prior (hot spots) location(s) for the related landscape value. For the rivers and wetlands LCA Figure 2 represents the locations having scores equal to or over the interpolation weight for each of 12 landscape values total scores for the area.

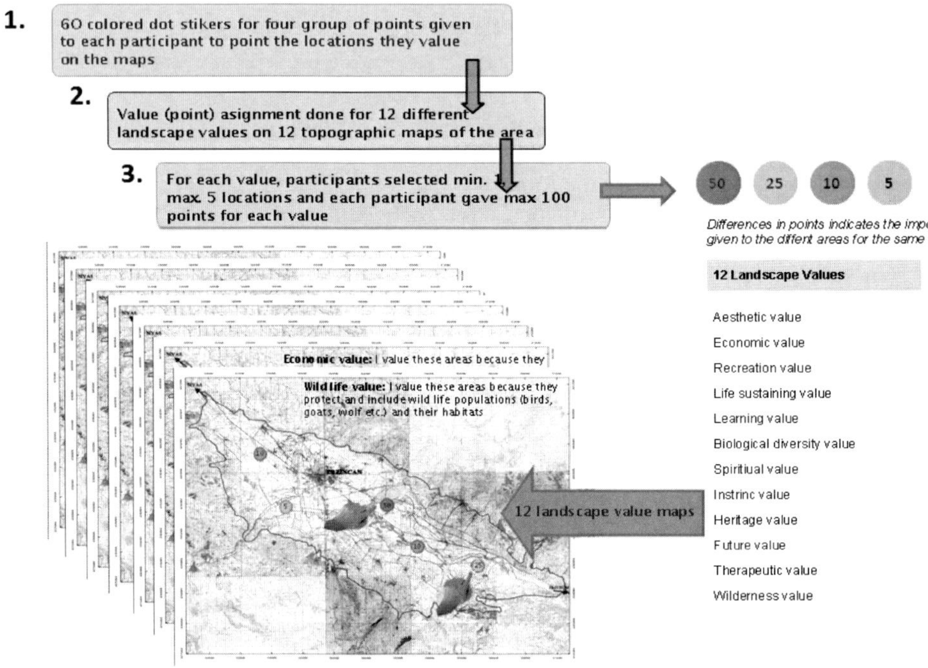

Fig. 1: Steps for landscape values mapping in the workshop held for Karasu River and its surroundings in Erzincan

3 Results

It was generally observed that the workshop participants had a common opinion about Karasu River had a "life giving" value to the people living in Erzincan province at large.

All judgments and opinions about the landscapes the people live in were generated in the 12 landscape values maps, prior values and hot spots map at the end of the mapping process. According to the *prior landscape values and hot spots in the rivers-wetlands LCA map* (Fig 2). Mertekli area located on the south-east bank of the Karasu River and Kemah Gate area on the south-west of the Erzincan Plain, where the River leaves the Plain and wetlands around these two locations were valued as prior locations and as hot spots for biological diversity and wilderness values for locals. The Sölperen Stream, flowing in a small valley while connecting to the Karasu River from the north-east was also valued as another prior location and hot spot for wilderness value in the area for local stakeholders.

Having scored a very high general average score for all of the 12 landscape values in the rivers-wetlands LCA, Ekşisu Wetlands which is an important natural area on national levele was valued as one of main prior location and hot spot. Işıkpınar (Vasgirt) Stream which is entering to the plain from the north-east location in a deep valley also had a high average score thus was qualified as prior locations and hot spots by mapping stakeholders.

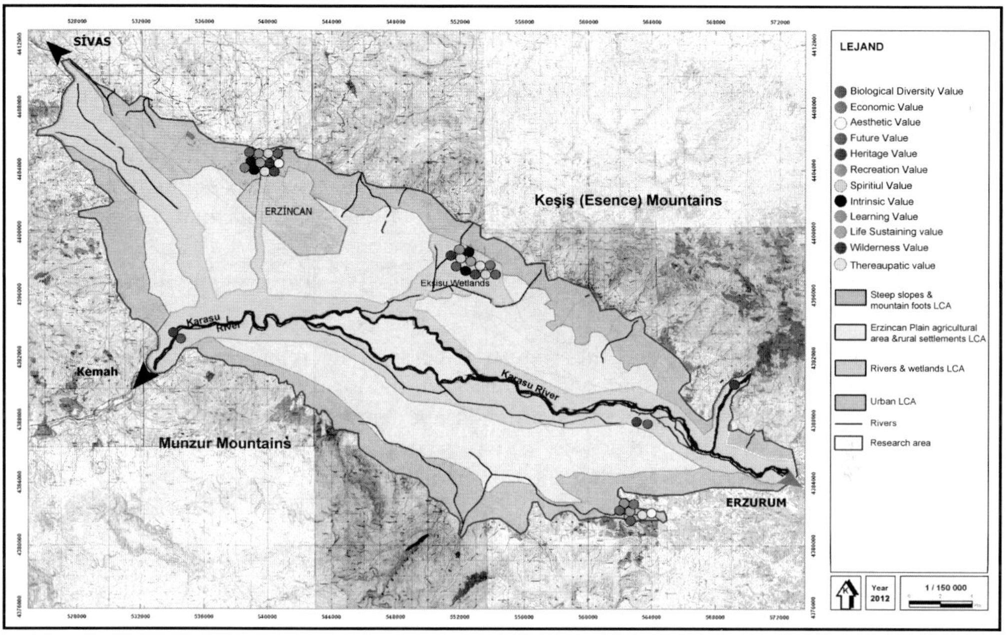

Fig. 2: Prior landscape values and hot spots in the rivers-wetlands LCA (Karasu River, Erzincan)

As far as the therapeutic, learning value, life sustaining value, economic value, biological diversity value, aesthetic value and recreation values were concerned, Girlevik Waterfall was identified as another prior and hot spot in the LCA.

4 Conclusions and Outlook

This research attempts to suggest a procedure that contributes to the active participation of local stakeholders in the landscape planning and decision making processes by involving the stakeholders' judgments and knowledge of the landscape they live in and interact with.

Holding workshops provided an opportunity for the participants to hold dialogs and discussions, share their opinions, values and judgments about the landscapes they live in. The mapping results showed important areas for particular values and expressed support for using landscape mapping in the formulation of common values and common objectives among stakeholders. Although the mapping results showed the common important areas for each landscape value, these results have been shared with the participants in another stakeholder meeting and the participants were totally agree with the results so with the evaluations' of each other's. In addition, participants of the mapping process use their words to convey information into spatial forms by showing their judgments on the map. This is an important output of the mapping process because it provides inclusion of views, opinions and judgments of locals in the planning process and creates an output for communication among planners, bureaucrats, local stakeholders and lay people. Therefore, landscape value mapping can be included in the diagnosis phase of landscape planning as an essential step to start a collaborative process. Analyzing the landscape values of stakeholders for the subject landscape in the diagnosis phase gives the opportunity to identify the common values, common important areas, conflicting values and conflicting activities in the landscape. Such a process improves the problem definition for the subject landscape and thus, improves the reality of landscape planning objectives as well as the overall effects of the participatory planning process. Moreover, integration of conflict analysis related with the current and proposed activities in the subject landscape, and the landscape values, to the diagnosis phase is another issue needed to be considered in a collaborative landscape planning process.

The mapping results suggest that the Karasu River and its related water resources and wetlands, provide present and future opportunities for the local people. Therefore, decisions related to these water resources and related landscape features in the area should be highly regarded during the planning processes to avoid conflict or negative response from the community during the planning implementation phase.

On the other hand, the mapping results only provided information that was based on local communities' daily use of the accessible landscape features. Other features that were inaccessible or not constantly used by the locals were not sufficiently considered irrespective of the value they might have. Extensive workshops that accommodate these unconsidered landscapes may work and may solve this bias and thus contribute to improve landscape value mapping. In addition to this, recording the basic information about the workshop participants' is another issue to be considered for the analysis and assessments regarding the different evaluations of different stakeholders.

Another noted issue is the similarities in the context and/or in the meanings of some landscape values example; biological diversity and wilderness value, future value and heritage value, recreation value and therapeutic value. Further studies are needed for the clarifications on these landscape values.

5 Acknowledgement

This research was conducted in the context of "*Water Resources Management and Definition of Landscape Quality Objectives Within the Scope of Collaborative Landscape Planning: Karasu River (Upper-Euphrates Basin-Erzincan) Case*" titled project (Project no: 110Y285) financed by The Scientific And Technological Research Council of Turkey (TUBITAK), CAYDAG Committee.

References

Alessa, L., Kliskey, A. & Brown, G. (2008), *Social-ecological hotspots mapping: a spatial approach for identifying coupled social-ecological space.* In: Landscape and Urban Planning, 85 (1), 27-39.

Brown, G. & Raymond, C. (2007), *The relationship between place attachment and landscape values: toward mapping place attachment.* In: Applied Geography, 27, 89-111.

Grimble, R. & Wellard, K. (1997), *Stakeholder methodologies in natural resource management: a review of concepts, contexts, experiences and opportunities.* In: Agricultural Systems, 55, 173-193.

Innes, J. E. & Booher, D. E. (2010), *Planning with complexity: an introduction to collaborative rationality for public policy.* 1st edition. London/New York, Routledge, Taylor & Francis Group, 1-30.

Luz, F. (2000), *Participatory landscape ecology – a basis for acceptance and implementation.* In: Landscape and Urban Planning, 50 (1-3), 157-166.

Margerum, D. R. (2002), *Collaborative planning building consensus and building a distinct model for practice.* In: Journal of Planning Education and Research, 21, 237-253.

Oliveira, R. & Dneboská, M. (2004), *From Landscape Perception Until Public Participation. How Long Is The Way?* Paper presented at International Conference "From Landscape Knowledge to Landscaping Action", Bordeaux, France.

Overall, M. P. (2005), *Toward a theory of collaboration for teachers and librarians.* School Library Media Research, 8. http://www.ala.org/ala/mgrps/divs/aasl/aaslpubsand journals/slmrb/slmrcontents/volume82005/theory.cfm.

Simão, A., Densham, J. P. & Haklay, (Muki) M. (2009), *Web-based GIS for collaborative planning and public participation: An application to the strategic planning of wind farm sites.* In: Journal of Environmental Management, 90 (6), 2027-2040.

Zhu, X., Pfueller, S., Whitelaw, P. & Winter, C. (2010), *Spatial differentiation of landscape values in the Murray River Region of Victoria, Australia.* In: Environmental Management, 45 (5), 896-911.

Collaboration of (Landscape Architecture) Students to Raise Consciousness of Trees

Anna CZINKOCZKY and Gyorgy SZABO

1 Introduction

Most of the features that have the greatest influence on our environment, including the climatic effects of urban vegetation (vegetation index) and other geographic features can be obtained in an enormous amount using remote sensing technologies. However, these indices cannot be directly perceived by humans, hence cannot be understood by ordinary or local people concerning the biological value of their environment. Ordinary citizens are not sensitive to these abstract scientific values, hence the gap between the quality and amount of data obtained by the sensors and the data comprehended and understood by local people is rapidly growing.

Information technology helps us to visualize and familiarize with the landscape and provide mass information about the environment. However, the increased usage of computers lessens the time one actually spends with real natural environment. Students are being criticized about spending most of their time in front of computers instead of observing nature or doing outdoor activities.

2 Material and Methods

2.1 IT knowledge of students

A recent study (DOOR et al., 2012) conducted at several US universities revealed that college students have rather advanced computer skills (approx. 60 % of first-year students have good IT skills, ~25% have excellent skills and only about ~15% of students have poor IT skills). This study also points out that the IT skills of students is about 15% higher than that of the faculty skills. It can be concluded that the increased use of computers did result in better IT skills – but according to the same report – most university professors do not think that the increased use of IT technology has any positive impact or beneficial for the quality of education and it results in a deeper understanding of the subject.

The goal to enhance the quality of Landscape Architecture education – obviously would be to use expertise of IT skills of the students to enhance their fundamental understanding of the complex issues of landscape development, but this process does not happen automatically.

I think a new approach is necessary to use digital knowledge to support traditional landscape studies, and since most students are more familiar with the digital world than with real nature – we should use the digital world to understand our physical and natural world better.

This is a rather inverse effect compared to the traditional learning process (that starts from natural experience and becomes more abstract and digital). The current trend should be based on the students' advanced digital knowledge (primary skill) to enhance their knowledge and concentration on real-world issues (secondary skill).

Fig. 1: A typical student of Landscape Architecture

The primary learning usually comes from personal experience and observation of the local environment. This mostly occurs in elementary schools. From personal and local experience, the students get broader and extended knowledge about the regional and global environment.

Fig. 2: From local observation

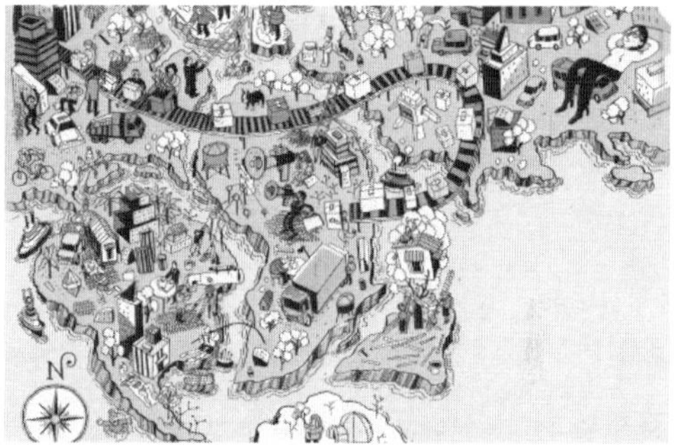

Fig. 3: Making regional connections

2.2 New approach with IT

It is specific to the generation of the digital age as students are becoming older and more mature, their connection to the real physical world decreases and the virtual environment will take an increased proportion of their lives.

Instead of fighting with this tendency, we rather have adopted to the knowledge acquisition methods of the young generation. In order to incorporate global knowledge to understand our personal and local environment the curriculum of landscape architecture must be reshaped and rearranged.

The use of computers can offer several new methods to have a broader perspective of a complex phenomena. Assignments still should start from the surrounding world and should be related to it, but the conclusions and findings can be more general both in spatial and temporal dimensions. The scope of the problem may involve larger geographic regions and

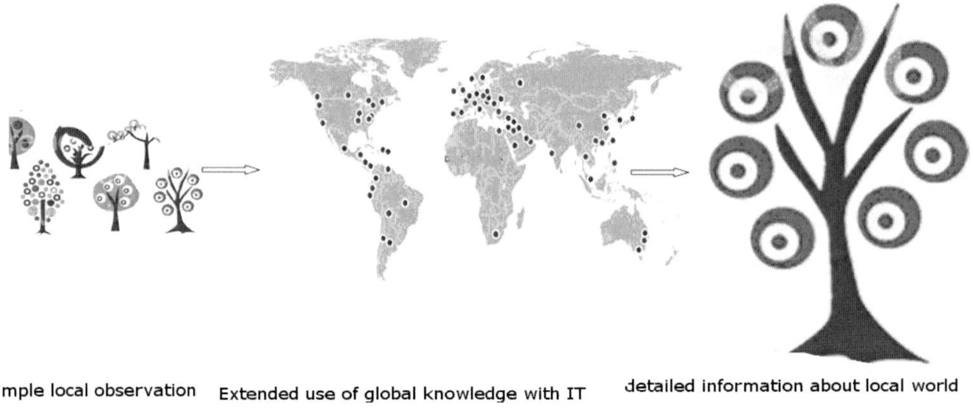

Fig. 4: Process of shared knowledge

we may ask questions about expected future situation (prediction). Modeling can serve to estimate future scenarios (temporal modelling) or spatial distribution change of the habitat zone (BEDE-FAZEKAS & CZINKÓCZKY, 2012).

The choice of the given IT platform is arbitrary. A special website can be set up, but it is possible to use even one of the most common social media website (i.e. Facebook), because the students are highly interested in social communication and spend a lot of time in this digital environment. The platform should not be designed rigorously, the students will naturally find the most convenient features for communication. Considering the shared information strategy of Web 2.0, shared knowledge is rather common (Wikipedia, Facebook groups dealing with certain topic, etc.). Even there are several open and free

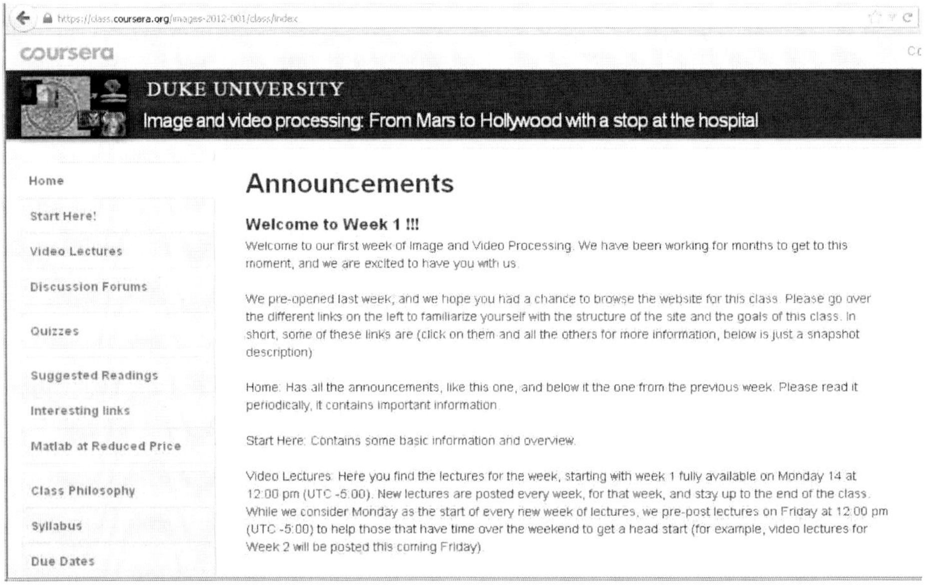

Fig. 5: A platform for a coursera.org web-class

international web-classes https://www.coursera.org/ where one can register for courses offered by the most prestigious universities for free of charge.

This website has a complete course material with discussion forums and on-line quizzes. It might be possible to offer a similar solution for landscape planning but it is rather time consuming and expensive to develop and our goal is not necessarily to offer a worldwide on-line course, but rather exchange ideas, data and communication of the practical issues. It is impossible foresee and predict all platforms and topics for collaboration, but starting with some practical examples certainly can increase the chance for further cooperation and sharing of ideas. As a lecturer of Mathematics for Landscape Architecture students, I have noticed, that even the future experts of landscape design have no clue about quantitative description about very basic and simple primitives of urban environment, such as the biological values of trees or bushes, their volumes, canopy areas or carbon dioxide absorption capability.

I gave an assignment to my first year Landscape Architecture students in Math to record a tree on their street, determine its height suggest that every student of Landscape Architecture is actively involved in a project called „my personal tree" where he/she gives a full description of a tree located in a city (including a picture, location with GIS coordinates, marking on Google Maps, identifying the type, approximate age, volume, canopy or other relevant characteristics).

3 Why Should LA Students Work Together and Who Should Organize the Collaborative Landscape Architecture Projects?

Students – like most people – follow trends or act together mainly because they see examples on TV, media, or just simply join certain campaigns or publicly organized events (critical mass, Earth day, etc.). Besides the regularly coordinated activities, there are other spontaneously organized gatherings and actions related to other public issues often for the purposes of entertainment, satire, and artistic expression. A famous example for this is the flash mob. Young people like unusual things, prefer surprises or actions that have not been seen before. That's why it is important to bring up new approaches and coordinated projects in science that are not standard, yet relevant for the young generation.

It is not feasible that media or other public campaigns will initiate project related to specific scientific field of landscape architecture. Popularity will not bring up every important issue, so it is the responsibility of the schools, universities or other international organizations (ECLAS, DLA conferences, etc.) to initiate and coordinate such wide scale cooperation. Things should be decided in a broad platform and narrowed down and tailored according to the actual school curriculum. In the present paper we are not proposing universal and everlasting projects, just we would like to mention a few samples, complex approaches of the cross- curricular cooperation that is primarily based on the IT technology.

4 Sample Project of Personal Tree

I found it to be a fun assignment for students to select a favorite plant in their local area (my Personal Tree) and try to estimate several complex attributes of it. In a case of a tree, the volume (amount of wood) can be important. The mathematics assignment asked to substitute to an approximation formula

$$v = (p_1 + p_2 \cdot d \cdot h + p_3 \cdot d + p_4 \cdot h) \cdot (\frac{h}{h-1.3})^k \cdot (\frac{d^2 h}{10^8}) \tag{1}$$

using specific parameters of a given tree.

Estimation of the height of a tree using a picture taken by a student

height of tree
approx. 4 times of the height of student = 4 x 1.76 m
7m

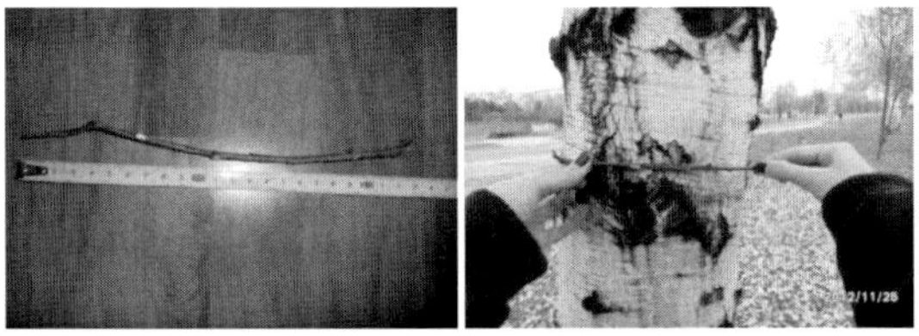

Fig. 6a: Estimation of a tree height **Fig. 6b, 6c:** Measuring diameter of a tree

The assigment asked for a complete documentation:

Fig. 7: Displaying the site on Google Earth with place marker

For the given type of tree (*populus alba*) the parameters of the volume formula (to be found in a table):

$p_1=3184.75$ $p_2=-0.34$ $p_3=10.21$ $p_4=247$ $k=3$

Substituting the values into the formula (1), we get: tree volume is: $0.236m^3=\underline{\underline{236000cm^3}}$

4.1 Habitat-range- mapping and personalization

The sample project above describes only one possible way how digital knowledge can be used to enhance the information obtained from personal world. There are some further possibilities to extend this problem of „getting to know my personal tree". It is possible to ask the students to display the habitat zone of the tree (by googling the habitat information from Wikipedia, displaying it on the map and making a real-life connection with someone who lives at the most distant part of the earth and shares the same attributes of living nearby a similar tree.) These projects may sound somewhat childish and not very scientific, but fun increases motivation and may create memorable events.

4.2 Temporal modeling

Besides recording certain physical parameters it is also important to include an attribute that shows the physical condition of the tree. Moreover, since the green value protection and maintenance is important if something is not adequate (such as the critical root zone or tree protection zone is blocked so that the tree cannot receive sufficient amount of rain due to the urban pavement), should be listed.

The first step is to create a regional database with predefined attributes which has to be designed by an appropriate group of experts and the second step is to combine the regional databases into national or larger scale units.

The maintenance of the database is also important, specially for the endangered objects. I think with this project we could enhance the public awareness of the biological importance and values of trees and with joint effort it will be possible to preserve the current urban vegetation and increase the quality of it even further. As a methodology, I suggest using mobile technology, smart phones that can be used as cameras for image capturing, data transmission.

And to add a trendy platform for communication , the most common social networks (FB or Twitter) could be added to make it suitable for students' communication and exchange of ideas.

One can argue that these platforms may not be permanent or scientifically rigorous , but they only serve as a gateway to connect participants. Moreover, the landscape related public projects do not need to be exclusively designed for experts. The wider the range of participators, the higher the social impact that can be achieved. Our final goal is to raise public awareness and activity towards environment consciousness and nature preservation and increase public participation in community related „green" issues.

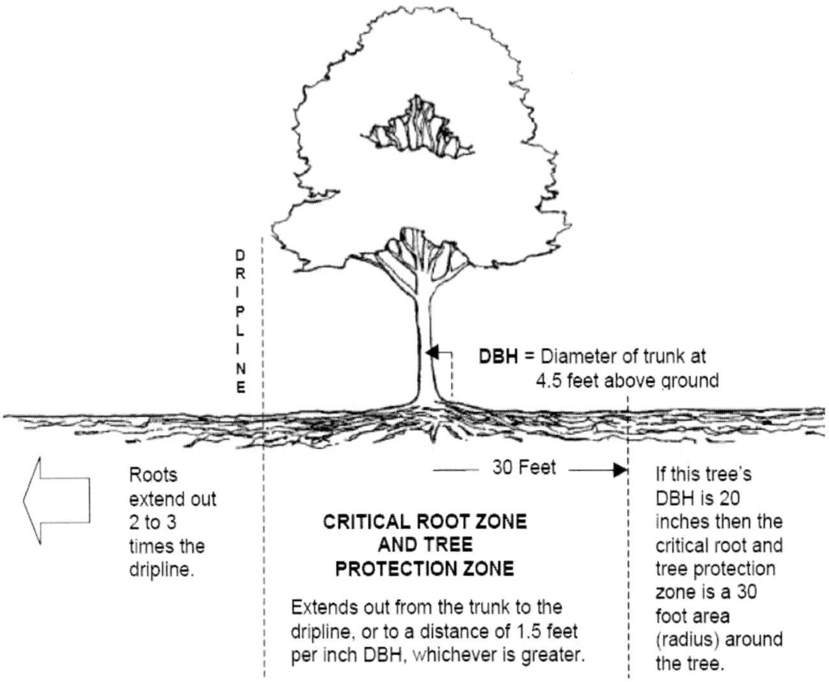

Fig. 8: Critical root zone of an urban tree

I suggest to develop an open source **tree – or landscape values database** which is shared and maintained by several universities and have a joint project of a year.

It is impossible to monitor every tree in a large city, but the most frequented downtown areas certainly deserve extra attention. It has happened many times in Hungary that the city has planted valuable mature trees and they died very quickly do to poor design of pavement or inadequate care. This deterioration can be prevented by regular monitoring of the condition and updating the tree database (i.e. every semester). The database can be used to query similar trees in different countries or regions and find common solutions for

Key aspects to consider (physical parameters) in case of a tree:

- the location (address and Geo-coordinates);
- species;
- size;
- age;
- overall health (scale 1- 10);
- recent changes of pavement or landscape;
- soil;
- Will roots pose a problem to sidewalks or driveway?
- Is it a high maintenance tree that will require frequent spraying for insect or disease control?
- Does it drop messy fruit pods or seeds?

- How adaptable is it to environmental changes?
- Is it in danger or critical situation? (scale 1- 10 = most critical situation)
- If in danger, what measures should be taken? (pavement, soil change, watering etc.)

The database can be filtered and queried and immediate measures should be taken to save the most endangered trees.

5 Critical Issues to Be Considered

5.1 Roots covered by pavement

If a large area around the tree is paved, the surface should be porous to allow water and air to penetrate. If brick or flagstone is used, joints should not be mortared. When pavement is nonporous, an opening should be left around the trunk of the tree. This opening should be approx. 2 m in diameter for small trees. The opening should be larger for mature trees unless the roots extend beyond the pavement into uncovered soil.

5.2 Interdisciplinary cooperation

When a database is set up, one probably immediately assumes an IT related task. It is indeed, but setting up a professional tree – database is more than just a routine IT procedure. Needless to say that it is essential to use relevant attributes of trees, which can correspond to the healthy urban habitat conditions, and can detect climatic changes or risks as well. For the best result, Landscape Architecture and Dendrology experts can join IT experts in finding the database structure and platform that can manage the tree related data.

Landscape Architecture students could take part in data acquisition, checking urban sites which are suitable for obtaining data, and setting up a framework for regular maintenance. Maintaining a database and providing valid data is a long-term task, so yearly student surveys can be organized to keep the data up-to-date.

IT involvement is not necessary a long term condition, it is primarily important at the beginning phase or when a different platform / system is used, but this may not happen regularly.

5.3 Realization of the database

The actual physical database could be realized in a LAMP-software stack (LINUX – Apache – MySQL – PHP) environment. This article does not intend to discuss the scope of technical details, it rather wants to introduce the concept only. The actual programming environment is beyond the scope of this paper.

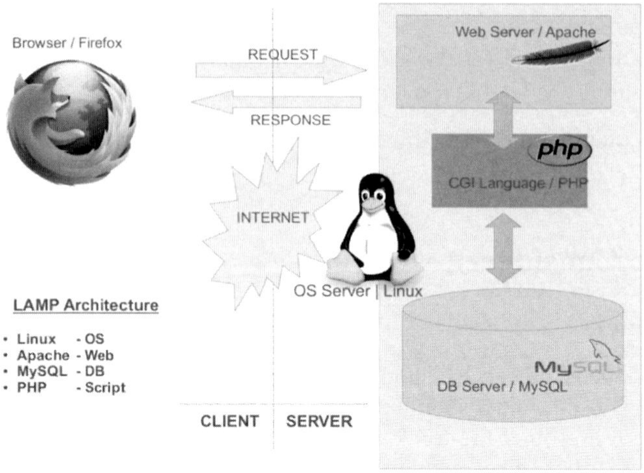

Fig. 9: LAMP environment (http://en.wikipedia.org/wiki/File:LAMPP_Architecture.png)

6 Summary

The aim of this paper is to raise the awareness of Landscape Architecture students concerning their local environment and to connect them with other students with different nationalities facing similar problems. Joint projects can be more enjoyable and fun by pointing to the similar problems of Landscape Architecture in different countries. This project is aimed to show our local environment in a larger perspective, hence it might serve as changing or reshaping our attitude towards nature using IT and scientific methods. The joint cooperation's can be performed at all levels of Landscape studies, i.e. it is possible to have BSc, MSc or Ph.D. level cooperation.

References

Bede-Fazekas, A. & Czinkoczky, A. (2012), *Visualization of the climate change with the shift of the so called Moesz-line*. DLA, 2012.

Brown, K. & Jennings, T. (2003), *Social consciousness in landscape architecture education: Toward a conceptual framework*. In: Landscape Journal, 22 (2), 99-112.

Brown, K. (2002), *Landscape architecture and social responsibility: Emerging concepts from a study of practice*. PhD Diss. University of Massachusetts, Amherst.

Doore, B. E., Fairman, J. E. & Doherty, D. (2012), *Study of Faculty and Student Perceptions of and Experiences with Technology in Post-Secondary Courses at Maine's Seven Public Universities*. University of Maine, Orono, Maine. DOI: 10.1034/j.1600-0889.1992.t01-1-00001.x.

Rabham, D. C. (2009), *"Crowdsourcing" the public participation process for planning projects"*. In: Planning Theory, 8, 242.

http://en.wikipedia.org/wiki/File:LAMPP_Architecture.png.

eGovernmentAR – Bildbasierende Kommunikationsmodelle als Schlüssel zum barrierearmen partizipativen eBeteiligungsverfahren

eGovernmentAR – Image-Based Communication as the Key to Barrier Free E-Participation Processes (English summary)

Lisa ROCKMANN, Simon ADLER und René KRUG

Summary

The public participation process concerning buildings of public interest can be reformed using new technologies. Current participation processes show deficits in communication and providing of information. Within the joint research project ArchKM-AR from Anhalt University and the Fraunhofer Institute IFF Magdeburg, a prototype for an interactive Augmented Reality Application with the objective of simplifying communication between experts and lay persons is being developed. The paper first defines the general task of current participation and communication processes and then shows the opportunities given by new communication technologies. The functional focus of this approach is applying Augmented Reality in combining 3D Models with the camera image of mobile tablet PCs.

Using the description of the communication process, the paper also describes the opportunities of Augmented Reality Technology. The application is intended to support E-Government processes. Governmental administrations may provide these digital interactions on the internet in order to provide participation services to citizens more easily and with less barriers.

1 Vorwort

Bilder können durch Neue Technologien eine Reformierung der Bürgerbeteiligung erzielen. Probleme der Kommunikation und Informationsvermittlung aktueller Beteiligungsverfahren und die Integrationsstrategie des eGovernment bildet die Grundlage für eine interaktive Augmented-Reality-Anwendung (AR = erweiterte Realität), die prototypisch in einem kooperativen Forschungs- und Entwicklungsprojekt zwischen der Hochschule Anhalt und dem Fraunhofer IFF in Magdeburg entwickelt wird. Ziel ist die Entwicklung einer vereinfachten Kommunikation zwischen Experten und Laien.

2 Hintergrund

Regelmäßig werden Bauprojekte auf vielen Planungsebenen von Bürgerprotesten oder nachträglicher öffentlicher Unzufriedenheit begleitet. Zugleich zeigen sich die beteiligten Akteure: Behörden, Projektträger sowie planende Disziplinen regelmäßig davon überfordert.

Gespräche mit Vertretern der planenden Disziplinen im Rahmen des Forschungsprojektes *ArchKM-AR* eröffneten ein aktuelles Beispiel einer Planung mit umfangreicher negativer öffentlicher Präsenz. Der Platz *Am Steintor* sowie anliegende Bereiche in Halle (Saale) wurden im Rahmen des Projektes *STADTBAHN Halle* aus städtebaulichen Gründen überplant. Dieses Vorhaben verursachte ein Aufbegehren in der Bevölkerung. Eine Bürgerin äußerte sich zur Vorgehensweise der Stadt: „[...] Es wirke so, als habe man die Planung, in der ein Erhalt des Brunnens vorgesehen wurde, nur gemacht um so zu tun, als nehme man die Meinungen der Bürger auf. Sie fordert, dass sich die Bürger mehr inhaltlich in die Planungen einbringen können.“[1] Als Reaktion auf die allseits präsente Problematik veröffentlichte Ende 2012 Bundesverkehrsminister Peter Ramsauer das vom *Bundesministerium für Verkehr, Bau und Stadtentwicklung* (BMVBS) erarbeitete *Handbuch für eine gute Bürgerbeteiligung: Planung von Großvorhaben im Verkehrssektor*. „Die Auseinandersetzungen um neue Straßen, Gleise oder Landebahnen zeigen nicht erst seit *Stuttgart 21*, dass wir bei Großprojekten ein besseres Miteinander von Politik, Wirtschaft und Bürgern brauchen.“[2]. Die Notwendigkeit einer besseren Bürgerbeteiligung ist dabei auch auf kleinere Projekte im urbanen Raum und deren Umgebung übertragbar. Im Projekt *ArchKM-AR* werden Methoden zur Verbesserung der Bürgerbeteiligung durch bildbasierende Kommunikation untersucht, deren Erkenntnisse im Folgenden aufgeführt werden.

3 Kommunikation

Um die Vorteile von Visualisierungen für die Bürgerbeteiligung aufzuzeigen, werden die gegenwärtigen Standards der Kommunikation vorgestellt.

3.1 Kommunikationsmodelle

Die klassischen Kommunikationsmodelle nach Shannon & Weaver, Satir, Berne, Schulz von Thun und Watzlawick berücksichtigen unterschiedliche Faktoren, die eine Kommunikation beeinflussen und ermöglichen. Sie lassen sich jedoch auf die grundlegende Bedingung nach Shannon & Weaver reduzieren: Jede Art der Kommunikation bedarf eines Senders, einer verschlüsselten Nachricht und eines Empfängers. Detailliertere Denkweisen der Theoretiker ziehen zudem die nonverbale Kommunikation (Metakommunikation wie Mimik und Gestik) sowie jegliche Art der Wahrnehmung mittels Sinnesorganen (u. a. Erwartungen, Werte) in Betracht. Diese Aspekte dienen meist der Erfassung der zwi-

[1] HalleSpektrum: „Steintor-Umbau: Bürger wollen Bäume und Brunnen“. Stand: 22. Juli 2012 (abgerufen am 13.02.2012).

[2] Bundesministerium für Verkehr, Bau und Stadtentwicklung: „Ramsauer stellt Handbuch für gute Bürgerbeteiligung vor“. Stand: 06.11.2012 (abgerufen am 08.02.2013).

schenmenschlichen Beziehung und individuellen sozial-psychischen Ausprägungen im Kommunikationsprozess.

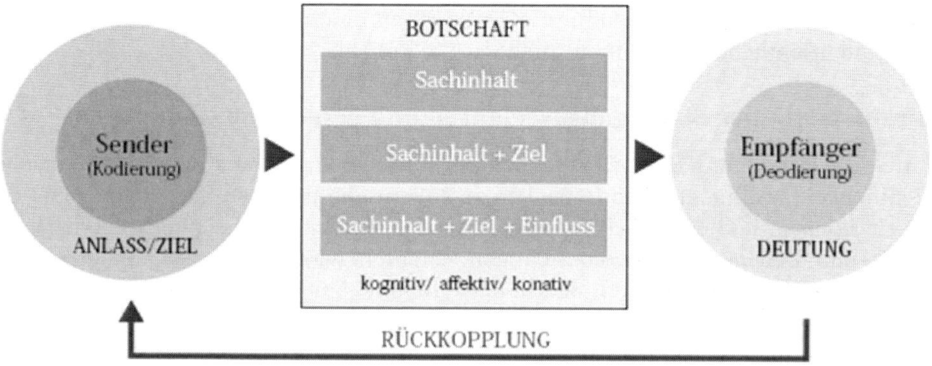

Abb. 1: Vereinfachtes Kommunikationsmodell (Quelle: KRUG, R. (2008): Architektur Kooperations-/Wissensnetz: Wissens- und Innovationstransfer in internetbasierenden Architektur-Kompetenznetzwerken, S. 77)

Bei der Kommunikation zwischen zwei Parteien kommt es nicht nur auf die Information und den Prozess ihrer Übertragung an, sondern vor allem auf deren Entschlüsselung. Kann der Empfänger die Informationen in ihrer Sinnhaftigkeit und Korrektheit nicht verstehen, ist die Kommunikation fehlgeschlagen. Dieser Umstand wird in den klassischen Kommunikationsmodellen nicht berücksichtigt. Es ist demnach erforderlich, dass Sender und Empfänger sich auf den jeweiligen Gesprächspartner einstellen und eine einheitliche Sprache finden. Diese Voraussetzung beruht hierbei nicht auf der zwischenmenschlichen Beziehung, sondern auf dem vermuteten und häufig unterschiedlichen Wissensstand der Gesprächspartner. Ein Experte sollte sein Vokabular und die Abstraktion seiner Ausführungen deshalb an das mentale Modell seines Gesprächspartners anpassen.

3.2 Experten-Laien-Kommunikation

Die sozial-psychischen Ausprägungen im Kommunikationsablauf einer Bürgerbeteiligung können vernachlässigt werden, da eine Experten-Laien-Kommunikation vorliegt. Dies bedeutet, dass den Kommunikationspartnern in der Regel von vornherein feste Rollen zugeschrieben sind. Bezogen auf die Bürgerbeteiligung hat der Planer die Rolle des Experten und die Aufgabe das Fachwissen an den Bürger (den Laien) zu vermitteln, damit dieser die Planung nachvollziehen und beurteilen kann. Der Bürger soll hierbei in die Lage versetzt werden, den Planungsstand mit seinen eigenen alternativen Ideen zu vergleichen, um diese eventuell wieder gezielt einer am Fachwissen orientierten Betrachtung zu unterziehen. Der Bürger ist häufig selbst ein Experte, da im Kontext einer Bürgerbeteiligung insbesondere Anwohner über bedeutendes Wissen über das entsprechende Planungsgebiet verfügen, das sich durch Alltagserfahrungen und den Austausch mit anderen Bürgern entwickelt hat.

Um diese Expertise zu nutzen, ist frühzeitig eine transparente und intensive Bürgerbeteiligung bei öffentlich wirksamen Projekten anzustreben.

4 Bürgerbeteiligung

Die derzeitigen Standards in der Bevölkerungsbeteiligung werden seit Kurzem von den Neuen Technologien revolutioniert. Im Folgenden wird die Effizienz herkömmlicher Beteiligungsmöglichkeiten untersucht und mit eGovernment, einer internetbasierenden Bürgerbeteiligung an öffentlichen Verwaltungsprozessen, verglichen.

4.1 Problematik der Beteiligungsverfahren und gesetzliche Grundlagen

„Eine effektive Beteiligung der Öffentlichkeit bei Entscheidungen ermöglicht es einerseits der Öffentlichkeit, Meinungen und Bedenken zu äußern, die für diese Entscheidungen von Belang sein können, und ermöglicht es andererseits auch den Entscheidungsträgern, diese Meinungen und Bedenken zu berücksichtigen; dadurch wird der Entscheidungsprozess nachvollziehbarer und transparenter, und in der Öffentlichkeit wächst das Bewusstsein für Umweltbelange sowie die Unterstützung für die getroffenen Entscheidungen.“[3] Das Europäische Parlament und der Rat der Europäischen Union beschreiben so die Vorteile einer Bürgerbeteiligung, wonach die Präsentation von Informationen den bestimmenden Faktor einer erfolgreichen Partizipation darstellt.

Angaben des Statistischen Bundesamtes[4] zeigen, dass das tägliche Zeitbudget einer vollzeiterwerbstätigen Person im Schnitt durch die Erfüllung der Grundbedürfnisse, wie Erwerbstätigkeit, private Verpflichtungen und Freizeitbeschäftigung erschöpft ist. Insbesondere, wenn eine persönliche Betroffenheit fehlt, liegt ein geringes Interesse für die Beteiligung an Bauprojekten vor. Außerdem sind „ortsübliche“[5] Bekanntmachungen und anschließend ausgelegte Informationen, wie sie im §3 BauGB genannt sind, in der Regel für den Bürger als Laien oft nicht nachvollziehbar.

Die Bekanntmachung mittels Aushängung förmlicher Schreiben an öffentlichen und sozialen Einrichtungen wie Sporthallen, Vereinshäusern und Verwaltungsgebäuden etc. ist zudem nicht mehr zeitgemäß und deckt nur einen geringen Teil der Zielgruppen ab, da weite Teile der Bevölkerung mittlerweile moderne Medien als Informationsquelle und Kommunikationskanal nutzen. Neue Informations- und Beteiligungsmöglichkeiten, wie mit AR können für diese Nutzergruppen eine passende Alternative darstellen.

Im Baugesetzbuch §4a wird auf den zusätzlichen Einsatz elektronischer Informationstechnologien hingewiesen, die auf der Verfügbarkeit des Internets beruhen. Damit können Informationen webbasierend, bzw. durch virtuelle soziale Netze weitläufig publiziert werden.

Planungsdaten in derzeit gebräuchlicher, digitaler Form sind für den Laien allerdings inhaltlich nur schwer zugänglich. Pläne, Bilder, Perspektiven und umfangreiche Textinformationen können unter Umständen nicht korrekt in Bezug zur IST-Situation eines Vorhabens in Deckung gebracht und somit ggf. fehlinterpretiert werden. Zusätzlich dazu können

---

[3] Europäische Union (Hrsg.), EG-RL über Öffentlichkeitsbeteiligung 2003/35/EG v. 26.5.2003. ABl. Nr. L156, 17 (abgerufen am 13.02.2012).

[4] Statistisches Bundesamt: DE Statis: Zeitverwendung (abgerufen am 06.02.13).

[5] Baugesetzbuch in der Fassung der Bekanntmachung vom 23. September 2004 (BGBl. I S. 2414), zuletzt geändert durch Artikel 1 des Gesetzes vom 22. Juli 2011 (BGBl. I S. 1509); §3 Abs. 2.

fehlende oder unverständliche Informationen, z. B. zu den Hintergründen bereits getroffener Entscheidungen, Missverständnisse erzeugen. Um dies zu verhindern, sind weitere Erklärungen des Experten an den Laien erforderlich, wobei die Erarbeitung eines angemessenen Schreibens zur Initiierung dieser Kommunikation oft zu aufwendig ist.

Selbst in öffentlichen Informationsveranstaltungen, in denen der direkte Kontakt zwischen Laie und Experte möglich ist, ergeben sich Fehlschlüsse. Da dem Laien eine intensive Auseinandersetzung mit den dargebotenen Informationen zeitlich nicht möglich ist und persönliche Eindrücke und Erfahrungen mit der Planung nicht hinreichend abgeglichen werden können.

Im Ergebnis lassen sich drei zentrale Barrieren in der Öffentlichkeitsbeteiligung feststellen. In der sukzessiven Abfolge des Partizipationsvorganges besteht die Schwierigkeit darin, alle Bevölkerungsvertreter über eine Planung und deren Datenveröffentlichung zu informieren, und das für die Beteiligung notwendige Interesse zu wecken. Außerdem liegt aufgrund fehlenden Fachwissens oft eine schlechte Informationsverarbeitung beim Laien vor. Und drittens besteht zwischen Laien und Experten häufig nur eine knappe Kommunikation oder sie fehlt sogar ganz.

Durch die Beseitigung dieser Hindernisse ist eine verbesserte Öffentlichkeitsbeteiligung im Sinne der EU möglich. Sie räumt der Öffentlichkeit das Recht ein „[...] Stellung zu nehmen und Meinungen zu äußern, wenn alle Optionen noch offen stehen und bevor die Entscheidungen über die Pläne und Programme getroffen werden [...].“[6] Doch dazu müssen die Planungsvorschläge von den Bürgern verstanden werden und Rücksprachen mit den planenden Disziplinen ermöglicht werden. An dieser Problemstellung knüpft das Forschungs- und Entwicklungsprojekt ArchKM-AR an.

4.2 eGovernment

eGovernment stellt den von Behörden und öffentlichen Einrichtungen angebotenen internetgebundenen Service für Dienste und zur Informationsvermittlung dar.

Nach Jahren der Forschung und Erprobung wird die internetbasierende Partizipation in Deutschland derzeit nur vereinzelt angewendet. Im Gegensatz zu den Vereinigten Staaten von Amerika, Großbritannien oder den skandinavischen Ländern, die auf internationaler Ebene als Vorreiter in der Erprobung verschiedener Formen der E-Partizipation gelten, sind die Entwicklungen für eine umfangreiche (E-)Kooperation hierzulande noch weitgehend unvollendet. „Da es sich dabei um eine anspruchsvolle Form der Beteiligung handelt, kommt sie nur selten zum Einsatz. Die Nutzung von Informations- und Kommunikationstechnologien für diesen Zweck ist noch weitgehend experimentell.“[7] Bundesländer und Kommunen verwenden zur innovativen Partizipation für Stadt- und Raumplanungen regelmäßig die *Online-Konsultation*. Exemplarische eGovernment-Angebote sind *ELSTER*, eine Anwendung für Online-Steuererklärung, die *IHK-Online-Akademie*, sowie das *Stand-*

---

[6] Europäische Union (Hrsg.), EG-RL über Öffentlichkeitsbeteiligung 2003/35/EG v. 26.5.2003, ABl. Nr. L156, 18, Art. 2, Abs. 2b (abgerufen am 13.02.2012).

[7] Albrecht, S. et al., „E-Partizipation – Elektronische Beteiligung von Bevölkerung und Wirtschaft am E-Government“, S. 5.

ort-Informations-System Bayern (SISBY) und die *Digitale Signatur,* ebenfalls von der *Industrie- und Handelskammer* (IHK).

4.3 Zielgruppen für eGovernment

Studien zur Zielgruppenforschung im eGovernment zeigen eine deutliche Abhebung potenzieller Nutzergruppen zum Bevölkerungsdurchschnitt. Insbesondere, in den älteren Bevölkerungsschichten ist häufig eine ablehnende Haltung feststellbar. Außerdem weisen erwerbstätige Nutzergruppen mit höherem Bildungsstand und Einkommen eine häufigere Nutzung des Internets auf. Bei Personen, die sich diesem Profil nicht zuordnen lassen, liegt hingegen ein potenzielles Nutzungsdefizit vor.

Eine entsprechend einfache Gestaltung der eGovernment-Systeme vorausgesetzt, kann durch eGovernment eine große Bevölkerungsschicht adressiert werden, da neben der generellen Internetverfügbarkeit lediglich ein allgemeines Verständnis und Interesse für moderne Medien erforderlich ist.

Die Charakterisierung der Bürger erfolgt anhand folgender Nutzertypen[8]:

- „digitale E-Government Pragmatiker,
- mobile E-Government Pragmatiker,
- prädigitale E-Government Desinteressierte,
- digitale (E-)Government Skeptiker,
- digitale (E-)Governemnt Unkundige".

5 Bildbasierende Kommunikation mit Augmented Reality

Das Potenzial erfolgreicher Wissenskommunikation durch die Verwendung von Bildern ist der zentrale Schwerpunkt, des Forschungsprojektes *ArchKM-AR*. Der Einsatz von Augmented Reality kann in der ePartizipation, speziell beim eGovernment, neue Möglichkeiten der Informationsvermittlung eröffnen. Deshalb sollen nun die Anwendung der AR und ihre Wirkung auf den Partizipationsprozess dargestellt werden.

5.1 Augmented Reality

Zur Unterstützung von Kommunikationsprozessen werden häufig Bilder eingesetzt. Im Bereich der Architekturvisualisierung werden diese Abbildungen auf Basis dreidimensionaler Planungsdaten erzeugt und stellen einen räumlichen und möglichst realistischen Eindruck des Bauvorhabens dar. Die digital generierten Bilder zeigen zu Werbezwecken jedoch häufig eine idealisierte Umgebung, sodass sie vom Laien oft als manipulativ wahrgenommen werden.

Bei der Augmented-Reality-Technologie werden Videobilder der Realität mit künstlich generierten Bildinhalten überlagert. AR-Visualisierungen werden bisher vor allem im Marketing eingesetzt, indem die gedruckten Informationen auf Werbeträgern, wie bspw. Prospekten oder Produktverpackungen, durch virtuelle 3D-Inhalte erweitert werden.

---

[8] eGovernment MONITOR 2012: „Nutzertypen: E-Government" (abgerufen am 13.02.13).

In der Produktion wird die AR außerdem zur industriellen Werkerassistenz eingesetzt. Hierfür werden Werkern zusätzliche Informationen zu Arbeitsprozessen in ein kopfgetragenes Display (HMD) eingeblendet.

Bei einer AR-Anwendung besteht die Herausforderung darin, die generierten Inhalte entsprechend der realen Gegebenheiten auszurichten. Heutige mobile Endgeräte wie Smartphones oder Tablets bieten hierfür neben der erforderlichen Kamera auch zusätzliche Sensoren sowie ein ausreichend großes Display.

AR-Marketinganwendungen benötigen meist nur einen geringen Abstand der Kamera zum Zielort der Überlagerung, sodass das verwendete Printmedium nahezu bildfüllend durch die Kamera erfasst wird. Mittels Bildverarbeitung werden Bildmerkmale im Videobild gesucht und mit zuvor trainierten Merkmalen, z. B. Markern verglichen. Da die Position der trainierten Merkmale auf dem Printmedium bekannt ist, kann eine Korrespondenz hergestellt und die aktuelle Position und Orientierung (Pose) des Printmediums berechnet werden.

In der Architektur ist die Überlagerung in Kamerabildern des Außenraums erforderlich. Die Beleuchtung ist hier umwelt- und wetterbedingt heterogen und der Abstand zum Ort der Überlagerung beträgt mehrere Meter, damit ein Planungsvorhaben an seinem Planungsort augmentiert werden kann. Hierfür ist die Verwendung kontrastreicher und strukturierter Schwarz-Weiß-ID-Marker (Fiducial Marker, Abb. 2) erforderlich.

Abb. 2: ID-Marker (A) – wird in Bildinhalten effizient detektiert. (B) Sein Innenbereich ermöglicht die Identifikation und Bestimmung der Orientierung. Interaktive Überlagerung eines Gebäude-3D-Modells am Ort des Bauvorhabens (C). (Quelle: IFF Magdeburg, 2012)

Mobile Endgeräte verwenden einen automatischen Weißabgleich und komprimieren das Kamerabild, wodurch hochfrequente Bildinhalte verlustbehaftet sind. Durch den Autofokus, der verwendeten Kameras, der sich zum Teil nicht auf eine feste Brennweite festlegen lässt, werden ggf. Objekte im Vordergrund fokussiert. Durch diese Ursachen kann die zur Identifikation erforderliche Bildschärfe bei großen Abständen u. U. nicht erreicht werden.

Wird ein Marker erkannt, erfolgt die Bestimmung seiner Pose zur Kamera. Ist der Marker im Kamerabild zu klein, kann seine Pose nicht numerisch stabil bestimmt werden, wodurch die Überlagerung zittert oder kippt. Zur korrekten Überlagerung in der Zielentfernung müsste dazu ein sehr großer Marker (1-2 m Kantenlänge) verwendet werden, der jedoch auf Baustellen oder in Bereichen des öffentlichen Interesses nicht praktikabel ist. Im Projekt *ArchKM-AR* wurde eine Methode entwickelt, bei der stattdessen vier Marker (30 cm Kantenlänge) genutzt werden, die eine Ebene aufspannen und sich in vorhandene Baustellenschilder integrieren lassen. Sie können im Kamerabild bei einer Entfernung bis zu 20 m erkannt werden und ermöglichen durch ihren Verbund eine stabile Bestimmung der Pose (Abb. 2). Durch die zusätzliche Verwendung der in mobilen Endgeräten vorhandenen Inertialsensoren kann die Pose weiter stabilisiert und auch verfolgt werden, sollten die Marker kurzzeitig nicht im Kamerabild enthalten sein.

Die prototypische Anwendung soll mit Hinblick auf eine Verwendung in planerischen Disziplinen den Beteiligten auf Experten- und Laienebene die Möglichkeit bieten, Planungsprojekte eigenständig zu untersuchen. Die Darstellung der Vorteile der Anwendung erfolgt in Kapitel 5.2.

5.2 Potenzial von Augmented-Reality für die Bürgerbeteiligung

Die Verwendung von AR bietet dem Bürger verschiedene Vorteile, die ihm das Beteiligungsverfahren erheblich erleichtern und somit sein Interesse enorm steigern können.

Vor-Ort-Aspekt
Dieser Punkt beschreibt den Effekt, bei dem der Betrachter aufgrund der Anwesenheit am Planungsort bewusst oder unbewusst jegliche Umwelteinflüsse (u. a. Wind, (Straßen-) Lärm, Licht und Gerüche) wahrnimmt. Des Weiteren fließen persönliche Empfindungen zur Umgebung in die meist intuitive Bewertung der Planung mit ein. Der Betrachter kann außerdem Anlass oder Notwendigkeit sowie Eckpunkte und Rahmenbedingungen der Planung in der Umgebung feststellen.

Zeit-Aspekt
Im Gegensatz zu bekannten Beteiligungsverfahren mit termingebundener Auslegung von Informationen oder öffentlichen Veranstaltungen hat der Nutzer die Freiheit, den Zeitpunkt sowie Dauer und Häufigkeit der Betrachtung individuell zu bestimmen. Die Beteiligung erscheint dadurch weniger gezwungen und amtlich.

Perspektiven-Aspekt
Dem Betrachter wird mittels der AR-Anwendung eine natürliche und ihm vertraute Perspektive auf das Planungsobjekt geboten.

Dimensionen-Aspekt
Durch die zwangsläufige Anwesenheit am Ort der Planung und dem genannten Perspektiven-Aspekt werden dem Betrachter die Dimensionen der Planung im Maßstab 1:1 präsentiert. Er erfährt somit eine exakte Vorstellung der Größe des Planungsobjektes.

Bewegungs-Aspekt
Der Betrachter hat die Möglichkeit das Gebiet unter Vorbehalt der lokalen Begebenheiten und Zugangsrechte eigenständig zu begehen und in seiner Ganzheit zu erfassen.

Interaktivitäts-Aspekt
Der Nutzer ist gezwungen sich das Planungsobjekt eigenständig zu erschließen, d. h. ihm werden keine Perspektiven oder Ansichten vorgegeben. Der Betrachtungswinkel wird selbstständig bestimmt.

Informations-Aspekt
Die AR-Technologie stellt Informationen auf die Weise dar, die dem Menschen als natürlichste und einfachste erscheint. Die visuelle Wahrnehmung ist der grundlegende Fakt zum Verständnis eines Objektes. Mittels AR wird eine Überlappung von Fachinformationen (3D-Objekt) und der Realität erzeugt. Es bedarf keiner aufwendigen Entschlüsselung der Fachinformationen. Zudem können Zusatzinformationen mittels Text, Bild, Video und Audio in die Visualisierung eingebunden werden.

Kommunikations-Aspekt
Je nach Zielstellung der Visualisierung kann die Software zu einer zeitgleichen Kommunikation neben der Betrachtung verwendet werden. Diese eröffnet dem Nutzer aufgrund der ungezwungenen Atmosphäre, die die Anwendung erzeugt, eine schnelle und einfache Alternative zum förmlichen Briefverkehr der bisherigen Beteiligungsverfahren. Die Kommunikation kann ähnlich den zeitgemäßen internetbasierenden Messenger-Systemen gestaltet werden und würde damit eine breite Nutzergruppe ansprechen.

5.3 Kommunikationsmodell mit Augmented Reality

Schon in der jüngsten Vergangenheit wurde festgestellt, dass der Kommunikationsprozess zwischen Experten und Laien (Wissenschaft und Öffentlichkeit) in vielen Bereichen nicht als einseitige Informationsvermittlung angesehen werden kann. Vielmehr wird sie als Aushandlungsprozess beschrieben (HESSLER, 2004). Die Kommunikation stellt eine gemeinsame Entwicklung von Wissen dar, wobei Wissenschaftler ihre Erkenntnisse an das Vokabular und Verständnis der Öffentlichkeit anpassen müssen, um einen effektiven Austausch zu ermöglichen. Diese Vorstellung sollte als Vorbild für die Bürgerbeteiligung bei Planungsprojekten dienen. AR bietet für die Kommunikation zwischen Experten und Laien eine erhebliche Erleichterung (siehe Abb. 3).

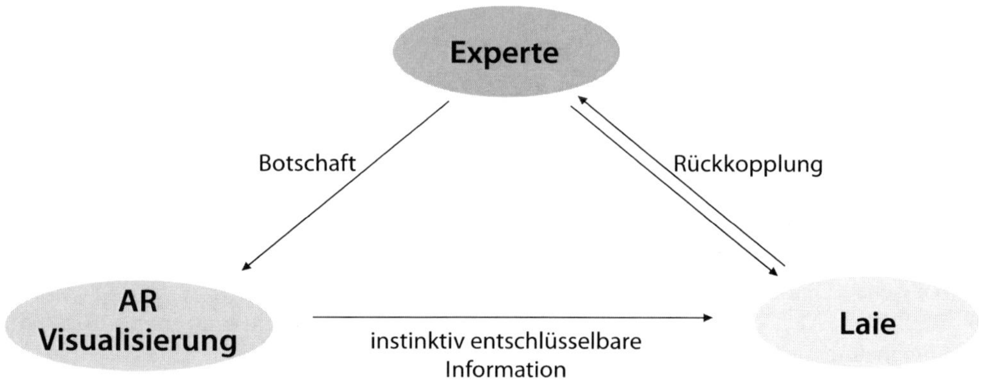

Abb. 3: Kommunikationsmodell mit Augmented-Reality (Quelle: eigene Darstellung, 2013)

Die Augmented-Reality-Technologie ist in der Lage die kritische Entschlüsselung von Fachinformationen durch den Laien deutlich zu befördern. Mit ihrer Hilfe besteht damit die Aussicht, dem Bürger eine vereinfachte Kommunikation mit Experten zu ermöglichen, so sein Verständnis für Projekte und Planungsvorhaben zu erhöhen und sein Interesse an der Beteiligung öffentlicher Projekte potenziell zu steigern. Voraussetzung dafür ist lediglich die Begeisterung für die Anwendung der Technologie sowie dem vereinfachten Umgang mit Planungsvorschlägen innerhalb partizipativer Beteiligungsverfahren.

Literatur

Abawi, D., Bienwald, J., Dorner, R. (2004), *Accuracy in Optical Tracking with Fiducial Markers: An Accuracy Function for ARToolKit.* In: Proceedings of the 3rd IEEE/ACM International Symposium on Mixed and Augmented Reality (ISMAR), 260-261.

Albrecht, S., Kohlrausch, N., Kubicek, H., Lippa, B., Märker, O., Trénel, M., Vorwerk, V., Westholm, H. & Wiedwald, C. (2008), *E-Partizipation – Elektronische Beteiligung von Bevölkerung und Wirtschaft am E-Government".* Studie im Auftrag des Bundesministeriums des Innern, Bremen, 2008, S. 3-6. http://www.ifib.de/publikationsdateien/ifib-zebralog-e-partizipation-kurz.pdf (14.02.13)

Anderson, D., Kernchen, S., Adler, S., Mecke, R. (2012), *Anwendung von Mixed Reality im Produktmarketing.* Fachtagung – Digitales Engineering zum Planen, Testen und betreiben Technischer Systeme. Tagungsband 15. IFF-Wissenschaftstage, 207-217.

Becker, J., Bergener, P., Fielenbach, K., Fuchs, P., Herwig, S., Karow, M., Niehaves, B., Räckers, M. & Weiß, B. (2008), *E-Inclusion: Digitale Integration durch E-Government.* Studie im Auftrag des Bundesministeriums des Inneren. Münster. http://udoo.uni-muenster.de/downloads/publications/2011.pdf (08.05.2013).

Bromme, R., Jucks, R., Rambow, R. (2004), *Experten-Laien-Kommunikation im Wissensmanagement.* In: Reinmann, G. & Handl, H. (Hrsg.), Der Mensch im Wissensmanagement: Psychologische Konzepte zum besseren Verständnis und Umgang mit Wissen. Göttingen, Hogrefe, 176-188.

Bundesministerium für Verkehr, Bau und Stadtentwicklung (Hrsg.) (2012), *Handbuch für eine gute Bürgerbeteiligung: Planung von Großvorhaben im Verkehrssektor.* Berlin, 5-14, http://www.bmvbs.de/cae/servlet/contentblob/81212/publicationFile/65799/handbuch-buergerbeteiligung.pdf (17.01.2013).

Europäische Kommission (2010), *Europäischer eGovernment-Aktionsplan 2011-2015: Einsatz der IKT zur Förderung intelligent, nachhaltig und innovativ handelnder Behörden.* Brüssel, 2-5. http://www.it-planungsrat.de/SharedDocs/Downloads/DE/Entscheidungen/5._Sitzung/Anlage_EU%20E-Government%20Aktionsplan_kommentiert_GS%20IT-PLR.pdf?_blob=publicationFile (14.02.13).

Europäische Union (Hrsg.), *EG-RL über Öffentlichkeitsbeteiligung 2003/35/EG v. 26.5.2003.* ABl. Nr. L156, 17-18. http://www.bmu.de/fileadmin/bmu-import/files/pdfs/allgemein/application/pdf/rl200335eg_oeffent.pdf (13.02.2012).

Hahnzog, S. (Hahnzog Organisationsberatung), *Ausgewählte Aspekte der Kommunikationspsychologie: Kommunikationstheorien.* http://www.hahnzog.de/organisationsberatung/lexikon#Kommunikationspsychologie/ (05.02.2013).

Havag, *Stadtlinie I.* http://www.havag.com/news-infos/projekte-havag/Stadtbahn/stadtbahnlinie-1 (12.02.2013).

Heßler, M., Hennig, J. & Mersch, D. (2004), *Visualisierung in der Wissenskommunikation*. Explorationsstudie für das BMBF, Berlin. http://www.sciencepolicystudies.de/dok/explorationsstudie-hessler.pdf (04.02.2013).

Initiative D21, ipima: *eGovernment MONITOR 2012*. http://www.egovernment-monitor.de/startseite.html (14.02.2013).

IHK Niederbayern, *E-Government*. http://www.ihk-niederbayern.de/innovation-umwelt/kommunikationstechnologie-e-government/e-government/ (13.02.2013).

kikidan, *Kommunikationsmodelle im Vergleich, Arten der Kommunikation,* http://www.kikidan.com/artikel/kommunikationsmodelle.html (05.02.2013).

Statistisches Bundesamt: *DE Statis: Zeitverwendung*. https://www.destatis.de/DE/Zahlen Fakten/GesellschaftStaat/EinkommenKonsumLebensbedingungen/Zeitverwendung/Tabellen/ZeitVerwRentnerVollzErwerb_ZBE.html (06.02.2013).

Invited Contributions

Visual Project Analysis for an Office Park – Evaluation of Method Used, Comparing Before, Simulation and After

Beurteilung der Visualisierung eines Büroparks für die Belange der Umweltplanung – Vergleich der Simulation mit dem gebauten Projekt

Erich BUHMANN

Introduction / *Einführung*

For the evaluation of the impact of building a large-scale office park of 640,000 square meters on the surrounding landscape, a landscape visualization study was performed. The Campeon office park is the new headquarters for Infineon Technologies AG. The visual landscape evaluation study was done as part of the environmental impact assessment for the legal planning *(Bebauungsplan)*. The scale of the study is 1:5,000 to 1:2,000. These scales represent the level of preliminary planning of the development. The intention of this study was to communicate the dimension of planed development and the concern to visual aspects of the overall design.

The changes were visualized in the urban planning from potentially representative user points. The need for renewal of the legal planning enabled the environmental planner to evaluate the visualization method initially used. After a period of ten years, the original simulated scenes could be compared with the current landscape.

Fig. 1:
Project site as non-accessible agricultural land in 2001 prior to Development / *Die nicht zugängliche Ackerfläche im Jahre 2001 vor der Bebauung der Fläche*

Fig. 2:
Office park in spring, 2011 as open-to-the-public landscape park / *Im Jahre 2011: Büropark als frei zugänglicher Landschaftspark*

This paper is concerned with describing the method used originally for the visual simulation of landscape change, the evaluation of the method in comparing the simulations with the built situation, and finally discussing future methods utilizing 3D GIS tools.

Für die Beurteilung der Beeinträchtigung des Landschaftsbildes durch den Neubau der 64 Hektar umfassenden Unternehmenszentrale „Campeon" der Infineon Technologies AG, wurde auf der Basis des städtebaulichen Entwurfs die Veränderung von repräsentativen Nutzerstandorten im Rahmen der Umweltverträglichkeitsprüfung für den Bebauungsplan visualisiert. Die Fortschreibung des Bebauungsplans nach zehn Jahren ermöglichte dem Umweltplaner den Vergleich der damals gewählten Methode der Projektvisualisierung mit der tatsächlichen Änderung des Landschaftsbildes.

In diesem Beitrag soll die ursprünglich gewählte Methode der Landschaftsbildprojektvisualisierung auf der Ebene des Bebauungsplans (verkürzt als B-Plan bezeichnet) im Maßstabe 1:5.000 bis 1.2.000 dokumentiert werden. Im Anschluss wird die Aussagekraft der Simulationen hinsichtlich zukünftiger vergleichbarer Projektvisualisierung durch den Vergleich mit der gebauten Situation ausgewertet. Abschließend soll diskutiert werden, wie heute 3D GIS Instrumente hier ergänzend eingesetzt werden sollten.

2 Method of Visualization / *Wahl der Visualisierungsmethode*

Scale and LOD / *Maßstab und Visualisierungsebene*

The Level of Detail (LOD) and the appropriate visualization techniques have to be considered during the conception phase of project visualization. Landscape models can be derived from different data sources for the different levels of details. In the discussed office campus project, the midsize-scale between the following three levels of detail (LOD) had been applied.

Bei der Konzeption einer Studie zur Bewertung der zu erwartenden Änderung des Landschaftsbildes wird zunächst unter anderen der Grad der Detaillierung LOD und die damit im Zusammenhang stehende Wahl der Visualisierungstechnik und die Datengrundlage berücksichtigt werden. Die Visualisierung auf der Ebene des B-Planes im städtebaulichen Maßstab entspricht dem mittleren LOD.

LOD 1: Regional Planning Scale: Far Distance / *Regionalplanungsmaßstab* Scale / *Maßstab* 1:200,000 – 1:25,000

- Landform: Radar data, 90 meters by 90 meters from SRTM (Shuttle Radar Topography Mission)
- Vegetation: Remote sensing satellite data, such as LANDSAT 80 m by 80 m / *Fernerkundungsdaten von erdbeobachtenden Satelliten*
- Software: ArcGIS, ArcGlobe, ERDAS, Google Earth et al.
- Observer location: from above (helicopter) / *Beobachtungspunkt: größere Höhe*

LOD 2: Community Planning and Zoning Scale/Middle Distance, *Städtebaulicher Maßstab, Vorentwurf* Scale / *Maßstab:* 1:25,000 – 1:1,000

- Landform: Topography from the state surveying authority, 10 m by 10 m
- Vegetation: Orthophotos (DOP 20) from the state surveying authority
- Architecture: ATKIS Geometry, 3D CAD of the involved engineering companies
- Software: ArcGIS, ArcGIS CityEngine, Autodesk Infrastructure Modeler, 3D Studio Max, Photoshop et al.
- Observer Location: From above, up to 20 meters above the ground and eye level
- Model: GIS model of scenarios, data provided by architect, hydrologists, environmental planer et. al.

LOD 3: User Planning Scale/Near Distance / *Maßstab Entwurf und Werkplanung* Scale / *Maßstab* 1:1,000 – 1:10

- Landform: Topography from the state surveying authority, 10 by 10 m performed with an airborne scanner, 1 m. by 1 m, or LIDAR data
- Vegetation: 3D Object
- Architecture: ATKIS Geometry, 3D CAD of the involved engineering companies
- Software: Lenne 3D, ArcGIS, 3D Studio Max, Photoshop et al.
- User Location: Eye level
- Model: GIS model data provided by architect, hydrologists, environmental planer et al.

The architect provided the 3D model of the preliminary planning. According to the code for legal zoning, the details included the footprint for the building, the lake, the main open park area, the building height, the clear definition of roof gardens on top of the buildings and the general colour scheme. It did not include textures or any additional details.

Der Architekt hat das 3D-Stadtmodell seines Vorentwurfs mit den wesentlichen im B-Plan festzulegenden Inhalten zur Visualisierung der Änderungen des Landschaftsbildes für die Integration in das 3D-Modell für die Visualisierung zur Verfügung gestellt. Mit dem 3D-Modell wurden die Gebäudegrenzen und die Höhe der Gebäude sowie die Abgrenzung des geplanten Sees digital übergeben. Weiterhin wurden die Festlegungen auf vegetative Dachbegrünung und die grundsätzliche Farbgebung der Fassaden nachrichtlich übernommen. Weitere Texturen und Detaillierungen wurden in diesem Maßstab gezielt nicht berücksichtigt. Die Lage der Grünflächen übernahm der Umweltplaner aus dem Grünordnungsplan, der von ihm selbst bearbeitet wurde.

General Workflow of Simulation / *Die wesentlichen Bearbeitungsstufen der Landschaftsbildsimulation*

For the first Visual Landscape Evaluation of the large scale office park, the following principal Workflow for the visualization on LOD level 2 was adapted:

Preparation of Visual Landscape Evaluation

- Define objectives of visualization
- Select the appropriate visualization method
- Decide with client on scope of study
- Refine visualization method according to scope of study

Visual Landscape Visualization of Images Using 3D Modeling

- Choose method for fixing the proportion of proposed design (defining fixed points)
- Capture high resolution images with fixed points
- Build 3D model of preliminary design
- Locate the 3D model according to the fixed points in the 2D image
- Complete the 2D simulation with 3D rendering, applying 3D libraries and final rendering of 2D image

Visual Landscape Evaluation

- Compare "Before" and "After", quantify and evaluate the visual impact
- Discuss the effect of visual impact of preliminary design and suggest visual improvements

Evaluation of Visual Landscape Simulation After Construction

- Compare images of visualisation after construction
- Suggest further changes and additions to improve and maintain visual quality
- Suggest improvements for further Visualisation of Projects

Für Landschaftsbildvisualisierung des großflächigen Büroparkes im städtebaulichen Maßstab wurde folgender "Workflow" gewählt:

Vorbereitung der Landschaftsbildbewertung

- *Definition der Zielsetzung für die Visualisierung*
- *Auswahl der geeigneten Visualisierungsmethode*
- *Abstimmung des Visualisierungsumfanges mit dem Auftraggeber*
- *Anpassung der Visualisierungsmethode nach der Abstimmung mit dem Auftraggeber*

Visualisierung von Landschaftsbildern mit der Unterstützung von 3D Modellierung

- *Auswahl der Methode um die Fixpunkte des geplanten städtebaulichen Entwurfs in das Foto einzupassen*
- *Aufnahme von hochauflösenden Bestandsbildern mit den Fixpunkten*
- *Modellierung des 3D-Models des Vorentwurfs*
- *Einfügung des 3D-Models mithilfe der Fixpunkte in das 2D Image*
- *Komplettierung der 2D-Simulation mit „3D-Rendering", Anwendung von 3D-Bibliotheken und abschließende Bildbearbeitung der 2D-Simulationen*

Bewertung der Landschaftsbildsimulationen

- *Vergleich der „vorher"- und „nachher"-Bildpaare hinsichtlich der quantitativen Änderungen der einzelnen Landschaftselemente und verbal argumentative Bewertung der geplanten Änderung*
- *Diskussion der Effekte der Änderung des Landschaftsbilds des derzeitigen Vorentwurfs und gegebenenfalls Ableitung von Vorschlägen für die Reduzierung der Eingriffe in das Landschaftsbild*

Bewertung der Landschaftsbilder nach der Bebauung

- *Vergleich der Bilder des gebauten Zustandes mit der Situation vor der Bebauung und mit den Simulationen des Vorentwurfs*
- *Ableitung von Vorschlägen hinsichtlich der weiteren Entwicklung zur Erhaltung eines möglichst hochwertigen Landschaftsbildes*
- *Ableitung von Vorschlägen für die Verbesserung von Visualisierung von Projekten*

Ethic Principals by SHEPPARD / *Die Prinzipien für Landschaftsbildsimulation nach SHEPPARD*

According to Sheppard (1982), Appleyard (1977) had already suggested the following "public" criteria for judging a "good" simulation:
It should be realistic, accurate, comprehensible, evaluable, engaging, flexible, and cheap.

These criteria were developed further by Sheppard. In order to prove the extent of changing landscapes to the client and/or to the public, the landscape modeler has to follow the ethical principles of Sheppard (1989) listed below in order to show objectiveness:

Neben der Festlegung des "Levels" und des "Workflows" sollen die allgemeinen gutachterlichen Regeln für die Landschaftsbildvisualisierung, wie diese sehr früh schon von Appleyard (1977) und dann zuletzt von Sheppard (1989) weiterentwickelt wurden, eingehalten werden:

- Representative Simulation / *repräsentativ*
- Accurate Simulation / *akkurat*
- Credible Simulation / *glaubwürdig*
- Comprehensive Simulation / *ausreichend und umfassend*
- Bias-Free Simulation / *vorurteilsfrei*
- Defensible / *vertretbar und angemessen*
- Engaging / *engagiert und verbindlich*
- Accessible / *nachvollziehbar*
- No "sale effects" such as animation etc. / *ohne „Verkaufseffekte"*

3 Landscape Visualization / *Landschaftsbildsimulation*

Selection of Representative Vistas / *Auswahl repräsentativer Nutzerstandorte*

Before development, the planning area was farmland and within the context of recreation was accessible mostly by way of the road Zwerger Allee. A high user frequency could also be noted from the direction of the Fasanenpark train station. Furthermore, the property could be briefly viewed by drivers on the A8 autobahn heading out of Munich. Additionally, children playing on the hill in a playground to the south could look to the area of the future Campeon office park.

Based on these perspectives, the following six viewing areas were selected to compare "before and after" views

- S1 View of office park after entering from the autobahn bridge
- S2 View from the hill the existing playground
- S3 View from Fasanenpark train station (S-Bahn)
- S4 View from Zwerger Allee (I) after entering from Fasanenpark residential area
- S5 View from Zwerger Allee (II) at the level of the future office site
- S6 View from A8 autobahn

Vor der Bebauung war das Planungsgebiet eine Ackerfläche und im Rahmen der Naherholung vor allem von einem Weg, der Zwerger Allee, erlebbar. Eine hohe Betrachter-Nutzerfrequenz weist das Gelände weiterhin von der S-Bahnhaltestelle „Fasanenpark" aus auf. Weiterhin konnte das Ackergrundstück von den stadtauswärts fahrenden Autofahrern von der Autobahn A 8 für einen kurzen Moment wahrgenommen werden. Einen weiteren Überblick über den künftigen Büropark Campeon können spielende Kinder von dem Spielhügel im südlich angrenzenden Spielplatz erhalten.

Es wurden daher folgende sechs repräsentative Standorte für Besucher für den Vergleich von „vor-nachher" ausgesucht:

- *S1 Blick vom vorhandenen Büropark nach Eintritt über die Autobahnbrücke*
- *S2 Blick vom Spielhügel des vorhandenen Spielplatzes*
- *S3 Blick vom S-Bahnhof Fasanenpark*
- *S4 Blick von der Zwerger Allee (I), nach Eintritt vom Wohngebiet Fasanenpark*
- *S5 Blick von der Zwerger Allee (11), in Höhe der zukünftigen Bebauung*
- *S6 Blick von der Autobahn A8*

Fig. 3: Selecting representative user vistas towards the 640,000 square meter development (about 750 meter by 900 meter) from S1 to S6, and selection of view directions and view angle / *Auswahl der Nutzerstandorte, der Blickrichtung und des Blickwinkels*

Dimension and Selection of View Angle / *Auswahl der Blickrichtung und des Blickwinkels*

The human eye is capable of observing a landscape within a view angle ranging from 180° to 200°. As a rule however, the view is focused mostly within the range of 60° to 100°. For the visualization of individual pictures we therefore used a view angle of 84° which was taken from a 35 mm format. If a wider view angle had been used, the horizon would have been shrunk. With the lateral view angle of 84°, a representative view angle for landscape visualization of individual pictures was chosen.

From every point where a picture was made, the view was always to the middle of what would become the office park so that from every potential user point the view was directed toward the site of the future office park.

Das menschliche Auge ist in der Lage, Landschaft bis zu einem Blickwinkel von etwa 180° bis 200° wahrzunehmen. In der Regel fokussiert sich der Blick jedoch auf einen Bereich der Wahrnehmung zwischen 60° und 100°. Für die Einzelbildvisualisierung haben wir daher einen mittleren Bildwinkel von 84° genommen, der sich aus einer 35-mm-Kleinbild-Brennweite im Querformat ergibt.

Ein weiterer Bildwinkel hätte die Höhe der visualisierten Gebäude und Anlagen optisch gedrückt. Mit dem seitlichen Blickwinkel von 84° ist ein repräsentativer Blickwinkel für die Landschaftsbildvisualisierung von Einzelbildern gewählt worden. Als einheitliche Blickrichtung wurde von jedem Bildstandort aus die spätere Mitte der geplanten Bürobebauung anvisiert. Somit wird von jedem Besucherstandort aus der Blick direkt auf die spätere Bebauung gerichtet.

Precision of Location / *Maßstabstreue und Genauigkeit*

All points where the photos were taken from, as well as the 18 outer corners of the planned buildings, were surveyed by a surveying office according to location and height, and marked with wooden poles. For the photos, a sunny late summer day with a clear view to the Alps was picked. The photos were processed as fine grained negatives and then scanned in a professional lab. They were then burned onto PRO PHOTO CDs. The high resolution led to a data volume of 72 MB per photo file. This high resolution allowed for elements with a height of 10 cm to be recognized from a distance of one kilometer. In the method used the incorporation of the edges of the planned buildings could be, depending on the distance, placed with a precision of between 20 and 50 centimeters.

Consequently, the 3D computer model of the buildings and the planned site (elevation model of sound barrier, etc.) could be continually mounted with more points, true to scale. For the visualization of the planned trees, a height of 8 to 10 meters was used which approximated the height of the woody plants after 10 to 15 years of growth.

The planning scale of the preliminary design from 2001 was simulated. For the landscape simulation, the proportions of the planned buildings and elements such as trees and terrain modeling were exactly rendered onto 3D computer models. The individual textures of the architecture and the detail true individual features were not stressed so as to keep the proportions of the landscape in the foreground.

In the Munich area, Foehn weather conditions allow for views of the Alps from many locations. This view is also possible from Zwerger Allee. For this reason, the Alps panorama from S4 was again presented in detail as can be seen in the enlargement.

Alle Fotostandorte und 18 äußere Eckpunkte der späteren Gebäude wurden von einem Vermessungsbüro nach Lage und Höhe vermessen und dauerhaft mit Holzpfählen markiert. Für die Aufnahmen wurde ein sonniger Spätsommertag mit „Föhnblick" auf das Alpen-Panorama gewählt. Die Aufnahmen wurden mit feinkörnigen Negativen durchgeführt und dann in einem professionellen Labor gescannt und auf PRO PHOTO CD gebrannt.

Fig. 4: Marking the building corners on site using 2 meter poles at all images taken of the existing condition "before" in order to fix them accurately in the digital model / *Markierung aller Gebäudekanten mit 2 Meter hohen Rohren, damit diese dann später genau in das digitale Modell eingepasst werden können*

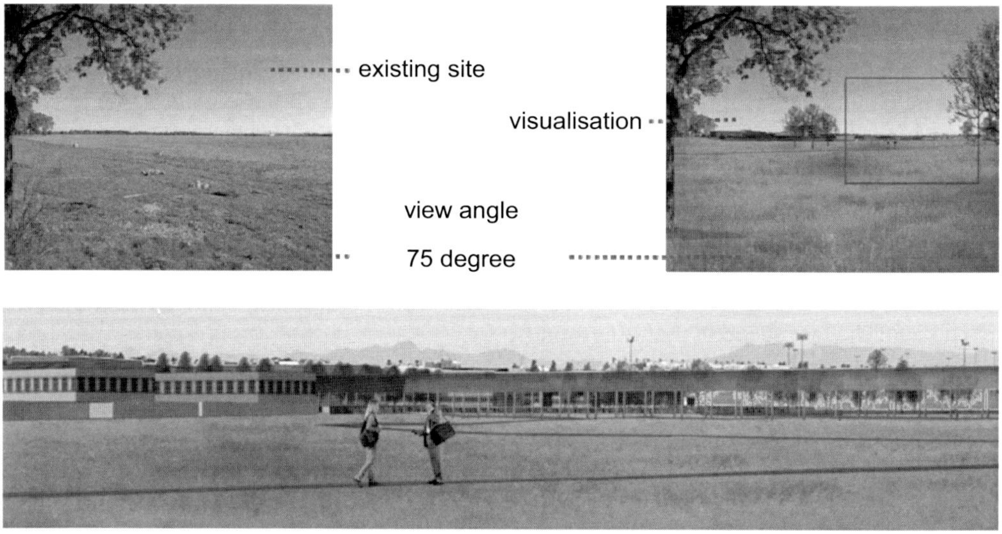

Fig. 5: Detail showing the Alpine panorama / *Detailvergrößerung mit Alpenpanorama*

Fig. 6: The sequence described in the Workflow / *Bearbeitungsschritte wie unter "Workflow" beschrieben:*

a) High resolution image of existing situation / *Hochauflösendes Bestandsbild*
b) 3D CAD model scaled to high resolution image by the marked fix points of the building corners / *Fixpunkte der Gebäudekanten werden im Bild festgehalten*
c) Rendering using 3D libraries / *Einsatz von Rendering und 3D Bibliotheken*
d) Final photoshop work / *Abschließende Bildretuschen*

Die hohe Auflösung führt zu einer Datenmenge von 72 MB je Bilddatei. Diese hohe Auflösung erlaubt die Erkennung von 10 cm großen Elementen in einer Entfernung von einem Kilometer. Die Einpassung der Raumkanten der Gebäude konnte in der gewählten Methode entfernungsabhängig in einer Genauigkeit von 20 bis 50 cm erfolgen.

Das 3-D Computermodell der Gebäude und des geplanten Geländes (Höhenmodell des Lärmschutzwalles etc.) konnten damit immer mit mehreren Punkten maßstabsgerecht „eingehängt" werden. Für die Visualisierung der geplanten Bäume wurde eine Höhe von 8-10 m verwendet, was einem Zustand in etwa 10-15 Jahren nach der Pflanzung von Großgehölzen entspricht.

Simuliert wurde im städtebaulichen Maßstab der Stand der Entwurfsplanung im Jahre 2001. Für die Belange der Landschaftsbildsimulation wurden die Größenverhältnisse der geplanten Gebäude und der raumwirksamen Elemente, wie Bäume und Geländemodellierung, in einem 3D-Computermodell exakt berechnet. Für die Darstellung von Einzeltexturen der Architektur und auf detailgetreue Einzelausstattung wurde in diesem städtebaulichen Maßstab auf die detailgenaue Wiedergabe verzichtet, um die Proportionen der landschaftlichen Einbindung an sich in den Vordergrund zu stellen.

Bei Föhnwetterlagen kann man in der Münchner Landschaft an vielen Stellen die Alpenkette erleben. Dieser Anblick ist auch von der Zwerger Allee her erlebbar. Deshalb wurde vom Fotostandort 4 (S4) das Alpenpanorama noch einmal deutlich herausgearbeitet, wie dies in der Vergrößerung sichtbar wird.

The evaluation of the visualization was discussed with the client and the architectural team. The visualization stressed the main opportunity for visual sensitivity of the urban design: two building free corridors, the "inner campus" and the 120 meter wide corridors towards the exiting development in the west to allow for air circulation from the rural areas in the south and to protect the vista to the Alps. The need for air circulation and the value of the vista resulted in building heights of only 4 to 5 floors (see simulation figure 5 and image of build structure in figure 2). Only two higher buildings in the north had been planned to indicate the more traditional image of a headquarters. The simulation was used for public meetings with the city to communicate the dimension of the built development.

After 10 years, the update of the legal planning permit for completing the not yet built plots gave the opportunity to compare the landscape simulation of the original design with the mostly completed office park.

Die Simulation der städtebaulichen Planung auf der Ebene des Bebauungsplanes wurde mit dem Auftrageber und den Architekten diskutiert. Die Simulationen machten die Bedeutung der freien Korridore in Nord-Südrichtung und die relativ niedere Bebauung von 4 von 5 Stockwerken für die Luftzirkulation und die Erlebbarkeit des Naturraumes im Süden von München deutlich. Diese planerische Zielsetzung konnte mithilfe der gutachterlichen Visualisierungen im Rahmen der öffentlichen Beteiligungen vorgestellt werden.

Im Rahmen der Fortschreibung des B-Plans nach 10 Jahren konnte nun diese Simulation mit dem weitgehend fertiggestellten Büropark verglichen werden.

4 Evaluation of Method Used Compared With Built Situation / *Bewertung der für die Landschaftsbildvisualisierung verwendeten Methode durch Vergleich mit der gebauten Situation*

2001

Simulation

2011

Fig. 7: View 1: comparing "before"– "visualization"– "built, after 10 years" / *Blick vom Standort 1: "vorher" – "nachher" – „gebaut nach 10 Jahren"*

2001

Simulation

2011

Fig. 8: View 2: comparing "before" – "visualization"– "built, after 10 years" / *Blick vom Standort 2: "vorher" – "nachher" – „gebaut nach 10 Jahren"*

2001

Simulation

2011

Fig. 9: View 3: comparing "before" – "visualization"– "built, after 10 years" / *Blick vom Standort 3: "vorher" – "nachher" – „gebaut nach 10 Jahren"*

Fig. 10: View 4: Quantitative comparison "before" – "visualization"– "built, after 10 years" / *Blick vom Standort 4: Quantitativer Vergleich "vorher" – "nachher" – „gebaut nach 10 Jahren*

Sample of Evaluation after Development / *Quantitative Auswertung nach Landschaftselementen der erfolgten Bebauung im Vergleich mit Ausgangsbestand und Visualisierung des Vorentwurfs*

	Before		Visualization		Built (2011)		±%
Sky	216	30%	129	18%	144	20%	+2%
Tree+ Shrubs	144	20%	173	24%	158	22%	–2%
Grassland	216	30%	324	45.1%	216	30%	–15.1%
Field	144	20%	0	0%	129	18%	+18%
Building	0	0%	87	12%	58	8%	– 4%
Paths	0	0%	7	0.9%	15	2%	+1.1%
Pixels	720	100%	720	100%	720	100%	

Tab. 1: Development of distribution of main landscape elements for view 4: comparing "before" – "visualization" – "built, after 10 years" / *Quantitative Auswertung der Anteile wesentlicher Landschaftselemente für das Bild des Ausgangsbestandes der Visualisierung des Vorentwurfs und der Situation nach 10 Jahren (mit der noch nicht vollständigen umgesetzten Bebauung)*

The set of images of the original site before, of the simulation, and of the condition after 10 years of growth were compared with a raster analysis relative to the amount of sky, vegetation, grassland, field, buildings and paths.

Result: The chosen method of landscape visualization very well described the dimension of the changing of the landscape. The decision, based on the dimension on the project that all views from the surrounding potential user points were towards the theoretical middle of the project was also plausible when repeating the pictures at a later time. Additionally, all user points were so accurately described in the initial study that making photos of the built condition from the same locations could be carried out with great assurance.

Differences to the simulation could be observed in the foreground through the growth of shrubs, or by changes in detail which in part were outside of the project, or by buildings which have not yet been built. Through the comparison it became clear that two planned building parcels have not yet been built on so that in these areas the amount of building in the image counts then it appears in the simulation.

One can reach the conclusion that for a project with a circumference of approximately 3,300 meters, six points for visualization are clearly too few to satisfactorily describe the changes of the landscape for any possible long term analysis. For this undertaking, before and after visualizations of the previous and future user points of the surroundings would have had to be done approximately every 300 meters.

Google Earth allows for a good, generally accessible documentation of the development of the project. After searching for *Munich Campeon* and activating the icon *historic pictures,* the project can be followed from farming use in 2001 until today, including the period of construction and the current addition of the breeding pond for European Green Toads in the north-east corner in 2012. This time -travel using archive areal photos, very clearly shows the dramatic change to the landscape. In this case the change led to a landscape valorization from intensively farmed land to an extensive park for the entire district, lucrative work places in information technology, as well as a habitat for a large range of species. The judgment of the changes to a landscape through a project is often very focused on the effects to certain species. Here, however, we must admit that the overall effect presented by the simulations of the creation of the office park nowhere matched the dimension of increase in value of the area for all parties concerned.

Die Bildinhalte der ursprünglichen Bestandsphotos, der Simulation und des Zustandes nach 10 Jahren wurden mit einer Rasterauswertung nach dem Anteil von Himmel, Gehölze und Bäume, Wiesen, Gebäude und Wege verglichen.

Ergebnis: Die gewählte Methode der Landschaftsbildsimulation hatte die Dimensionen der Änderungen des Landschaftsbildes grundsätzlich sehr gut beschrieben. Die Entscheidung aufgrund der Dimension des Projektes von den umgebenden Wegen jeweils den Blick in die theoretische Mitte der Projektes zu legen, ist auch bei der Wiederholung der Bilder zu einem späteren Zeitpunkt plausibel. Weiterhin waren alle Besucherstandorte in der Ausgangsstudie so eindeutig beschrieben, dass die Bilder des gebauten Zustandes mit einer großen Sicherheit von der gleichen Stelle aus aufgenommen werden konnten.

Unterschiede zur Simulation ergeben sich jedoch im Vordergrund durch Wachstum von bestehenden Gehölzen oder durch Änderungen von Details, die zum Teil außerhalb des Vorhabens liegen, oder durch noch nicht gebaute Gebäude. Im Vergleich wurde deutlich, dass etwa 20 Prozent der Fläche bei dieser Prüfung nach 10 Jahren noch nicht bebaut waren.

Es ist daher fest zuhalten, das bei einem Umgriff von ca. 3.300 Metern sechs Visualisierungsstandpunkte deutlich zu wenig sind, um die Änderungen des Landschaftsbildes des Gesamtprojektes auch für eine etwaige Projektfortschreibung zufrieden stellend beschreiben zu können. Für dieses Vorhaben hätten im Durchschnitt etwa alle 300 Meter eine „Vorher-Nachher"-Visualisierung von den ehemaligen und zukünftigen Besucherstandorten der umgebenden Wege erfolgen müssen.

Eine gute allgemein zugängliche Dokumentation zur Entwicklung des Projektes ermöglicht Google Earth. Nach der Standortsuche „München Campeon" kann bei der Aktivierung der Anzeige „Historische Bilder" das Vorhaben von dem intensiven Acker im Jahre 2001 bis heute, auch mit dem im Jahre 2012 ergänzten Ersatzlaichplatz für Wechselkröten im Nordosten, verfolgt werden. Diese Zeitreise mit den historischen Luftbildern macht noch einmal die dramatische Änderung dieser Landschaft deutlich. Die Änderung in diesem Fall führte zu einer Landschaftsaufwertung von einer intensiven Ackerfläche zu einem extensiven Volkspark für den gesamten Stadtteil, lukrativen Büroarbeitsplätzen für eine Reihe von IT-Mitarbeitern und einem großen Artenspektrum. Die Beurteilung der Änderung des Landschaftsbildes eines Projektes drückt im Verhältnis zur meist sehr fokussierten Betrachtung der einzelnen betroffenen Arten ein breites Spektrum an Synopsen aus. Dennoch sei hier eingeräumt, dass die Gesamtwirkung der mit den Simulationen dargestellten Schaffung eines offenen Büroparkes selbst vom Gutachter nicht in der gesamten Dimension der Wertsteigerung für alle Nutzergruppen vorstellbar war.

5 Future Methods Utilizing 3D GIS Zools / *Zukünfige Möglichkeiten mit der Verwendung von 3D-GIS*

At this time of real time modeling, when GIS works on all devices, when LIDAR data is at our fingertips, when data services are likely to be offered all over Europe and when GIS is finally making the attempt to get 3 dimensional, the described method of using high resolution photographs and a 3D wire frame model to generate realistic 2D images appears to be outdated. However, the basic workflow is not so very different today; all processes are faster, and there are definitely more 3D image libraries available, and many of the processes are now known routines to many young professionals in landscape architecture. During this visualisation we had been very pleased that the imaging expert Guido Liesecke, who as a trained landscape architect, understood our intention for bias free work. Today we would additionally have an interactive model of the entire office park, but we still would need detailed visualizations from representative user points. It is the attempt of the author to prepare this interactive virtual version using 3D GIS and continue monitoring the development of this exciting office park of Maier Neuberger Architekten, Munich, Rainer Schmidt Landscape Architects Munich and gtl Kassel.

Im Zeitalter von Echtzeitmodellierung, wenn Applikationen von GIS für alle Gerätearten („Devices")verfügbar sind, wenn LIDAR-Daten in GIS integriert werden können, wenn Kartendienste und -services bald auch in Europa angeboten werden und in einer Zeit in der GIS zunehmend auch 3-dimensional wird, scheint der beschriebene Ansatz einer mit 3D-Modell unterstützten 2D-Landschaftsbildsimulation technologisch überholt. Das Novum dieses Beitrages ist somit, dass die Aussagekraft dieses klassischen Ansatzes hier am gebauten Projekt geprüft werden konnte. Um gutachterliche „Vor – Nachher"-Bildpaare zu generieren, ist der grundsätzlich Workflow auch 12 Jahre später vergleichbar. Alle Prozesse und der Datenaustausch sind einfacher und schneller, 3D-Bildbibliotheken sind kein Expertenwissen mehr und der Umgang mit den Softwarefunktionen ist für viele der jüngeren Kolleg/innen aus der Landschaftsarchitektur zur Routine geworden. Während der ursprünglichen Visualisierung waren wir dem studierten Landschaftsarchitekten Guido Liesecke, der sich in seinen ersten Berufsjahren auf Visualisierung spezialisiert hatte, dankbar, da er unserem Anliegen vorurteilsfreie Landschaftsbildvisualisierungen zu erzeugen sehr aufgeschlossen war. Heute sollte parallel im städtebaulichen Maßstab das gesamte Modell interaktiv zur Diskussion und Entscheidungsfindung herangezogen werden. Zielsetzung des Autors ist es daher ein solches interaktives virtuelles 3D-GIS für das laufende Monitoring der Entwicklung dieses vorbildlichen Büroparkes von Maier Neuberger Architekten München, Rainer Schmidt Landschaftsarchitekten München und gtl Kassel zu entwickeln.

References / *Referenzen*

Appleyard, D. (1977), *Understanding Professional Media: Issues, Theory, and a Research Agenda.* In: Human Behavior and Environment, ed. by Altman, I. & Wohlwill, J. F. New York, Plenum Press.

ATELIER BERNBURG (1991), *Umweltbericht nach Paragraph 2a BauGB (Umweltverträglichkeitsprüfung UVP). Anlage D: Landschaftsbildsimulation von repräsentativen Besucherstandorten.* Im Auftrag der MoTo GmbH, Unterhaching.

Buhmann, E. (1994), *EDV in der Landschaftsbildsimulation. Technische Möglichkeiten. Design.* In: Garten und Landschaft, 10/1994, 31-32.

Buhmann, E., Nothhelfer, U. & Pietsch, M. (Eds.) (2002), *Trends in GIS and Virtualization in Environmental Planning and Design.* Proceedings at Anhalt University of Applied Sciences 2002. Heidelberg, Wichmann.

Buhmann, E., Paar, P., Bishop, I. D. & Lange, E. (Eds.) (2005), *Trends in Real-Time Visualization and Partizipation.* Proceedings at Anhalt University of Applied Sciences 2005. Heidelberg, Wichmann.

Buhmann, E. & Pietsch, M. (2008), *Interactive Visualization of the Impact of Flooding and of Flooding Measures for the Selke River, Harz.* In: Buhmann, E., Pietsch, M. & Heins, M. (Eds.), Digital Design in Landscape Architecture 2008. Proceedings at Anhalt University of Applied Sciences 2008. Heidelberg, Wichmann, 152-162.

Buhmann, E. & Pietsch, M. (2008), *Interaktion mit digitalen Landschaften, Szenarien und Modellen.* In: Garten + Landschaft, 3/2008, 35-38.

Ervin, S. & Hasbrouck, H. (2001), *Landscape Modeling: Digital Techniques for Landscape Visualization.* McGraw-Hill.

Hoppenstedt, A. & Stocks, B. (1991), *Visualisierung von Landschaftsbildveränderungen.* In: BfANL (Hrsg.), Landschaftsbild – Eingriff – Ausgleich. Handhabungen der naturschutzrechtlichen Eingriffsregelung für den Bereich Landschaftsbild, Bonn.

Jessel, B., Fischer-Hüftle, P. & Jenny, D. (2004), *Erarbeitung von Ausgleichs- und Ersatzmaßnahmen für Beeinträchtigungen des Landschaftsbildes.* BfN – Bundesamt für Naturschutz, Bonn.

Krause, C. & Buhmann, E. (1989), *Bildsimulation für Projektanalysen / Video imaging: Improvement for visual project analysis.* In: Garten und Landschaft: Planen und Entwerfen mit dem Computer, 10/1989, 44-48.

Kretzler, E. (2003), *Improving Landscape Architecture Design Using Real Time Engines.* In: Buhmann, E. & Ervin, S. M. (Eds.), Trends in Landscape Modeling. Proceedings at Anhalt University of Applied Sciences 2003. Heidelberg, Wichmann, 95-101.

Mach, R. (2008), *Workflow for Interactive High-end 3D Visualization in Site Planning.* In: Buhmann, E., Pietsch, M. & Heins, M. (Eds.), Digital Design in Landscape Architecture 2008. Proceedings at Anhalt University of Applied Sciences. Heidelberg, Wichmann, 174-179.

Mach, R. & Petscheck, P. (2006), *Visualisierung digitaler Gelände- und Landschaftsdaten.* Berlin/Heidelberg, Springer Verlag.

Sheppard, R. J. (1982), *Predictive Landscape Portrayals: A Selective Research Review.* In: Landscape Journal, 1 (1), 9-14.

Sheppard, R. J. (1989), *Visual Simulation, A User's Guide for Architects, Engineers, and Planners.* New York, Von Nostrand Reinhold.

Sheppard, R. J. (1999), *Manipulation und Irrtum bei Simulationen. Regeln für die Nutzung der digitalen Kristallkugel.* In: Garten und Landschaft: Visual Landscape, 11/1999, 28-32.

Sheppard, R. J. (2012), *Visualizing Climate Change. A guide to Visula Commincation of Climate Change and Developing Solutions.* London/New York, Routledge.

Smardon, R. C., Palmer, J. F. & Fellemann, J. P. (1989), *Foundations for Visual Project Analysis.* New York, John Wiley Sons.

The Landscape Institute with the Institute of Environmental Management and Assessment (2002), *Guidelines for Landscape and Visual Impact Assessment.* Second edition. London/New York, Spon Press.

Data Sources / *Datenquellen*

Bayerisches Landesvermessungsamt, Orthofotos 1:10.000, Bild-Nr. 241 Bildflug 96106/1, Bild-Nr. 243 Bildflug 96109/1, AZ:DLZ-LB-2927.

C. Vohler Photogaphie, München, Bestandsfotos 2001.

Yangwa Lu, Bestandsfotos 2011.

Generating a Point Cloud from a Crowdsourced Photographic Survey

Ruben JOYE

1 Introduction

The use of point cloud data – a set of vertices in a three-dimensional coordinate system – is becoming more and more widespread. It's in particular a quick way of capturing and representing complex environments where measuring each little detail with traditional tools is impractical. Generally, point clouds are captured using sophisticated and expensive 3D laser scanners resulting in very accurate and extremely detailed results. LIDAR data for instance is a common example, already widely in use.

The aim of this paper is to show an alternative way to create a point cloud, based on a free-to use photogrammetric tool. The entire process from capturing the input photos, to turning them into usable point cloud data for use in AutoCAD or other programs is discussed here.

As a way to illustrate this method, the archaeological site of Sillyum in Antalya (Turkey) was chosen as a case study. One of the sites that was visited during the Le:Notre Landscape Forum, that took place in Turkey previous spring (18th – 21st of April 2012).

2 Sillyum as a Case Study

Different working groups were set up for the Le:Notre Landscape Forum in Antalya, focussing on various themes. I took part of the heritage working group supervised by my college Harlind Libbrecht (University College Ghent) and assisted by specialists Graham Fairclough (English Heritage), Karsten Jørgensen (Norwegian University of Life Sciences) and Diedrich Bruns (Kassel University).

Each of the working groups went on a different excursion related to their theme, we visited the archaeological site of Sillyum. Contrary to similar sites in the area like the overcrowded Pamphylian or Perge, Sillyum is barely known and far less documented. Combined with the distinct topography and the impressive ruins, some say it's more attractive than the better-known places (BEAN, 1968). Sillyum – like other heritage sites – is in many ways an interesting case study as point clouds are particularly useful for representing such complex environments.

Fig. 1: Le:Notre heritage group on the Sillyum site

After a thorough introduction to the site by the experts (Fig. 1), we split up along our particular interests. I took charge of a small group consisting of six people concentrating on documenting through photography. The idea was to get the most out of the hundreds of pictures that would be taken anyway. For this, we used the application Photosynth, developed and made freely available by Microsoft Labs (MICROSOFT, 2012). It maps photos relative to each other, resulting in a very user friendly and intuitive way of browsing a set of images. People unfamiliar with a certain site get a better impression as they can see the relation between each of the photos. It allows even further ways of use, since a point cloud can be derived quite easily from this data.

Our final aim was to generate a basic point cloud, solely based on photos taken collaborative with standard digital photo cameras. Our time on the Sillyum site was limited as we only had 3 hours, nonetheless we managed to apply this method on two smaller areas within this archaeological site and still found time to visit other parts of the site. We chose to focus on a palace close to the entry (Q) and the tower (D) (Fig. 2).

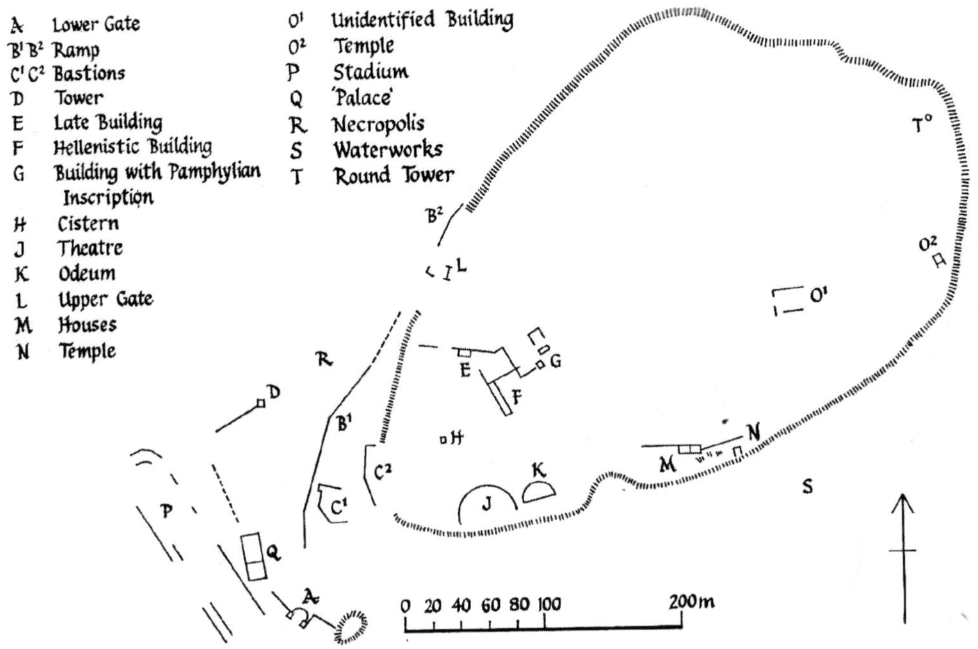

Fig. 2: Plan of Sillyum (BEAN, 1968)

2 Method and Results

2.1 Capturing the photographs

For Photosynth to allow to recognize the relative position of each photo, the following guidelines are important (PHOTOSYNTH TEAM, 2008):

- Panorama first, then move around: Start by taking a panorama of your scene, then move around and take more photos from different angles and positions. If you just do a panorama you won't end up with a good 3-D experience.
- Have enough overlap when shooting panoramas: Try for at least 50% overlap between photos.
- Shoot wide shots: Wide angle shots (photos taken from farther away, or with your camera's lens zoomed all the way out) reconstruct more reliably than closer shots. It's good to have close-ups, too, but you'll want to have good coverage of your subject with lots of nice overlapping wide shots.
- Move around the object capturing the object from all possible sides

In order to make sure that all of the photos taken by the participants of this experiment would be recognised, I made sure to cover the general landscape by taking numerous wide angle photos at various positions.

In total, P. Diehl, W. Dutch, I. Fasching, M. Kremmel, S. Rulinha and myself ended up taking 589 pictures of the palace area (Q), and 359 pictures in and around the tower (D).

Fig. 3: Palace area

Fig. 4: Tower

2.2 Processing the photos

A folder with +400 photos can quickly be more than one gigabyte in size, so a fast and stable broadband connection is needed to make this uploading process more bearable. It is also advised to reduce the number of pixels of the photos prior to feeding them in Photosynth. This may on the other hand result in Photosynth having more difficulty in matching all of the photos. So keep in mind not to reduce the file size too much. Somewhere between 2 – 4 megapixels seems most appropriate. This can be done prior to taking the photos by adjusting the resolution setting in the camera, or be done afterwards in a wide range of programs (e.a. Adobe Lightroom, Adobe Photoshop, Picasa …).

2.3 Creating the Photosynth

The Photosynth application is available for free, and can be downloaded from the Photosynth-website (MICROSOFT, 2012). A free Microsoft Live account needs to be created if you don't have one already. It is strongly advised to close all programs other than Photosynth while processing the photos, as the calculation and creation of the Photosynth is done on your own computer, and is a RAM-intensive process that might fail if the application runs out of memory.

The program is pretty straightforward and needs little explaining as it only consists of pointing the application to the photos that need to be used. Once the Photosynth has been calculated, the final step is an automatic upload of the photos that are part of the synth.

If all went well, you end up with a successful Photosynth tying all of your pictures together accessible on the Photosynth website (MICROSOFT, 2012) pressing the 'my Photosynths' button. Please note that this requires an up to date browser and the SilverLight plugin (MICROSOFT, 2013). A variable called 'synthy' displays how successful the program was at connecting your photos. The photos can be explored in many ways, the point cloud can also be visualized and navigated within the web browser itself (Fig. 5 and Fig. 6).

Fig. 5: The palace area in the Photosynth browser

Fig. 6: The tower in the Photosynth browser

2.4 Importing point cloud date into AutoCAD

Most AutoCAD-products can import point cloud data starting from version 2011. AutoCAD uses *.pcg as the default point cloud format, however a few other extensions can be opened.

For importing a Photosynth point cloud we've used the 'BrowsePhotosynth for AutoCAD' plug-in which is downloadable for free (WALMSLEY, 2010). Note that this requires having 'NET Framework 3.5 SP1' installed (MICROSOFT, 2008). Once the 'BrowsePhotosynth' plugin was installed, entering the command 'browseps' in the AutoCAD command line opens a web browser interface of the Photosynth website. Upon navigating to a specific Photosynth, this opens a thumbnail of the photos on the right hand side. Clicking this thumbnails opens the Point Cloud in AutoCAD and stores a local *.PCG file in the Mydocuments/Photosynth Point Clouds folder on the local computer.

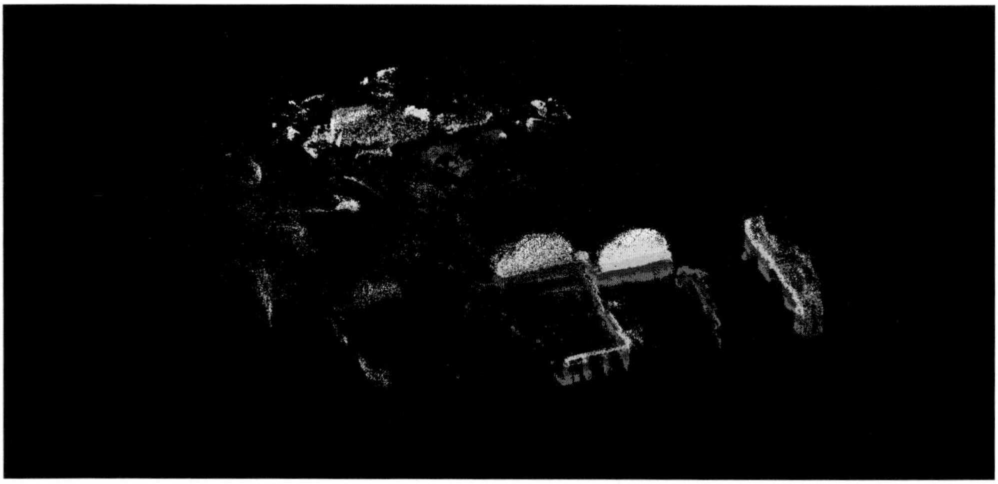

Fig. 7: Point cloud of the palace area classified by height

Using commands like 'scale' or 'align' in combination with a given reference length, this point cloud can be put to scale for further analysis and measuring. Using the 3D-orbit tool, the point cloud can be viewed in 3D, and the colour can also be set using height classify-cations (Fig. 7). In the recently released AutoCAD 2013 there are further tools to clip the point cloud in order to specify which points to display. The REVIT version of AutoCAD 2012 has a downloadable plug-in called 'point clouds feature extraction' downloadable for free at (AUTODESK LABS, 2012). Using the release 2011, it's most easy to store the point cloud as a *.DWG file and insert it in a new AutoCAD file as a XREF. Using the command xclip, the xref can be clipped to only show a specific part of the point cloud (e.a. front, back, side, section …) (Fig. 8 & 9). This allows for more easy measuring or tracing of the features.

Fig. 8: Orthogonal view of the tower and topography

Fig. 9: Two sections derived from the point cloud

If you prefer to use the point cloud in programs other than AutoCAD, a standalone exporter called 'synthexport' is also available for download (HAUSNER, 2010) and allows you to export to formats like *.obj, *.ply, *vrml, and *.x3d. Just like the 'BrowsePhotosynth for AutoCAD' plug-in, this also requires the 'NET Framework 3.5 SP1'.

3 Conclusions

The process of generating a point cloud solely based on photos is in most cases fairly simple using Photosynth. Results were achieved fast without much effort, even while the environment was very complex and highly detailed. In addition, it didn't require expensive (and heavy) equipment.

There are of course some important observations to be made that can limit the usability. For instance, Photosynth has a hard time dealing with repetitive patterns and complex occlusions. The high amount of detail in the heritage site therefor was actually a helpful factor. Making a Photosynth from a contemporary monotonous glass building would have proven much harder to accomplish. Currently the program is Windows-only and absolutely requires having a (fast) internet connection, both of which can be constraining factors.

Most importantly however, it is to be noted that a point cloud created with Photosynth will never have the same amount of detail and preciseness that you would get from a 3D laser scanner.

References

Autodesk Labs (2012), *Point Cloud Feature Extraction For Autodesk Revit.* http://labs.autodesk.com/utilities/scan_to_bim/ (23/04/2013).

Bean, G. E. (1968), *Turkey's southern shore: an archaeological guide.* Praeger.

Hausner, C. (2010), *SynthExport.* http://synthexport.codeplex.com/ (23/04/2013).

Microsoft (2008), *NET Framework 3.5 Service Pack 1.* http://www.microsoft.com/en-us/download/details.aspx?id=22 (23/04/2013).

Microsoft (2012), *Photosynth – capture your world in 3D.* http://photosynth.net/. (23/04/2013).

Microsoft (2013), *Silverlight.NET.* http://www.silverlight.net/ (23/04/2013).

Photosynth Team (2008), *Photosynth Photography Guide: Tips and Tricks.* http://blogs.msdn.com/b/photosynth/archive/2008/12/02/photosynth-photography-guide-tips-and-tricks.aspx (23/04/2013).

Walmsley, K. (2010), *October's Plugin of the Month: BrowsePhotosynth for AutoCAD.* http://through-the-interface.typepad.com/through_the_interface/2010/10/octobers-plugin-of-the-month-browsephotosynth-for-autocad.html (23/04/2013).

Heeding Its Lessons – How the New Field of Geodesign Can still Learn from Past Experiences

David L. TULLOCH and Grant F. WALTON

Invited Paper

Geodesign is the emerging area of study, research, and practice that combines design from disciplines like landscape architecture with science from fields like geography and environmental science and is facilitated through the newest innovations in geospatial technology (DANGERMOND, 2009; GOODCHILD, 2010; FLAXMAN, 2010). Concomitant with its status as a nascent field, geodesign has been the focus of special issues of journals, discussed in panels at conferences, featured in non-academic magazines and generated an annual standalone conference called the Geodesign Summit. A measure of the field anticipated stamina is that two American universities will begin geodesign degree programs in the Fall of 2013. Despite the energy and promise of the field, there are also a number of potential barriers to its success that will need to be overcome.

This paper presents a summary of two recent papers examining potential lessons for the emerging field of geodesign. One of the papers looks back to a similar time of convergence when many GIS innovations emerged from the landscape architecture program at Harvard in the 1960s and early 1970s. While many of the innovations from that period succeeded with exceptional impact, for years their use and integration was more apparent outside landscape architecture than within. The second paper considers experiences in regional design studios as a parallel to experiences by newly interested parties in geodesign. Both papers leverage their explorations as potential lessons for those contributing to the growth of geodesign today.

1 Learning from History

In, "Geographic information systems and landscape architectural design and scholarship: A source of heritage and tension" (TULLOCH 2012), lessons from the early years of GIS at the Harvard Graduate School of Design (GSD) are gleaned. Based heavily on first-hand eyewitness accounts, the paper describes how challenges faced by the early designer-innovators at GSD may apply to geodesign today.

Central to the experience at that time was the role played by Charles Harris, then the chair of landscape architecture, who hired Symap's creator Howard Fisher to found the Computer Graphics Laboratory. In a very short time, the GSD and the Lab became a hub of intellectual and innovative creativity attracting an impressive bevy of eventual GIS experts including Carl Steinitz, Nick Chrisman, Scott Morehouse, Tim Murray, Bruce Rowland, and Jack Dangermond.

Also attracted to this hub were the graduate students who went on to co-found the GIS company, Erdas: Lawrie Jordan, Steve Sperry, and Bruce Rado. As detailed by NIEMANN & NIEMANN (1994), these 3 students were part of a class at Harvard that sped the transformation by refusing to use the software (or pay tuition) until Harris found a way to make it more user-friendly. Over many years, the student body tested and improved the different versions of the nascent geospatial technologies through applied studio projects. While quite successful, these projects also provide insights into today's geodesign.

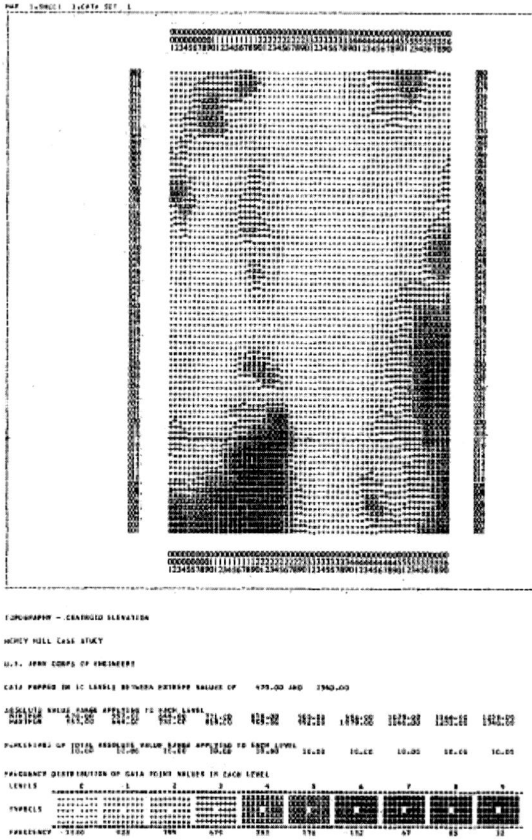

Fig. 1: The Computer Graphics Lab at Harvard used a project working for the US Army Corps of Engineers at Honey Hill (elevation map shown above) as a proof of concept for GIs as a scientific integration tool (MURRAY et al., 1971)

Because these software packages and geospatial practices were developed in a design school setting, it created opportunities to witness potential conflicts between design and science. The prominence of the program facilitated the rapid and widespread dissemination of the software to other programs; an undergraduate landscape architecture studio at Wisconsin (DANE ENVIRONMENTAL SYSTEMS EVALUATION, 1969) was already using Harvard's software in 1969 (Fig. 2). While GIS, then, and geodesign, now, hold promise in integrating these two perspectives, the struggles at Harvard's GSD also highlight areas where the two could create friction. Rather than seeing the science as a rigid barrier to creative design explorations, Harvard's early efforts show today's geodesign proponents ways that exciting solutions emerge when the two intersect.

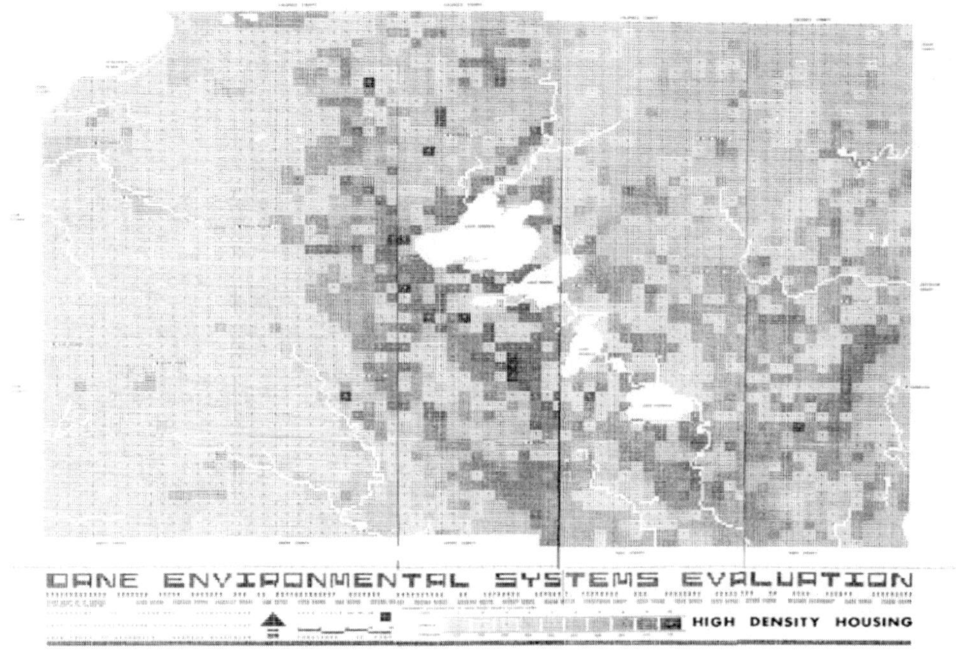

Fig. 2: One example of the rapid impact of Harvard's was the 1969 undergraduate landscape architecture studio taught at Wisconsin (DANE ENVIRONMENTAL SYSTEMS EVALUATION, 1969)

2 Learning from Students

The second paper, "Learning from Students: Geodesign lessons from the Regional Design Studio" (TULLOCH, 2013), turns to more recent work as a source of lessons. Pioneered by earlier landscape architecture professors like Phil Lewis at the University of Wisconsin and Ian McHarg at the University of Pennsylvania, regional design studios have long sought to create sweeping and creative visions for larger landscapes that are informed by information and science. In recent studios at Rutgers University, we can look students' collected experiences as a potential parallel to the experiences that new geodesigners may experience.

The primary examples used in the paper were studios investing design alternatives for: the Appalachian Trail at Bear Mountain in New York (Fig. 3); landscapes impacted by sea-level rise in Cape May County, New Jersey (Fig. 4); and the Stony Brook-Millstone Watershed in Central New Jersey. While each explored very different forces and uses, they all relied on the integration of GIS and various landscape sciences as important elements of their design process.

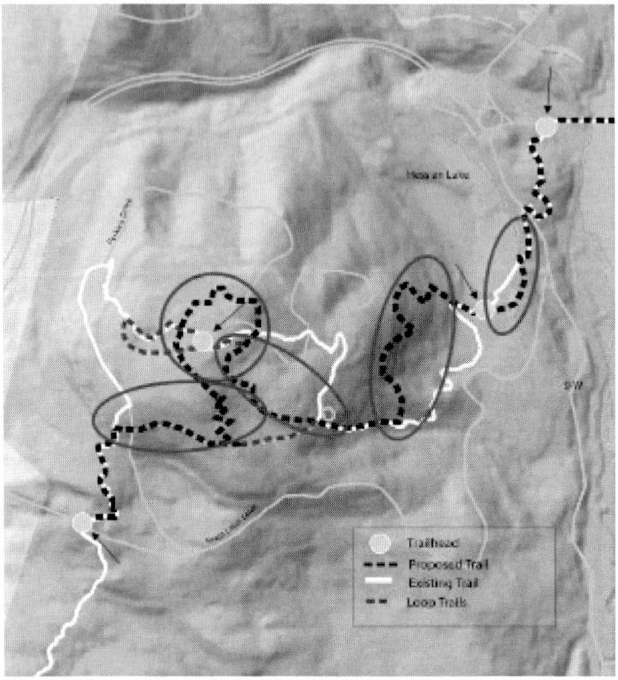

Fig. 3: Students redesigning a segment of the Appalachian Trail at Bear Mountain State Park in Harriman, New York were able to use GIS to guide potential designs and then analyze outcomes

Fig. 4: Students exploring design approaches for sea level rise impacted landscapes in Cape May County explored design solutions about education (above) as well as ecologically sensitive designs

Significant challenges come when, empowered by new technologies, the students encounter rather dramatic issues related to scale. Thinking large and small simultaneously is quite challenging, but recognizing failures in that seems even harder for many. With the larger scale and technology also comes vast quantities of information (and data) resulting in a form of information overload that often stops the student projects rather abruptly. Add to the quantity of information a frequently large quantity of stakeholders and participants, and it becomes clear that geodesign may require new tools and techniques for addressing these concerns (beyond the traditional use of multicriteria decision models).

For students in these studios, still learning design as a foundational skill, there often is a struggle that emerges between what they see as logic and design. STEINITZ (1979) characterized the importance of logic as part of "defensible design processes", but for new designers and geodesigners this can still be a challenge. Similarly, choices between creative or artistic expression and decisions grounded in science and information can be seen as a false choice, but one that geodesigners should anticipate.

3 Final Thoughts

These papers represent only a tiny part of a rapidly growing literature (see the geodesign bibliography at http://gisandscience.com/2009/08/13/geodesign-a-bibliography/). But they are meant to show that while geodesign is a new field built on new technologies, much about it is not. As a multidisciplinary field it will have difficult conflicts, maybe between scientists and designers, that add to the richness of the field but could also derail projects and conversations if we are not careful.

The paper focusing on students' lessons ends with a challenge for geodesign and landscape architecture writ large:

> For geodesign to grow and live up to its fullest potential, we need to be instilling a grand vision of geodesign in professionals and students as an expansive approach to serving society and improving the environment while empowering them to fulfill that vision. This is particularly true in urban planning and design, where these technologies bring with them the potential of new tools for shaping both old and new cities (TULLOCH, 2013).

Indeed, geodesign and digital landscape architecture require an educational response that reaches beyond today's students and out to practitioners and future designers of all stripes. But the challenge of making that happen will require overcoming the barriers we have learned of in history and studio.

4 References

Dane Environmental Systems Evaluation (1969), *Dane Environmental Systems Evaluation Project*. Report by the class, Landscape Architecture 651, University of Wisconsin-Madison.

Dangermond, J. (2009), *GIS: Designing Our Future*. In: ArcNews, 31 (2), 1, 6-7.

Flaxman, M. (2010), *Geodesign: Fundamental Principles and Routes Forward.* Talk at GeoDesign Summit 2010.

Goodchild, M. (2010), *Towards Geodesign: Repurposing Cartography and GIS?* In: Cartographic Perspectives, Special Digital Issue 2 (Fall 2010), 7-23.

Murray, T., Rogers, P., Sinton, D., Steinitz, C., Toth, R. & Way, D. (1971), *Honey Hill: A systems analysis for planning the multiple use of controlled water areas.* Department of Landscape Architecture Research Office, Graduate School of Design, Harvard University. http://www.iwr.usace.army.mil/docs/iwrreports/IWRREPORT71-9OCTOBER1971VOLUME1OF2.pdf and http://www.iwr.usace.army.mil/docs/iwrreports/IWRREPORT71-9OCTOBER1971VOLUME2OF2.pdf.

Niemann, B. J. & Niemann, S. (1994), *Revolutionaries in Action: Nick Faust, Lawrie Jordan, Bruce Rado, and Steve Sperry.* In: Geo Info Systems, 4 (2), 53-57.

Steinitz, C. (1979), *Defensible Processes for Regional Landscape Design.* In: Landscape Architecture Technical Information Series, 2 (1). Washington, D.C., American Society of Landscape Architects.

Tulloch, D. (2012), *Geographic information systems and landscape architectural design and scholarship. A source of heritage and tension.* In: Revue Internationale de Géomatique, Special Issue: Geodesign: From Theory to Practice, 22 (2), 169-184.

Tulloch, D. (2013), *Learning from Students: Geodesign lessons from the Regional Design Studio.* In: Journal of Urbanism: International Research on Placemaking and Urban Sustainability, DOI:10.1080/17549175.2013.765903.

Rapid Prototyping Techniques in Landscape Model Making

Maria Angelica GUTIERREZ GOMEZ

Invited Paper

Abstract

This study focuses on providing a reference of a practical implementation of Rapid Prototyping Techniques to generate a tangible 3D topographical representation of an existing landscape. It pretends to serve as an example to understand, compare, distinguish and execute a series of steps to develop similar exercises. Furthermore, it will expose the strengths and weaknesses found during the work with two of the most accessible machines throughout manufacturing centers, such as CNC router and Laser Cutter Machines, to be able to assess some of the relevant aspect as production time scale, costs, materials, required training and generally all the possible issues or findings that can come up in the process.

This hands-on experience follows the standard procedure of some of the Prototyping Laboratories available to students and practitioners at well-known European educational institutions, such as the RAPLAB at the ETH in Zurich and the Digital Manufacturing Center at The Barlett School of Architecture of UCL (University College of London).

1 Introduction

Models, as an essential tool of designers, have traditionally served to represent the actual and to explore what could be possible in the future. More than a merely representation, models suggest diverse spatial and structural solutions, as suggested by different authors such as Karen Moon, in her book "Modeling Messages: The Architect and the Model" Hence, the potential of exploring and developing models according to today's requirements and challenges, become an asset to understand and support design ideas. Especially when the landscape industry is everyday getting an increasing contribution from different fields of knowledge, such as engineering, urbanism, geology, cartography, architecture, etc, which are nowadays commonly communicated through a digital CAD language.

We are facing a time where material is enriched by information. Material and data are now interwoven to open up the gate for digital fabrication, especially when we deal with high speed of production that requires us to be more efficient. For diverse industries this has represented the need to adapt faster and more competitive design, manufacturing and communicative tools. Current sophisticated computer modeling techniques facilitate us to convert virtual information into analog products.

Despite the increasing availability and access of these technologies over the last decades, the used of them still concentrated in Product Design, Architecture, Industrial Engineering and Manufacturing industries.

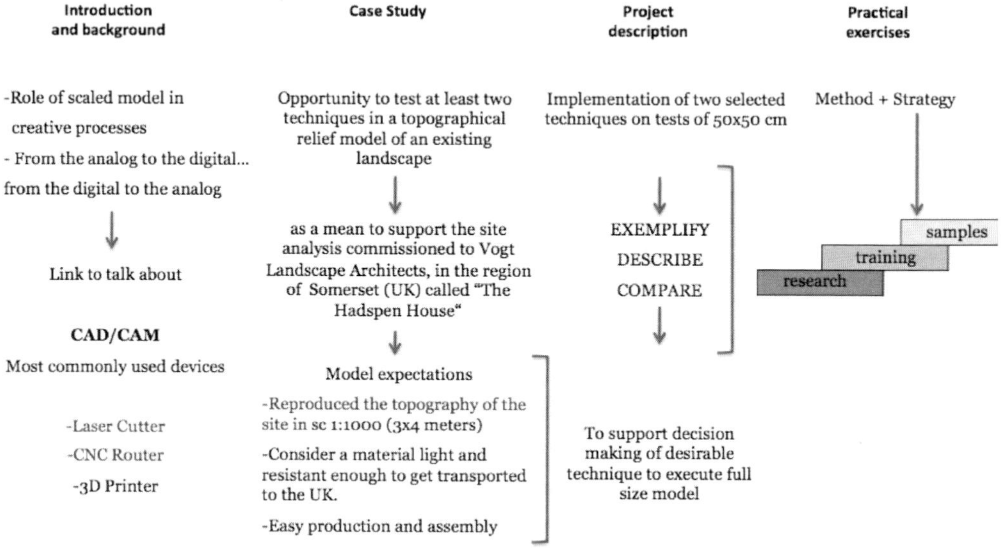

Fig. 1: Process diagram

2 Role of Analog Scaled Models in Creative Process

The author Albert C. Smith writer of "Architectural model as machine" says that a model is a machine for imagining, for developing the free association needed to develop new ideas. Based on this, one can recognize the transforming role of models from images into processes. Models are no longer considered static artefacts with a representational purpose. Smith explains in his book why models are important into the design process, by going through some historical examples of architectural models and their significance. He starts from the magic of the Egyptian scaled models that were thought to control nature and have some influence in the afterlife. Passing by the roman view of Vitruvius who understood and represented military camps, damns etc through small models of them. Or the Alberti's models during the renaissance that serve the builders when it came to the execution of the buildings, furthermore as the famous architect Spiro Kostof said "these models will enable us to weigh up repeatedly and examine, with the advise of experts, the work as a whole and individual dimensions of all the parts, and before continuing any further, to estimate the likely trouble and expenses".

Through similar historical journeys it is possible to appreciate how models were used as mechanisms of thinking about design and becoming communicative devices, being less ambiguous than drawings.

Models present themselves as containers of dynamic processes rather than static images that nowadays have been revolutionized thank to the influence of different disciplines taking part in a digital era with a very important role in design and fabrication.

3 Project Description

This practical exercise explores the implementation of two Prototyping techniques in the generation of a topographical relief model of an existing landscape. This specific experimentation is meant to support the site analysis carried out by Vogt Landscape Architects, where the intention was originally to detect potential view points on the open landscape located in Somerset (UK) called "The Hadspen House". The framework will be defined according to the expectations and model-making requirements of the real clients of the project.

Due to the lack of previous reference of similar large-scale models at Vogt Landscape Architects, it justifies the need of testing at least two different technical approaches to obtain diverse results suit-able to be adapted to the final model.

This hands-on experience pretends to generate a topographical representation of the same landscape location, based on the standard procedure of two of the most accessible machines through-out manufacturing centers, such as CNC milling machine and laser cutting machine. Thereby, it is possible to document and describe the subsequent achievements and failures of both processes, in order to obtain random results used as measurable tool to compare, analyze and evaluate the contribution of the selected techniques in this specific model making challenge.

In those terms, the goals are:

1. To exemplify, describe and compare a digital fabrication process in Landscape Architecture, by using at least two different technological approaches.
2. To implement and assess the process of milling and laser cutting when generating a topographical relief model.
3. To collect relevant aspects to support the election of a desirable technique to produce a full size model according to the model expectations.

Fig. 2a:
Contour lines

Fig. 2b: Both, the CNC router and laser cutter machine do require a Digital Terrain Model (DTM) to start any work. In this case, it has been generated by using AutoCAD Land Desktop to produce a triangular surface based on contour lines.

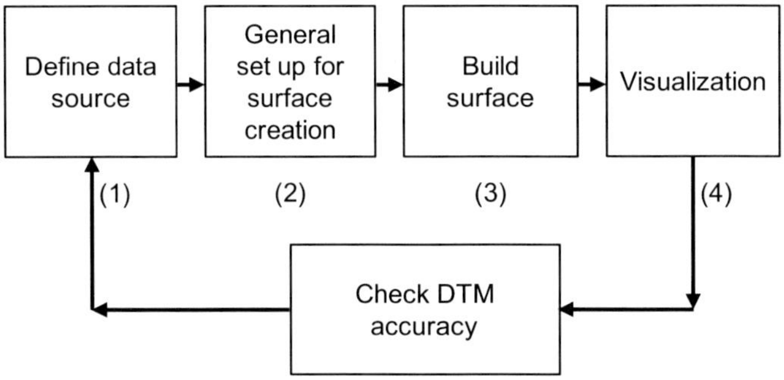

Fig. 3: AutoCad Workflow to generate DTM (Digital Terrain Model)

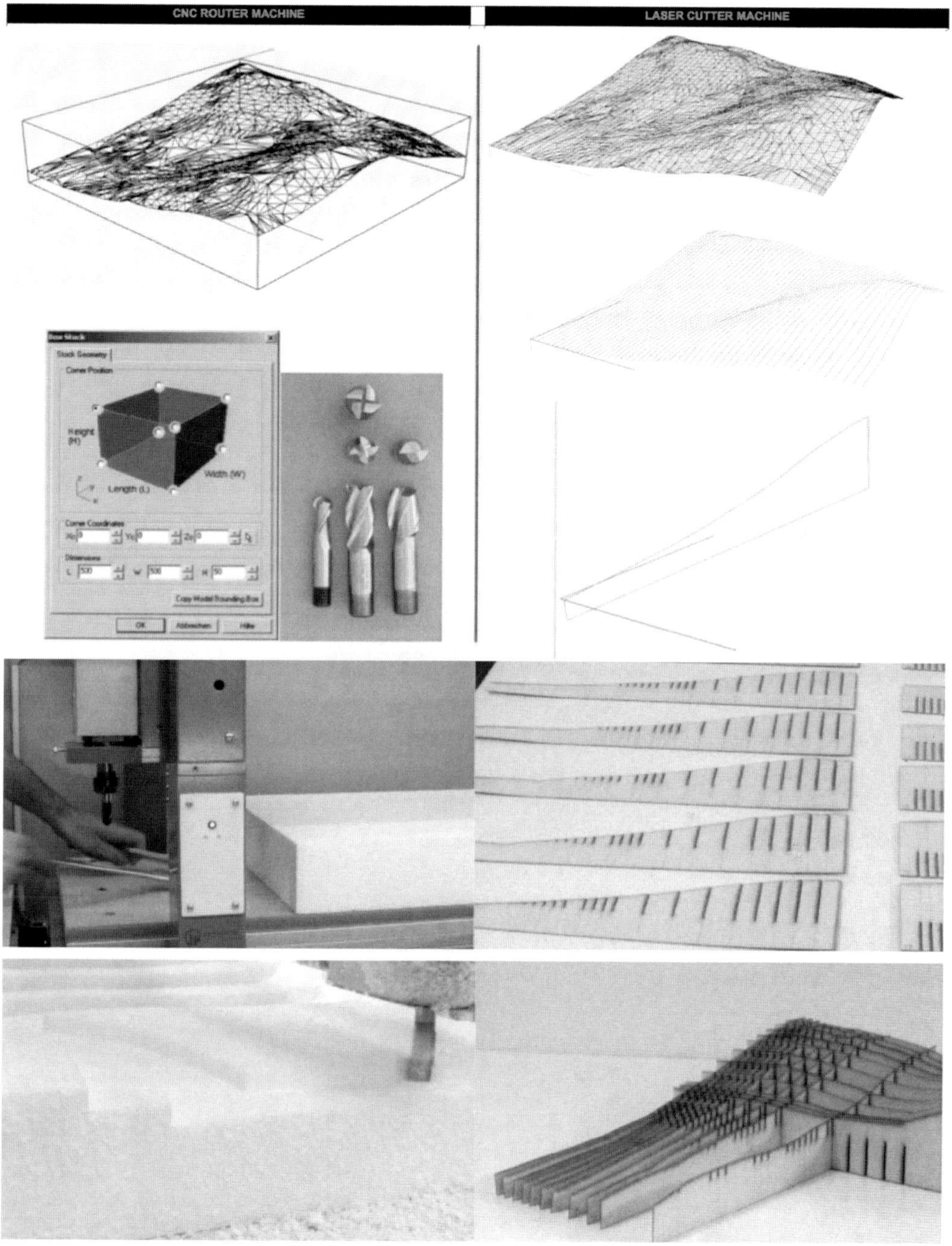

Fig. 4: Both, the CNC router and laser cutter machine do require a Digital Terrain Model (DTM) to start any work. In this case, it has been generated by using AutoCAD Land Desktop to produce a triangular surface based on contour lines.

4 Model Assessment

	CNC ROUTER MACHINE	
Technique	FORMING: Subtractive process , 3D remover. Mechanical operation	
Machine constraints	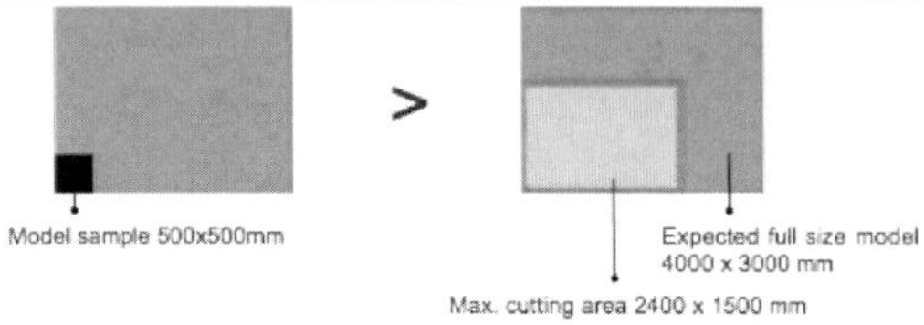	
Average Time	Model sample = 1- 1.5 hours	Full size model = 24 hours
Material	Foam	
Weight	50 Kg / m³	
Cost	20 CHF per hour	400-500 CHF = 24 hours
Training	- Digital files preparation is more time consuming . - General interface with the router is more complicated due to its mechanical nature. - Requires more maintenance and set up time.	
Job quality	- Range of smooth or rough finishes according to specific needs - Limited size and finishes of small details related to bits diameter - 2D and 3D engraving and cutting jobs - Depending on the type of job aasigned, one must stop and exchange the bits - Due to its large working area, few loose component are produced - Height limitations vary according to specific project, tool selection and path design - Direct contact with material may cause distortion or retooling	

LASER CUTTER MACHINE	
CONTOURING: Solids form by 2D profiles. Laser operation	
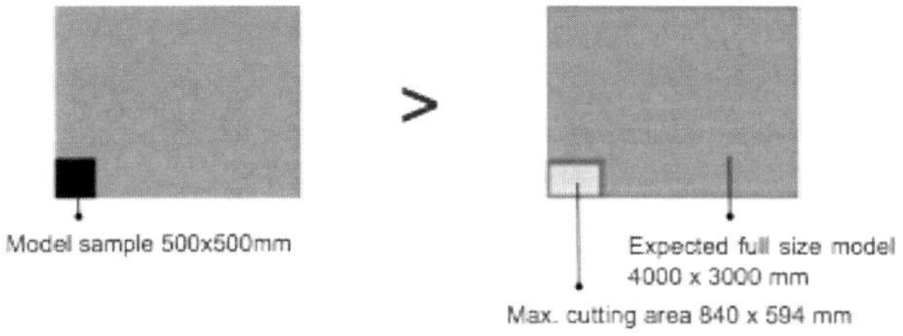	
Model sample = 1 hour (cutting job) 3-4 hrs (assembly)	Full size model = 14 hrs (cutting job) 6-8 hrs (assembly)
Cardboard	MDF
0.7 Kg / m³	700-750 Kg / m³
14,2 CHF (£10)	199 CHF (£140)
- Less sophisticated training required - Few simple operation steps and material set up - Simple system that demands less maintenance	
- Precise definition of cuts, fine edge finish - High definition cuts (90° angles, narrow slot width) - Only 2D engraving and cutting jobs - One continuous process, no need to interrupt in order to remove additional tool - Requires a strict and methodical collection of cut pieces - Height limitations of a model vary in relation to the material thickness - No direct contact with material that can cause distortion	

5 Conclusions

After comparing the results of the two model samples produced with different technologies and re-viewing the initial model making expectation, the following conclusion arise:

- Based on the desirable scale of 1.1000 of the designated landscape, both technologies (CNC router and Laser Cutter machine) are equally capable of producing an accurate topographical representation of the site, where only the range of finishes may vary. Nevertheless, for this specific case, the large scale of the model suggests that the more suitable tool is the CNC router due to its higher volume of production.
- In order to produce a model easy to transport and handle, the main aspect is the type of material to be used. Its consideration was obviously the low weight and available dimensions. Therefore, between the materials suitable for both machines, there were three selected: Styrofoam, Cardboard and MDF. From all of them the Styrofoam is highly recommended in terms of the large size of the model, which in combination with the extended working area of the CNC router could produce a full size model with a minimum amount of individual component.
- Regarding the different pieces that can potentially compose the model, is it true that a CNC router and a Laser Cutter can perform a very efficient production. What could make a big difference is the process of assembly, in which case, the first technology would be more convenient.
- The last relevant aspect relates to the expected period of time for completion. In this respect, both machines can meet the time frame expectations. The only aspect that should be taken into consideration is the available time slots allowed per person for each session in the corresponding Manufacturing Center.

In general, it is worth emphasizing that despite using very sophisticated machines with a very high level of accuracy, there is also a level of uncertainty; from the moment one manipulates the digital data until the final settings and material reactions in the output process. Rapid prototyping technique is a tool that is taking its place in landscape model making and also proffering solutions to issues of landscape development in the world as a whole.

References

Abruzzo, E., Ellingsen, E. & Solomon, J. D. (2007), *Models*. 306090 Books, Vol. 11.

Balmond, C., Davies, P., Jones, P. L., Ruy, D. & Sabin, J., *Cross-Catalytic Architectures*. www.slought.org.

Buhmann, E., Pietsch, M. & Kretzler, E. (Eds.) (2010), *Peer Review Proceedings of Digital Landscape Architecture 2010 at Anhalt University of Applied Sciences*. Wichmann.

Buhmann, E., Ervin, S. M. & Pietsch, M. (Eds.) (2012), *Peer Review Proceedings of Landscape Architecture 2012 at Anhalt University of Applied Sciences*. Wichmann.

Callicott, N. (2001), *The Pursuit of Novelty – CAD/CAM in Architecture*. The Architectural Press.

Eliasson, O. (2007), *An artist's book published by the Library Council of the Museum of Modern Art, New York.*

Ervin, S. M. (2009), *Digital: Analog: A Report from the Harvard Graduate School of Design.* ASLA, Digital Technology Newsletter, Fall 2009.

Fricker, P., Girot, C., Melsom, J. & Wemer, P. (ETH Switzerland), *3D Landscape Modeling and Planning Methods: From Reality to Virtuality and Back Again.* Teaching Experience within a Postgraduate Study Program in Landscape Architecture.

Glynn, R. (2009), *Digital Architecture: Passages Through Hinterlands.*

Gramazio, F. & Kohler, M., *Digital Materiality in Architecture.*

Groover, M. P. & Zimmers, E. W., Jr. (1984), *CAD/CAM: Computer-Aided Design and Manufacturing.* Prentice-Hall.

Gutierrez, M. A. (2012), *Rapid Prototyping Techniques in Landscape Model making.* Master Thesis at Anhalt University, Bernburg.

Hensel, M. & Menges, A. (2006), *Morpho-Ecologies.*

Iwamoto, L., *Digital Fabrications: Architectural and Material Techniques* (Architecture Briefs).

Kolarevic, B. & Klinger, K., *Manufacturing Material Effects: Rethinking Design and Making in Architecture.*

Lang, J., *A Model of the Design Process.* University of Pennsylvania.

Lee, K. (1999), *Principles of CAD/CAM systems.*

Manyoky, M., Wissen Hayek, U., Klein, T. M., Pieren, R., Heutschi, K. & Gret-Regamey, A. (ETH Switzerland), *A Concept for Collaborative Design of Wind Farms Facilitated by an Interactive GIS-based Visual-acoustic 3D Simulation.*

McMohan, C. & Browne, J. (1993), *CADCAM: Principles, Practice and Manufacturing Management.* 2nd Edition.

MIT OpenCourseWare (Massachusetts Institute of Technology), http://ocw.mit.edu.

Moon, K. (2005), *Modeling messages: The architect and the model.*

Moussavi, F., *The Function of Form.*

Reas, C. & McWilliams, C., *Form+Code in Design, Art, and Architecture* (Design Briefs).

Sheil, R. (Ed.) (2005), *Design through Making.* Wiley.

Smith, A. C. (2004), *Architectural Model as Machine.* Elsevier.

Steen, W. M. & Mazumder, J., *Laser Material Processing.* 4th Edition.

Topos – The International Review of Landscape Architecture and Urban Design. http://www.topos.de/

www.cadinschools.org.

www.data.org.uk – Design and technology association.

www.develop3d.com – Digital Magazine Technology for the Product Lifecycle.

Superkilen: Digital Mosaic

Martin REIN-CANO

SUPERKILEN

48 TIMER
4. april 2013

19. 20. & 21. APRIL 2013 · KULTURFESTIVAL
48
TIMER
NØRREBRO

Kulturfestivalen 48 Timer har infoområde på Den Røde Plads.

Hele Nørrebros festivalprogram: HER

Gemt i Nyheder | Kommentarer lukket

STEM PÅ SUPERKILEN
3. april 2013

Superkilen er blevet nomineret i AOK's Byens Bedste, i kategorien Byens Bedste Nye Initiativ

Her er et link til afstemningen: http://www.aok.dk/byens-bedste/byens-

Sider
» Nørrebros nye byrum – Superkilen
» Ansøgningsskema
» Formalia
» Sådan behandles ansøgninger
» Kontakt info
» Introduktion til Superkilen
» Projektbeskrivelse for Superkilen
» Partnerskabet for opførelsen af Superkilen
» Download logo
» Følg med via Facebook
» Kilebestyrelsens medlemmer
» Kommisorium
» Hvem byggede Superkilen?
» Link til projekthjemmesiden
» Kalender
» Lej Karmakanonen
» Superkilens historier app
» Superkilens historier online
» Ord om Superkilen

Kategorier
» Annoncer (4)
» Nyheder (26)

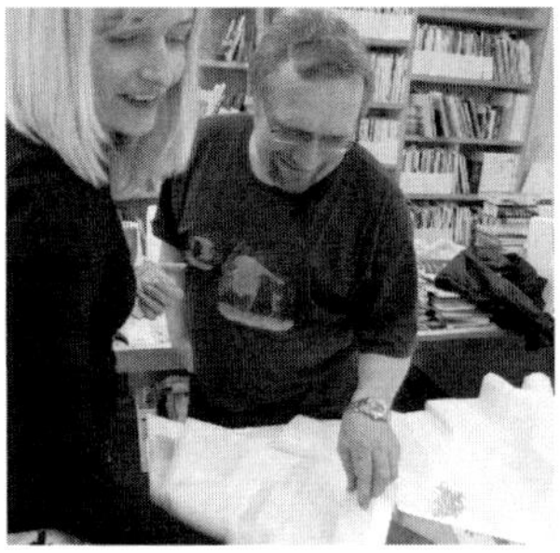

2 Digitally Sourcing the World

References and design in the global design age

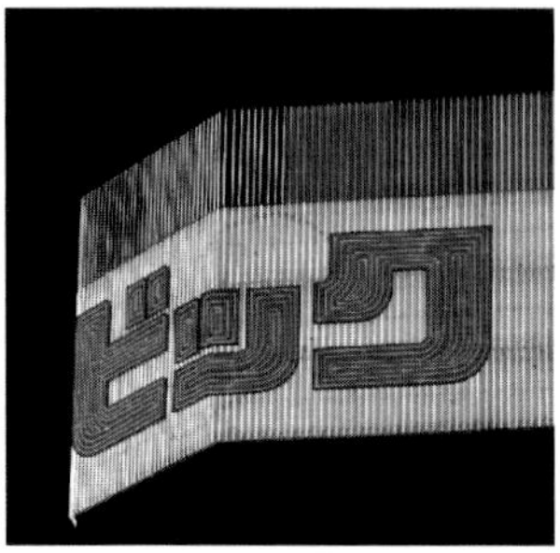

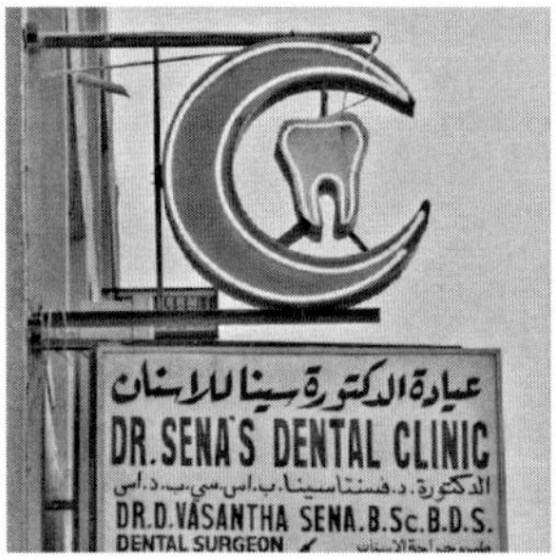

КРАСНАЯ
موقف باص
DeAngelis

3 Man-Made

People from all over the world make a place

4 Digital *Second Life*

Publication, virtual tour, app and Google

superkilen08.jpg
designboom.com Share
818 × 543 - superkilen urban park by BIG architects, topotek1 + superflex
Similar More sizes
Page 2
Superkilen-BIG-Topotek1-Superflex-9.jpg
inhabitat.com Share
728 × 485 - BIG's Quirky Kilometer Long Superkilen Park Celebrates Diversity in ...
Similar More sizes
Page 3

5 About a Digital Landscape

This project is a joy! It is not only original but stunning to behold. It is noteworthy for its aesthetic approach, which is straightforwardly artificial rather than pretending to be natural. (...) SUPERKILEN demonstrates the value of powerful visual and spatial moves while keeping connected to the realities of a contemporary multicultural context: the condition of many European cities. (...) The form of the project grew from a community participation process that led to a clear identity and enjoyable series of spaces. As a formal expression it breaks down into pieces and fits very nicely into the scale of the surrounding area as an interesting patchwork.[1]

Like a 'guerrilla approach', it has created an incredible amount of influence and changed the way people think about and consider projects.[2]

"(...) We weren't afraid of creating a tangled forest of symbols. What we wanted was a riotous mass of stimuli." (Bjarke Ingels)[3]

The juxtaposition and the assembly of the recognizable elements, shifted out of context or their place of origin, become a metonymy for the ethnic, social and cultural mix of 21st century global cities.[4]

A version of this park could be created anywhere, born using the same rules and methods, yet each version would be inherently different through participatory input. The landscape of Superkilen fits its urban area, and was built to enhance it. Nørrebro's complex situation – a demographic island of heterogeneous minorities in a homogeneous country – has forced new thinking for how a public space could be a participative element in urban culture. Superkilen is a precedent for a new way of engaging users in public spaces, in sharing cultures and in the aesthetics of 21st century parks.[5]

6 Project Dates

PROJECT:	Superkilen
PLACE:	Nørrebro, Copenhagen
TEAM:	TOPOTEK 1, BIG, Superflex
CLIENT:	City of Copenhagen and Foundation Realdania
TIME SCHEDULE:	Competition 2007: 1st prize; Planning 2007-2011; Realisation: 2010-2012

---

1 http://www.aia.org/practicing/awards/2013/regional-urban-design/superkilen/index.htm (24.04.13).
2 REIN-CANO (2013)
3 CIUFFI (2011)
4 METTA (2012)
5 BRIDGER (2013)

SIZE:	2,7 ha
WEBSITE:	http://www.superkilen.dk
APP:	http://superflex.net/tools/superkilen_app
AWARDS:	German Landscape Architecture Awards 2013 (recognition) red dot Award 2013 (Best of the Best) Civic Trust Award 2013 Archdaily- Building of the Year 2013 National AIA Award- American Institute of Architects (recognition) Architizer A+ Awards
NOMINATIONS:	Mies van der Rohe Awards 2013 Design Museum of the Year 2013 Årets Arne Award 2013 - Danish Architects' Association
PUBLICATIONS:	Newspapers / Magazines: 80 Books: 10 Websites: 86

References

Bridger, J. (2013), *Superkilen, Copenhagen*. In: anthos, 2/2013, 10-14.

Ciuffi, V. (2011), *A crowd of signs*. In: Abitare, 516 (10), 50-67.

Metta, A. (2012), *Urban & Design*. In: Paysage, 9/2011, 100-103.

Rein-Cano, M. (2013), Interview for the Red Dot Design Awards, 25. 03. 2013 (unpublished).

http://www.archdaily.com/286223/superkilen-topotek-1-big-architects-superflex/ (10.04.2013).

http://www.designboom.com/architecture/superkilen-urban-park-by-big-architects-topotek1-superflex/ (10.04.2013).

http://www.dezeen.com/2012/10/24/superkilen-park-by-big-topotek1-and-superflex/ (10.04.2013).

http://www.google.de (10.04.2013).

http://superflex.net (10.04.2013).

http://superkilen.dk (23.04.2013).

http://www.visitcopenhagen.de/Sehen-und-Erleben/Superkilen/382 (10.04.2013).

http://en.wikipedia.org/wiki/Superkilen (10.04.2013).

Photo Selection (Photographers in alphabetical order)

Eskord, Torben; Joosten, Hanns; Lindhe, Jens; da Schio, Maria

Geography – A Platform for Understanding

Jack DANGERMOND

Invited Paper, reprint from ArcNews, Fall 2012, page 7

At the 2012 Esri International User Conference, I shared the context that GIS professionals are working in today: living on a small planet; breathing the same air and becoming increasingly concerned about our future – our personal future, the future of our families and communities, even the future of life on the planet. The evidence suggests that our world is changing rapidly, with many trends that will be challenging for us personally and our organizations for the rest of our lives.

At the same time, we are living in an amazing time when scientific discovery and technological advancement are accelerating dramatically. We are making huge scientific discoveries and creating unfathomable volumes of data in the process. But these advances and the simple volume of data aren't enough. Clearly we need more integrated knowledge and ways to be able to make better decisions and create better outcomes. We need to harness our technology and our brain power to create a more sustainable future.

Geography

The role of geography is a platform for understanding the world. GIS is making geography come alive. It condenses our data, information, and science into a language that we can easily understand: maps.

These maps help us integrate and apply our knowledge. The same maps tell stories, stories about almost everything in our world. We need to better harness the power of GIS maps to engage everyone, telling the stories of what's happening to the world and creating maps that create a better future, a future with better outcomes.

I'm increasingly confident that our GIS community will do this. One reason is that GIS itself is advancing; it's getting more powerful and easier to use. It's evolving with lots of new capabilities. It's also moving to a new web/cloud based platform; one that will make GIS pervasive. GIS will evolve to a new level, creating "geography as a platform".

Re-imagining Our World

This new platform allows geographic knowledge to be widely shared, enabling widespread access and use of GIS.

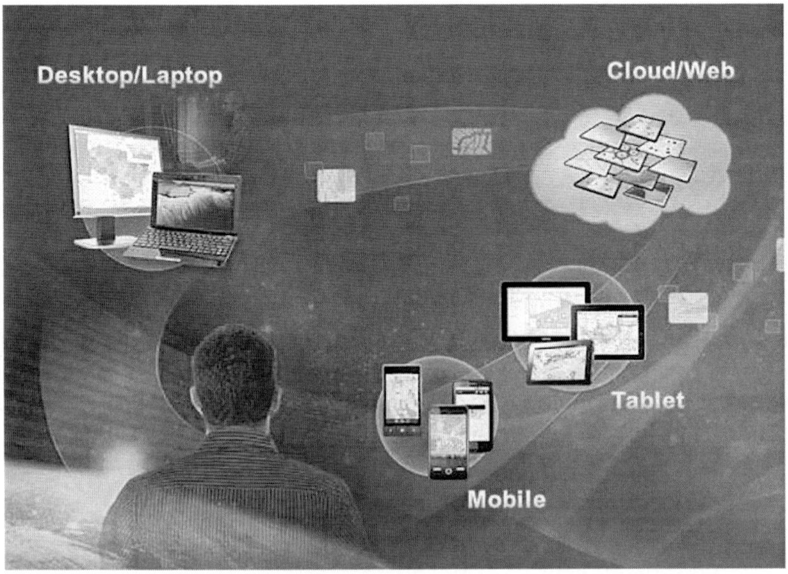

Fig. 1: Cloud GIS enables pervasive access, integrating traditional GIS with a whole new world of applications

At the same time, other trends, such as widespread measurement, big data, and ubiquitous computing, are advancing rapidly, including Software as a Service computing, device computing with lightweight and locationally aware applications, as well as supporting scientific exploration and innovation.

The convergence of GIS with these trends will enable us to integrate geographic knowledge into everything we do.

This new pattern integrates all types of geographic information – maps, data, imagery, social media, crowd sourced information, sensor networks, and much more.

Cloud GIS enables ubiquitous access and integrates the traditional work of geospatial professionals with a whole new world of GIS applications. It takes what have been relatively scarce commodities – stories and actionable geoinformation – and makes them abundant. Web maps provide the medium for integration and understanding and make this information widely accessible in simple forms. This widespread, easy access to geographic knowledge is what we mean by providing geography as a platform for understanding.

Cloud GIS changes the discussion, breaking down barriers between workflows, disciplines, and cultures.

A New Pattern

This new pattern breaks down the barriers between different workflows and disciplines and brings them together. This will enable us to better collaborate and share, as well as approach problem solving and decision making more holistically.

We are already starting to see organizations rapidly adopt this new pattern. The European Environment Agency, the United Nations Environment Programme, the World Bank, and many United States government agencies are adopting it. They are using cloud computing to support their own mission and, at the same time, sharing their knowledge with others. By sharing their geospatial knowledge in common cloud environments, they are creating a new kind of spatial data infrastructure.

Enabling the Platform

The sharing of geospatial knowledge will open our world and create a new level of understanding. As more organizations embrace this idea and adopt a culture of collaboration and sharing, the GIS community will benefit greatly. GIS practitioners will be able to do their work better and elevate the role of GIS in our organizations.

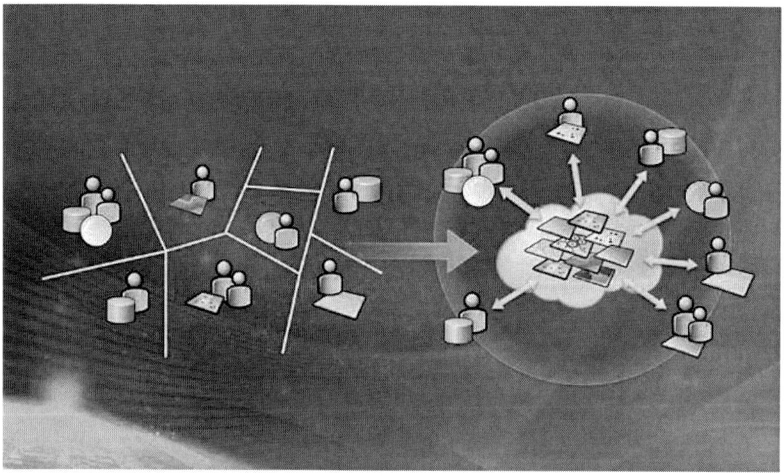

Fig. 2: Cloud GIS changes the discussion, breaking down barriers between workflows, disciplines, and cultures

Our work at Esri is about enabling our users to do their important work. We take that responsibility very seriously. I thank you for entrusting us to do that.

GeoDesign – Herausforderungen an einen verständigen Umgang mit GIS

GeoDesign – A Challenge to Improve Communicating of GIS Applications (English summary)

Hans-Georg SCHWARZ-V. RAUMER und Antje STOKMAN

Reprint of German Contribution to the Proceedings of 18. Münchner Fortbildungsseminar Geoinformationssysteme 8 – 11 April 2013, München, Chair: Prof. Dr. Thomas H. Kolbe, Technische Universität München and Prof. Dr. Matthäus Schilcher, Vorstandsvorsitzender Runder Tisch GIS e.V.; Koch, A., Bill, R. & Donaubauer, A. (Hrsg.): Geoinformationssysteme. Beiträge zum 18. Münchner Fortbildungsseminar 2013. Wichmann Verlag, Offenbach/Berlin, April 2013, p. 297-306.
Completed with English language summary for this publication.

Summary

Today geo-information is an omnipresent element of everyday life. Data collection „by the way“ and activity control „on the way“ is facilitated by apps, GPS, smart phones or satnav systems. Landscape planning and design is far from being pushed by the massive use of geo-information provided by geo-IT. Geodesign tries to bridge this gap without neglecting the necessity of creativity and communication in planning and design. But it has to tackle a lot of challenges and to overcome considerable obstacles with regard to conceptual rigor, technical restrictions and primarily concerning the organization of Geodesign as a collaborative process. Efforts can be seen in the development of man-machine-interfaces, and modeling tools. And with his „Geodesign Framework“ Steinitz (2012) has set a "landmark" which revises past experiences to create a comprehensive guideline for model based and collaborative planning and design. However at the core job level of Geodesign – as seen by the ESRI promoters of the subject – an iterative circle between Geo-data mapping, Sketching, GIS-Analysis, simulation model runs and visual impact representation has to be installed. The main methodical and technical problems arise from data availability, appropriateness of the used evaluation, process and impact models, restrictions from geo-processing facilities in GIS and computation time. Considering the organization of a collaborative workflow it is necessary that planners/designers accept restrictions from geo-spatial conditions, respect landscape function, are interested in modelling, and are aware of traps in model use. They have to have a clear idea of the target of a design. We suggest to (1) strictly respect the division of work between GeoIT/Geomatics, Geography/Landscape and Planning/Design experts and (2) to set high value on the communication process in-between them. Our article ends suggesting for a general structure of a Geodesign-Process and concludes that there have to be considered six questions before Geodesign has a chance to be implemented in planning practice.

1 Einleitung

Geoinformation hat Konjunktur. Wir orientieren uns mit Navigationshilfen und Smartphones, lokalisieren Standorte über GPS und Handy und wir können aus der „Wolke" immer und überall erfahren, wer oder was sich in unserer Umgebung gerade befindet. Geodatenerhebung „by the way" ist ebenso möglich wie die präzise Steuerung von Ereignissen „on the way".

Es liegt nahe, zu vermuten, dass diese Entwicklung an der Art und Weise, wie unsere gebauten und naturnahen Lebenswelten beplant und ausgestaltet werden, nicht vorbeigeht. Diese Vermutung weicht allerdings schnell der Skepsis, wenn man die Wünsche und Hoffnungen in Bezug auf eine digitale Landschaftsplanung mit dem vergleicht, was davon geblieben ist. Von der Idee der webbasierten Planung blieb nur (aber immerhin!) eine webbeschleunigte Planung mit webinformierten Planungsakteuren und -interessierten, die nunmehr schneller und umfassender über Sachlagen Bescheid wissen.

Auch im hier aufgerissenen Feld zwischen Technik und Geist geht es darum, erfolgreich Geoinformation in die Sphäre des Entwurfsprozesses und der Geokommunikation zu transportieren und dort adäquat und adaptiert zu integrieren. Und es wird am Ende des Papiers festzustellen sein, dass auch der Versuch, Geo-IT und Design als Geodesign zu vermählen mit nicht offensichtlichen Hindernissen belastet ist.

2 Wunsch und Wirklichkeit

Was versteht man unter dem Begriff „Geodesign"? Eine eher lapidar anmutende aber prominente Antwort bekommen wir zunächst aus der Esri-Welt. Danach wurde „Geodesign" als nunmehr mögliches Einsatzfeld von ArcSketch nach dessen Entwicklung bezeichnet[1]: die GIS-technische Aufbereitung und Bewertung einer Bierdeckelzeichnung. Die tatsächlich sehr umfangreiche Aufweitung dessen, was unter Geodesign subsumiert werden kann erfolgte dann über die seit 2010 in Redlands stattfindenden Geodesign Summits. Die Esri-Sicht auf Geodesign beschränkt sich dabei nicht auf einen Geo-IT-Blickwinkel. Vielmehr wird die Rolle von Geo-IT (GIS im weitesten Sinne) als Rückgrat einer geodatenbasierten räumlichen Denk-, Planungs- und Gestaltungswelt gesehen, die Partizipation und Kollaboration, Modellierung und Raumvisualisierung einbezieht und kreative mit rationalen Zugängen in der räumlichen Planung verbindet. Aus der Sicht von „Spatial Planning and Design" wird diese grundlegende Idee eines Geodesign durch das Harvard-Urgestein Carl Steinitz prominent befeuert. Er definiert Geodesign schlicht als „... process of changing geography by design" (STEINITZ, 2012, IX, im Vorwort).

Die bisher kommunizierten grundlegenden Idealkonzepte von Geodesign sind iterativ-zyklisch. Zunächst ist da die Idee einer interaktiven Arbeit mit GIS und Modellen, die die lineare Prozedur Eingabe-Verarbeitung-Ausgabe kurzschließt. Abbildung 1 stellt diesen Vorgang schematisch dar. Die Rolle des Einsatzes von Geodatenanalyse, von Modellen und Visualisierung ist nicht mehr auf einen unären und einmaligen Produktionsvorgang be-

---

[1] "Imagine if your initial design concept, scribbled on the back of a cocktail napkin, has the full power of GIS behind it ..." (DANGERMOND, 2009).

schränkt, sondern basiert auf im Idealfall schnellen Wiederholungen, die kreatives Entwickeln über Versuch und Korrektur ermöglichen. Die Entwurfsarbeit folgt dabei dem Paradigma, dass Rauminformation (auch in der ‚veredelten' Form komplexer Raumanalysen) einerseits und die Antizipation der Wirkungen von im Entwurf eingebrachten Raumveränderungen grundlegend notwendig für einen erfolgreichen Gestaltungsprozess sind.

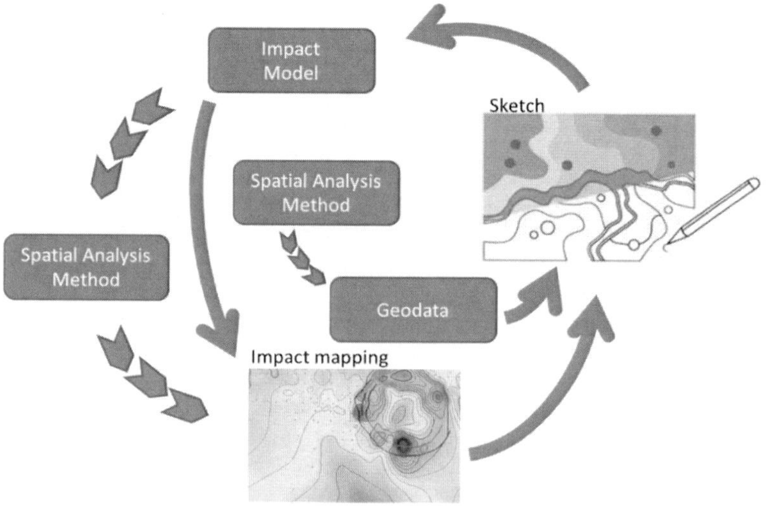

Abb. 1: Iterative Verknüpfung von kreativer Entwurfstätigkeit und geodatenbasierten Modellen (eigene Darstellung; Logos von Esri, Redlands/W. Xie and T. Brown, Alberta)

Dieses erste zyklische Geodesign-Konzept beansprucht im Kern die Einführung einer komplexen Mensch-Maschine-Interaktion. Steinitz hingegen betont die Zyklen, die (im Idealfall) ein GIS- und modellgestützter *Planungsprozess* durchlaufen sollte und bezeichnet diese als „Framework for Geodesign" (STEINITZ, 2012). Abbildung 2 verdeutlicht zunächst den übergeordneten Kommunikationszyklus eines Geodesignteams mit den an der Planung Interessierten. Steinitz sieht dann eine verständige Planung in der Abarbeitung von sechs Schlüsselfragen, die zu einem Methodenset führen, das im Ergebnis einen Planentwurf unterstützt, der (und das ist die kleine Raffinesse, die er vorschlägt) mit den gleichen Schlüsselfragen wiederum hinterfragt wird. Der vollständige Geodesignprozess durchläuft diesen Zyklus in Iterationen, die man als Scoping-, Methodenwahl- und Implementationsphase bezeichnen kann.

So reizvoll und sinnfällig Workflowzyklen sich ausdenken lassen, die Praxis steht vor einigen Problemen, die zuweilen und in Teilen unüberwindbar scheinen.

Kontext. Ein Geodesign-Workflow hat zunächst klar darzulegen, auf welcher „Produktlinie" dieser sich hinsichtlich Methodik, Technik, Ziel und Pragmatik bewegt. SCHWARZ-V.RAUMER UND STOKMAN (2012) unterscheiden hier zunächst nach der *Funktion* des Resultats. Informationstechnik und Prozessorganisation müssen sich danach richten, ob analytische, partizipative oder normative Zwecke mit der Gestaltungsaufgabe verbunden sind. Die Funktion des Produkts und deren bindenden Charakter, in etwa wie von DÜHR (2007, 58) in

Abbildung 3 angedeutet, bestimmt nicht nur die Wahl von Methodik und Technik, die im Geodesignprozess zum Einsatz kommt, sondern auch dessen kommunikative Ausgestaltung einschließlich eingesetzter Visualisierungsformen.

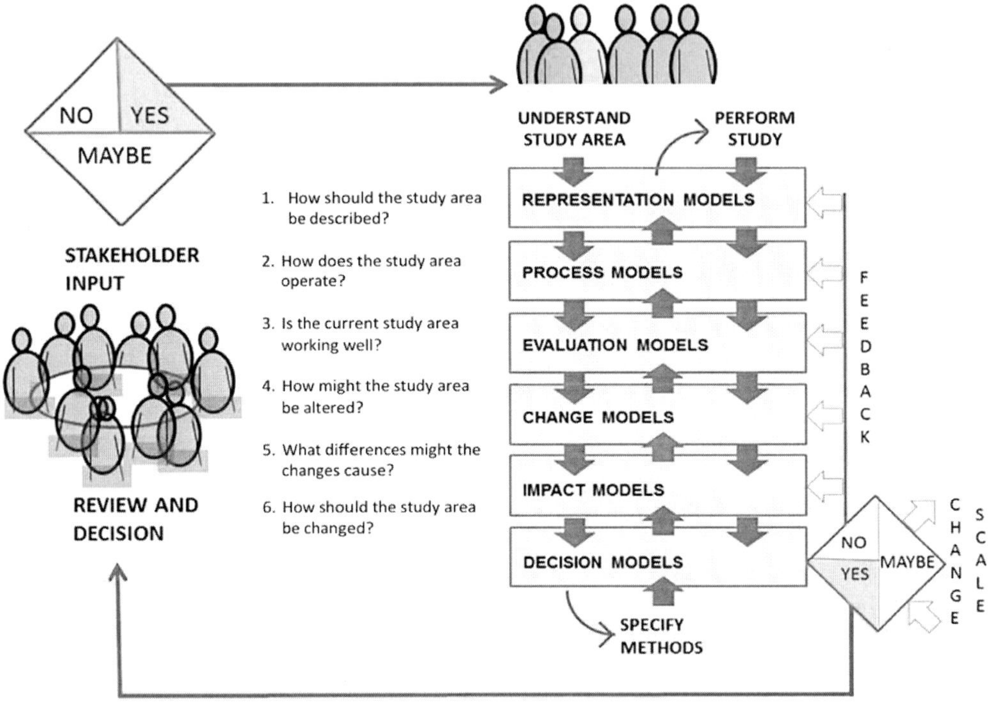

Abb. 2: Das „Geodesign Framework" (STEINITZ, 2012, 25)

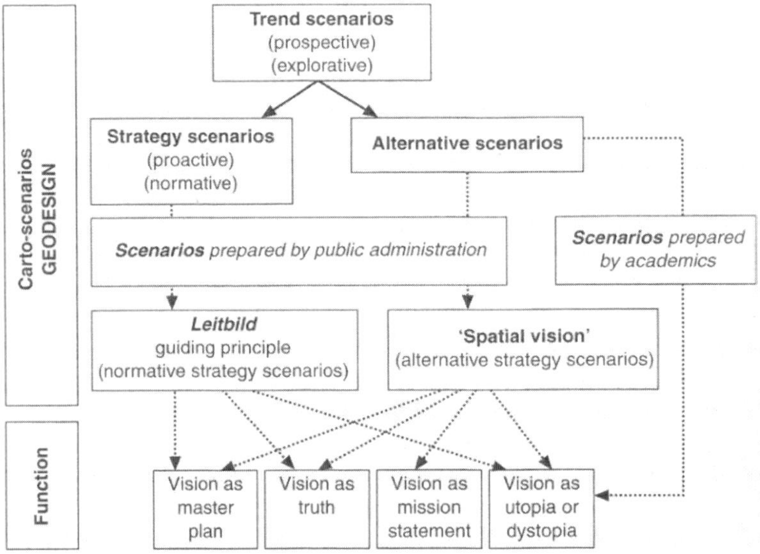

Abb. 3: Typen und Funktionen über Geodesign entworfener Zukünfte aus DÜHR (2007, 58)

Im Rahmen der Klärung kontextueller Fragen ist auch zu klären, ob die Qualität der eingesetzten Daten und Methoden der Entscheidungsrelevanz des Entwurfskontextes entspricht.

Problem Technik. Geodesign wartet mit dem Anspruch auf, komplexe räumliche Zusammenhänge in gestalterische Entscheidungszusammenhänge „on the fly“ einbringen zu können. Hohe Ansprüche an Input-Devices können hier mittlerweile zunehmend befriedigt werden. Hinsichtlich Benutzerschnittstellen hingegen ist Entwicklungsbedarf zu konstatieren. So bieten unsere GIS auch nach der legendären Erfindung von ArcSketch keine bequeme Funktionalität an, eine intuitive Skizze in einen operablen Geodatensatz zu transformieren. Eine weitere Herausforderung besteht in der Rechenkapazität, die (gemeinhin) für Geodesignvorhaben zur Verfügung steht. Um es kurz zu formulieren: das Rechentempo des Geoprocessing in ArcGIS ist nicht ausreichend, das Ergebnis komplexer Bewertungsmodelle mit einer Antwortzeit zu liefern, die einem Work*flow* entspricht. Und schier undenkbar ist die Verwendung von GIS-Geoprocessingfunktionalitäten, wenn es um die „high-speed“-Berechnung komplexer Prozessmodelle geht.

Problem Kollaboration. Die geschilderte Idealvorstellung eines Geodesign sieht vor, dass Planer/Entwerfer, IT-Fachleute, ggf. wissenschaftliche Spezialisten und andere Beteiligte vernetzt zusammenarbeiten. Über grundlegende Mechanismen und Formate, wie diese Zusammenarbeit optimal zu organisieren ist, gibt es kaum systematisch aufgearbeitete Erfahrungen. STEINITZ (2012) arbeitet allerdings Planungsfallstudien auf und dokumentiert Beispiele kollektiver Planung. http://geodesigneducation.com/ hingegen diskutiert ausbildungsbezogene Geodesign-Formate. Grundsätzlich bietet der Geodesign-Ansatz so ein weites, aber nicht einfaches Experimentierfeld. Der Schluss dieses Artikels macht hierfür einen Strukturierungsvorschlag.

Anspruch und Wirklichkeit. Eine Querschnittsanalyse dessen, was derzeit unter dem Label „Geodesign“ firmiert, ergab, dass die oben geschilderten Rahmenkonzepte nur partiell in den dokumentierten Projekten realisiert sind und dennoch mit viel Euphorie als „Geodesign“ kommuniziert und als methodisch-technische Innovation proklamiert werden. Aus den geschilderten Problemlagen ist dies natürlich verständlich und so ist es denn auch akzeptabel, wenn man „Geodatenbasiertes Entwerfen“, „wirkungsantizipierendes Entwerfen“, „kollektiv-partizipative Landschaftsgestaltung“ oder aber auch „Szenariovisionen“ als „Geodesign“ bezeichnet. Viele solcher „kleinen Lösungen“ bauen auf (meistens mehrkriteriellen) Eignungsbewertungen auf oder verknüpfen räumliche Simulationsergebnisse mit räumlichen Entscheidungsproblemen.

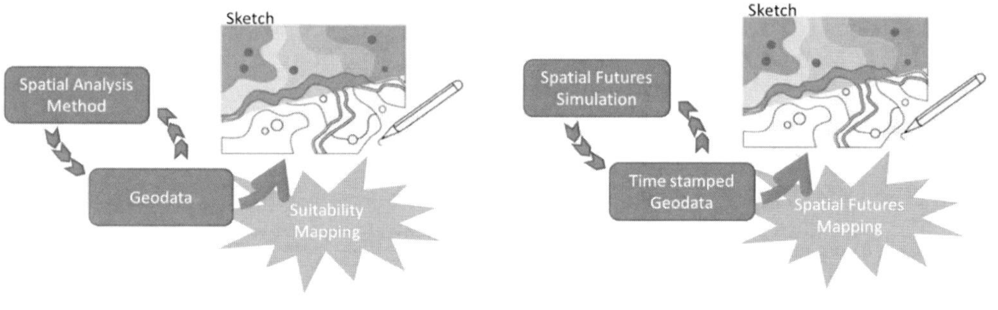

Abb. 4: Häufige „kleine“ Geodesign-Lösungen; vgl. Abbildung 1 (eigene Darstellung; Logo von Esri, Redlands)

3 Was ist ein Entwurf?

Das Entwerfen ist eine Form der Wissensproduktion, die eine evolutionäre menschliche Handlungsweise darstellt und uns dazu befähigt, komplexe Entscheidungen in Bezug auf „positiv bewertete, wirklichkeitsfähige Möglichkeiten" zu treffen (POSER, 2004). Im professionellen Bereich kombiniert der Entwerfer kontextabhängige, anwendungs- und wertorientierte Ausgangsbedingungen mit empirischen, ingenieurswissenschaftlichen Fakten, die in komplexen, logischen Entscheidungsketten zu einem Ergebnis geführt werden (SCHÖN, 1991).

Im Entwurfsprozess entstehen eine Vielzahl von Darstellungen und Notizen in Form von Texten, Berechnungen, Zeichnungen, Grafiken und Modellen, die als Mittel der Veranschaulichung und Kommunikation mit anderen Menschen dienen. In komplexen Entscheidungsketten werden Randbedingungen, Funktionsweisen, Plausibilitäten und Qualitäten von Entscheidungen, aber auch eventuelle Fehler überprüft, diskutiert und gegebenenfalls verbessert. Das Ergebnis eines Entwurfs können sowohl räumlich-konkrete Konstruktions- bzw. Gestaltungsvorschläge, aber auch konzeptionelle Strategien sein, die einen Handlungsrahmen festlegen, ohne das räumliche Ergebnis konkret festzulegen. Das Entwerfen ist also „ein integrierender Erkenntnisprozess", der in einem iterativen Prozess analytisches wie intuitives Verstehen und projektives Handeln verknüpft (VON SEGGERN et al., 2008).

Eine wichtige Grundlage stellt in diesem Zusammenhang PROMINSKI (2004) zur Verfügung, der auf der Basis aktueller Erkenntnisse aus Naturwissenschaften und Wissenschaftstheorie zeigt, wie der Prozess des Entwerfens bei der Entwicklung von Landschaft ökologische, ökonomische und soziale Aspekte integrieren und zu komplexen Lösungen führen kann. Er weist nach, dass Entwerfen eine eigenständige Erkenntnisform ist und eine dem Wissenschaftsmodus-2 (NOWOTNY et al., 2001) zuzuordnende wissenschaftliche Handlungsweise ist, die in der Lage ist mit Komplexität, Unsicherheit oder Wertkonflikten umzugehen.

Mit ihrer kontextuellen, transdisziplinären und anwendungsorientierten Ausrichtung suchen die „Modus 2"-Wissenschaften nicht mehr nach allgemein gültigen Prinzipien wie die „Modus 1"-Wissenschaften, sondern sind gekennzeichnet durch die "kontinuierliche Konfiguration und Rekonfiguration von Wissen, das auf einer temporären Basis in unterschiedlichen und heterogenen Anwendungskontexten zusammengefügt wird" (NOWOTNY et al., 2001). Im Gegensatz zu etablierten wissenschaftlichen Routinen und Methoden einzelner Disziplinen bringt der transdisziplinäre Modus 2-Forschungsansatz im Forschungsprozess unterschiedliche Akteure und Disziplinen zusammen, die jeweils unterschiedliche Sichtweisen und Methoden der Problembearbeitung mit sich bringen. Entsprechend müssen bezogen auf den jeweiligen Forschungsgegenstand spezifische „Forschungssettings" entworfen werden, die sowohl eine innovative, Disziplinen übergreifende Fragestellung als auch neue Methoden der Forschungsarbeit beinhalten.

Der hier zur Diskussion stehende iterativ-zyklische Geodesign-Prozess greift also die originäre Methodik des Entwurfsprozesses auf, kann und sollte als „Modus 2"-Forschungsansatz betrachtet und ausgestaltet werden.

4 „GIS 2.0“ oder „GISler 2.0“?

Brauchen wir neue GIS oder bessere Geo-IT-Bediener? Das „GIS for everyone“ gibt es nunmehr seit 20 Jahren (so wurde die erste mit Windows-Oberfläche ausgestattete Arc-View-Version betitelt). In der Landschaftsplanung ist dieses Werkzeug – mittlerweile, gefüttert mit weitreichenden Datenbeständen – bei der analytischen Hinterlegung von Regionalisierungen und Lokalisierungen von Schutz- und Nutzungswidmungen unverzichtbar geworden. Die gängige Arbeitspraxis passt dabei vor allem zum *rationalen* Entwurf, die Trennung von Analyse und Planentwicklung wird durch den Erfolg der Pragmatik perpetuiert. Und beides ist Rahmen der GIS-gestützten Planung und Gestaltung zu überwinden! Für den Weg dorthin sind vor allem zwei Anstrengungen zu unternehmen:

Die Implementierung einer Geodesign-Umgebung erfordert umfangreiche *methodisch-technische Vorbereitungen*. Zunächst bestehen diese in der Bereitstellung valider und ausführbarer Modelle. Die Einbindung in einen Geodesign-Prozess begründet nicht, auf die Erfüllung notwendiger Anforderungen an die Modellierung wie strukturelle Evidenz, plausible Sensitivität und akzeptable Validität verzichten zu können. Es gilt nach wie vor, dass gute Modelle einen hohen Datenbedarf erzeugen und falls der Versuchung, mit einfachen Modellen auszukommen, nachgegeben wird, muss dies kommuniziert, akzeptiert und bei der Bewertung des Ergebnisses berücksichtigt werden. Mit GIS lassen sich nur begrenzt aussagekräftige Modellergebnisse herstellen und diese sind meist nur auf einer Screening-Ebene angesiedelt. Beispielsweise lässt sich die veränderte Luftschadstoffimmission in einer Wirkungssimulation, die einen städtebaulichen Entwurf bewertet, nur über das Abarbeiten einer komplexen Modellkette und unter Verwendung spezialisierter Programme (ggf. als GIS-angebundene Erweiterung) realisieren.

Eine weitere technische Voraussetzung besteht in der Zurverfügungstellung operabler und angepasster Benutzerschnittstellen. Digitalisierung und Visualisierung müssen flexibel und den Anforderungen des Entwurfs entsprechend erleichtert vorbereitet sein, funktionierende CAD-Schnittstellen sind hier genauso notwendig wie Datenschnittstellen von und zu Modellrechnungen.

Wichtiger als die Bereitstellung einer technischen Geodesign-Umgebung ist aber die Herstellung einer funktionierenden *Arbeitsteilung*. Als grundlegende Konsequenz aus dem oben geschilderten ist zunächst festzustellen, dass Geodesign ohne die Einbindung eines Geoinformatikers und/oder Geomodellierers nicht möglich ist. Auch wenn die Ausbildung von Landschaftsarchitekten, Stadt- und Freiraumplanern eine solide GIS-Ausbildung beinhalten würde, Datenbankmanagement, Skript- und Modellcodierung oder das Aufsetzen von umfangreichen Geoprocessing-Modellen können kaum Bestandteile eines entsprechenden Syllabus werden. Die Voraussetzungen für die Entwurfsarbeit hingegen ergeben sich aus den Zielen eines Geodesignprojekts:

- Offenheit für kreativitätseinschränkende Restriktionen aus geographisch-räumlichen Gegebenheiten und räumlich-funktionalen Zusammenhängen,
- Verständnis für den Modellierungsprozess und die Begrenztheit der Aussagekraft von Modellergebnissen,
- eine klare Zielorientierung im Entwurf und
- weitgehende Akzeptanz, Entwurfsarbeit in einer ungewohnten technischen Umgebung zu leisten.

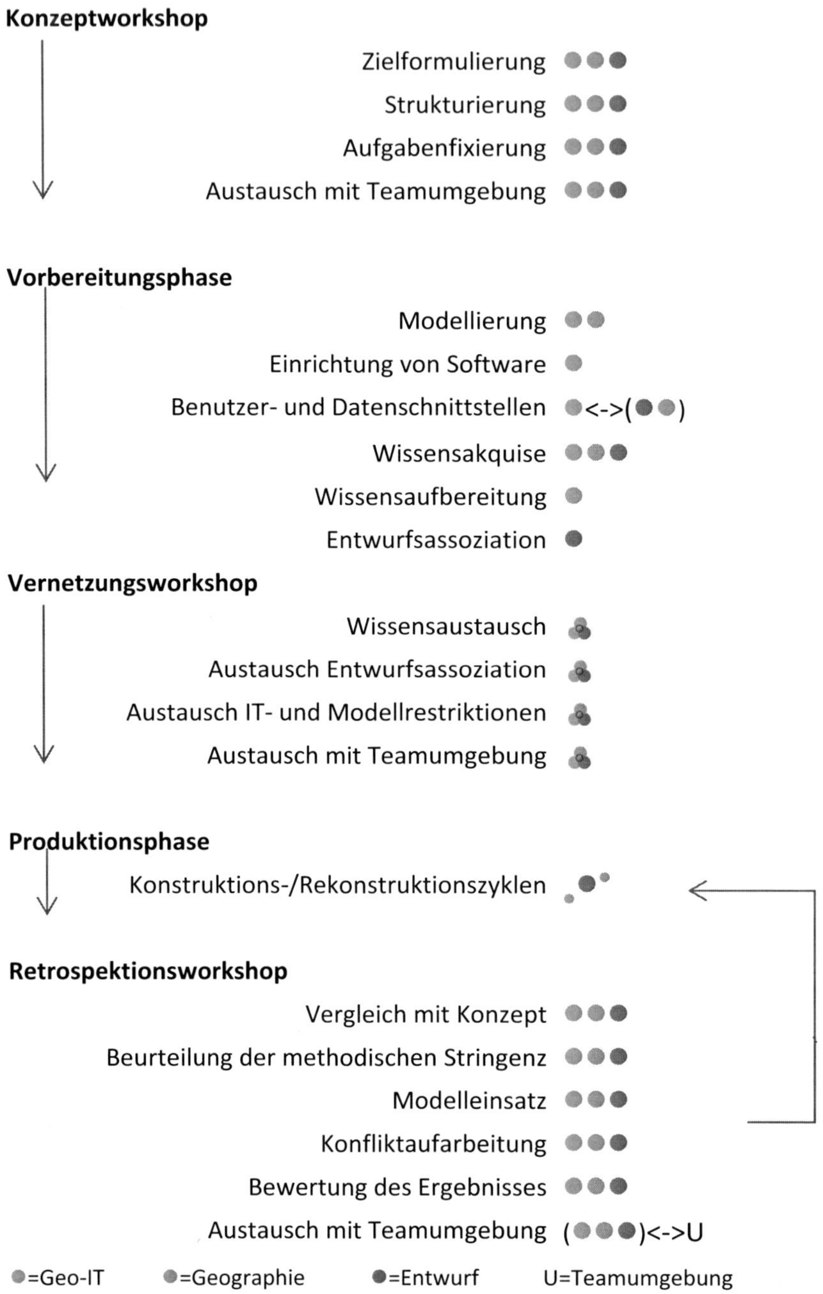

Abb. 5: Schema, Akteure und Aktivitäten eines Geodesign-Prozesses

Für die Zusammensetzung eines Geodesign-Teams ist es unabdingbar, dass in ihm mindestens drei Rollen vertreten sind, d. h. eine klare Zuordnung zu den Rollen „Geo-IT", „Geographie" und „Entwurf" getroffen ist (ggf. kann auch ein „Tandem" sich eine Rolle teilen und natürlich kann auch ein Wunderkind mit einem Solo aufwarten; größere Teams benötigen einen „Teamer" mit Koordinations-, Moderations- und Mediationsfunktion). Ein u. E. sinnvolles Konzept unterteilt die Arbeit eines Geodesign-Teams in zwei mit Workshops eingerahmte Phasen: Die „Vorbereitungsphase" wird in einem Konzeptworkshop vorbereitet und kann u. U. eine erhebliche Zeitspanne in Anspruch nehmen. Ein Vernetzungsworkshop bereitet die „Produktionsphase" vor. Der Workshop zielt auf ein größtmögliches Verständnis aller im Team Beteiligten hinsichtlich (1) der geographischen Gegebenheiten (Strukturen und Prozesse, landschaftsphysisch wie auch sozio-ökonomisch und soziokulturell), (2) der Entwurfsassoziation und Zielfokussierung und (3) der IT- und Modellierungsbesonderheiten. Wie auch der Konzeptworkshop bezieht der Vernetzungsworkshop die externen Beteiligten (Auftraggeber, Interessenvertreter, Experten, Interessierte etc.) mit ein. In der Produktionsphase findet schließlich der eigentliche Geodesignprozess statt. Idealerweise als Gruppenarbeit, in der die oben beschriebenen Rollenträger aus ihrer Warte aber synchron Konstruktions- und Rekonstruktionszyklen durchlaufen. Abbildung 5 ordnet den Workshops und Prozessphasen Aktivitäten zu und verdeutlicht, welche Aufgaben den drei Rollen „Geo-IT", „Geographie" und „Entwurf" zugeschrieben sind.

5 Schluss

Begriffe entstehen und bestehen dann, wenn ihre Funktion, Zusammengehörendes zu bündeln auf Sinn gebenden Bedarf trifft. Die ‚Catchphrase Geodesign' (SCHWARZ-V.RAUMER & STOKMAN, 2012) hat mittlerweile diesen Werdegang bewältigt und es wäre nicht gerecht, den Boom des Begriffs nur in der euphorischen Promotion durch Esri begründet zu sehen. Zwar gibt es Tendenzen, dass das Label inflationär vergeben und weit gedehnt wird (ähnliches hatte übrigens auch der Begriff „GIS" zu erleiden), dennoch *bündelt* „Geodesign" bislang nur umständlich umschreibbare Formen konstruktiver und lebensweltverändernder Anwendungen raumbezogener Informationstechnologien die schon lange gepflegt wurden.[2] Und das ist neu und gut. Endlich ein Schlauch für den Wein.

Wir haben uns mit diesem Beitrag bewusst auf die Entwurfsinterpretation des Begriffs Design gestützt. Gerade hier gibt es (geradezu überfällig) „Neuland" zu entwickeln. Und zwar hinsichtlich innovativer Entwurfsformate, die die Gestaltung von Lebensräumen mit Wissen und moderner Geo-Informationstechnologie verbinden, die aber auch in der Lage sind, die strenge Trennung zwischen Landschaftsplanung, Landschaftsarchitektur und Freiraumplanung aufzuheben (Geodesign als Konvergenzprogramm; vgl. hierzu STOKMAN & VON HAAREN (2010) sowie SCHWARZ-V.RAUMER & STOKMAN (2012)).

Die oben dargestellten Anforderungen an GIS-Spezialisten hierfür sind nicht trivial. Die technische Herausforderung stellt hohe Ansprüche an eine Ausbildung in räumlicher Modellierung, an Erfahrung in der Bereitstellung und Gestaltung von Benutzerinteraktion und an einen versierten Umgang mit Visualisierungstechniken. Um diesen Ansprüchen ent-

[2] Das Zitat „'We've been doing Geodesign for years' was a statement commonly overheard at the first Geodesign Summit held in January 2010" (ARTZ, 2010) bestätigt dies.

gegenzukommen, müssen unsere GIS-Plattformen auch weiterhin mit innovativen Werkzeugen weiterentwickelt werden. Mehr noch aber gilt es für die GIS-Welt, einer per se nicht-technischen und per Modus kreativen Arbeitskultur von Planern und Entwicklern zukünftiger Geographien im Geodesign-Prozess entgegenzukommen: mit einem ausgeprägten Maß an Bereitschaft zur Interaktion und mit einem aktiven und verständigen Willen zur Zusammenarbeit.

Schließlich und abschließend sei angemerkt, dass im Einzelfall hinsichtlich der Sinnhaftigkeit, einen Geodesign-Prozess anzustoßen – in Anlehnung an VAN DELDEN et al. (2004) – sechs Fragen im Vordergrund stehen müssen:

- Die *Struktur*frage: Wer ist überhaupt bereit, die etablierte Planungs- und Gestaltungskultur aufzubrechen?
- Die *Strategie*frage: Worin besteht der Nutzen der Vorgehensweise?
- Die *Verfügbarkeits*frage: Sind Qualität und Quantität von Daten, Kenntnissen und Modellen ausreichend?
- Die *Glaubwürdigkeits*frage: Besteht Konsens über Struktur und Annahmen der eingesetzten Modelle und werden deren Ergebnisse vertraut werden können?
- Die *Signal*frage: Trifft die Art der Ergebnisinformation der eingesetzten Methoden den gewünschten Informationsbedarf?
- Die *Kultur*frage: Wird der Endnutznießer des Resultats bereit sein, das Produkt zu akzeptieren und seine Entscheidungsfindung entsprechend anpassen?

Literatur

Artz, M. (2010), *Geodesign: Changing Geography by Design.* http://www.directionsmag.com/article.php?article_id=3435 (30.01.2013).

Dangermond, J. (2009), *GIS: Designing Our Future.* http://www.esri.com/news/arcnews/summer09articles/gis-designing-our-future.html (30.01.2013).

Dühr, S. (2007), *The visual language of spatial planning: exploring cartographic representations for spatial planning in Europe.* Abingdon, UK.

Flaxman, M. (2010), *Fundamentals of Geodesign.* In: Buhmann, E., Pietsch, M. & Kretzler, E. (Eds.), Peer Reviewed Proceedings of Digital Landscape Architecture 2010 at Anhalt University of Applied Science. Berlin/Offenbach, Wichmann Verlag, 28-41.

Nowotny, H., Scott, P. & Gibbons M. (2001), *Re-Thinking Science.* Cambridge, Polity Press.

Poser, H. (2004), *Entwerfen als Lebensform. Elemente technischer Modalität.* In: Kornwachs K. (Hrsg.), Technik – System – Verantwortung (= Technikphilosophie, 10). Münster, LIT, 561-575.

Prominski, M. (2004), *Landschaft entwerfen.* Berlin, Reimer Verlag.

Schön D. (1991), *The reflective practitioner.* Aldershot, Avebury.

Schwarz-v. Raumer, H.-G. & Stokman, A. (2012), *Geodesign – Approximations of a Catchphrase.* In: Buhmann, E., Pietsch, M. & Kretzler, E. (Eds.), Peer Reviewed Proceedings of Digital Landscape Architecture 2012 at Anhalt University of Applied Science. Berlin/Offenbach, Wichmann Verlag, 189-198.

Stokman, A. & von Haaren, C. (2010), *Integrating Science and Creativity for Landscape Planning and Design of Urban Areas*. In: Weiland, U. & Richter, M. (Eds.): Urban Ecology – a global Framework. Oxford, Blackwell Publishing.

Steinitz, C. (2012), *A Framework for Geodesign: Changing Geography by Design*. ESRI Press, Redlands.

Van Delden, H., Luja, P. & Engelen, G. (2004), *Final report on the development and application of the MedAction PSS as part of 'MedAction: Policies to Combat Desertificaton in the Northern Mediterranean Region'*. EU-DGXII, Brussels (contract EVK2-2000-22032).

Von Seggern, H., Werner, J. & Grosse-Bächle, L. (2008), *Creating Knowledge. Innovationsstrategien im Entwerfen urbaner Landschaften*. Innovation Strategies in Designing Urban Landscapes. Berlin, Jovis Verlag.

Verwendung von Fachsoftware im kommunalen Grünflächenmanagement

Stephan HERBARTH

Invited Paper

1 Einführung

Der Einsatz von Fachsoftware im kommunalen Grünflächenmanagement ist mittlerweile fester Bestandteil des täglichen Verwaltungsprozesses in fast allen Kommunen. Verwaltungsrechtliche Erfordernisse wie die doppelte Buchführung in Konten (DOPPIK) macht es in Zukunft notwendig, den Geld- und Materialeinsatz präziser zu steuern und zu kontrollieren. Trotzdem Fachsoftware seit Jahren verwendetet und deren Leistungsspektrum regelmäßig angepasst und ergänzt wird, existieren bisher keine Erhebungen über den deren Einsatz. Eine Umfrage unter Grünflächenämtern bzw. mit der Grünpflege anvertrauten Stellen lieferte Aussagen zum Einsatz, dem tatsächlich genutzten Abfragespektrum und zu den Nutzern der entsprechenden Programme. Aspekte des Nutzens der Software werden ebenfalls, aus Sicht der Nutzer, abgefragt. Es lassen sich vom bloßen Flächennachweis bis zur Kostenkalkulation verschiedenste Aufgaben mit der Software abarbeiten. Die Recherche zeigte, dass es bisher keine Erhebungen darüber gibt, mit welcher Software gearbeitet wird, wo es Vorteile gibt, wer mit der Software arbeitet und wie die zukünftige Bedeutung von Fachsoftware beurteilt wird.

2 Gegenstand der Untersuchung und Vorgehen

2.1 Fachsoftware

Vielfältigste Aufgaben und Anforderungen werden an Grün- und Freiflächen und damit letztendlich auch an das Management gestellt. Die Fülle der gesammelten Daten zu den Objekten und Flächen und den dazugehörigen Informationen erzeugt große Datenbestände. Diese zu vernetzen und zu bündeln, oder gar erst zugänglich zu machen, ist der fachliche Anspruch, der an die Software gestellt wird. Fachsoftware für das kommunale Grünflächenmanagement besteht in aller Regel aus zwei Komponenten – einer GIS-Anwendung und einer Datenbank mit Objekt- und Flächendaten. Aufgabe ist es, schnell und übersichtlich Informationen zu den Objekten und Flächen zu liefern. Ergebnisse und Daten können als Grundlage für die Kosten- und Leistungsrechnung gelten, als Arbeits- und Flächennachweis, als Argumentationsgrundlage für verschiedenste Diskussionen und Auswertungen und Weiteres dienen.

2.2 Bestandsanalyse

Informationen zum Einsatz von Grünflächenmanagementsoftware liegen für wenige Städte vor. So zum Beispiel für Berlin, Stuttgart und Bremen. Bisher unbeantwortet blieben jedoch Fragen, welche den mittelbaren und anwendungsbezogenen Einsatz in den Kommunen betreffen. Hier ist der Ansatz der Bestandsanalyse zu suchen.

Die Ermittlung der aktuell in der kommunalen Verwaltung verwendeten Software und deren Einsatzgebiete erfolgte über eine Onlineumfrage. Die Bereitstellung dieser Umfrage erfolgte über die GALK in 2012. 50 Antworten wurden anschließend ausgewertet. Schwerpunkte der Umfrage waren:

- Nutzung der Software in den Verwaltungen,
- die genutzte Software,
- das abgefragte Leistungsspektrum,
- Kaufüberlegungen.

3 Ergebnisse

Die gestellten Anforderungen an die Fachsoftware leiten sich aus den Aufgaben, welche von den Nutzern an diese gestellt werden ab. Eine nicht auf die Ziele der Kommune zugeschnittene Software hat wenig Bestand. Zu unterscheiden sind fachliche Anforderungen (Was muss die Software können?) und Aspekte, welche die Handhabung und Nutzung berühren, z. B. Bedienbarkeit. Die Erwartungshaltung kann beispielsweise rein betriebswirtschaftlich ausgerichtet sein, oder auch die Gesamtheit der Grün- und Freianlagen betrachten (Ästhetik, Kosten, Nutzbarkeit und Weitere) – dennoch lässt sich eine gemeinsame Schnittmenge formulieren.

3.1 Geschäftsprozess unterstützen

Die anfallenden Arbeiten müssen durch die Softwareanwendung unterstützt, erleichtert oder vereinfacht werden. Kosten- und Zeitersparnis sind hier sicher an erster Stelle zu nennen. Gesetze, Vorgaben, Richtlinien müssen beispielsweise in die Software eingearbeitet werden. Komplizierte Systeme und schwierig zu bedienende Software verfehlt den Sinn, welcher dem Einsatz bzw. der Entscheidung für Fachsoftware in aller Regel vorausgegangen ist – auch die Editierbarkeit bzw. die Einbindung eigener Aspekte in die Programme sollte mit möglichst wenig Aufwand möglich sein. Der hohe finanzielle und personelle Aufwand für die Einrichtung eines System muss sich in der Nutzungshäufigkeit- und -intensität widerspiegeln. Die Software sollte der Mehrzahl der Mitarbeiter zur Verfügung stehen und von diesen genutzt werden und nicht nur einem kleinen Kreis technisch Versierter. Voraussetzung dafür ist hier Nutzerfreundlichkeit und einfache Handhabung. Daten und durchgeführte Abfragen müssen als Nachweis und Arbeitsgrundlage dokumentierbar und in Karten und Plänen ausdruckbar sein.

3.2 Datenverwaltung- und -austausch

Die Kopplung der Fachdaten (mit Informationen zu den Flächen und Objekten) und der GIS-Daten macht eine kontinuierliche Datenpflege unumgänglich! Ungenaue Daten führen immer zu falschen Aussagen. Dateneingabe und -pflege sind sehr zeitaufwendig und kostenintensiv. Vor Ort aufgenommene Daten sollten unkompliziert in das vorhandene System überspielbar sein oder sich automatisch aktualisieren lassen. Um eine Mehrfachnutzung des Datenbestandes (intern wie extern) zu erleichtern, müssen die erzeugten Dateiformate für einen möglichst großen Nutzerkreis einlesbar sein. Besonders im Hinblick auf Datenaustausch mit anderen Abteilungen und der Fremdvergabe von Leistungen.

3.3 Vorteile der Software

Die Anschaffung von Software wird immer von dem Gedanken begleitet, Vorteile aus der Nutzung zu ziehen. Ob der Betrieb positiv und effizient empfunden wird, hängt maßgeblich von den Rahmenbedingungen und Kriterien wie Datenverfügbarkeit- und -qualität ab.

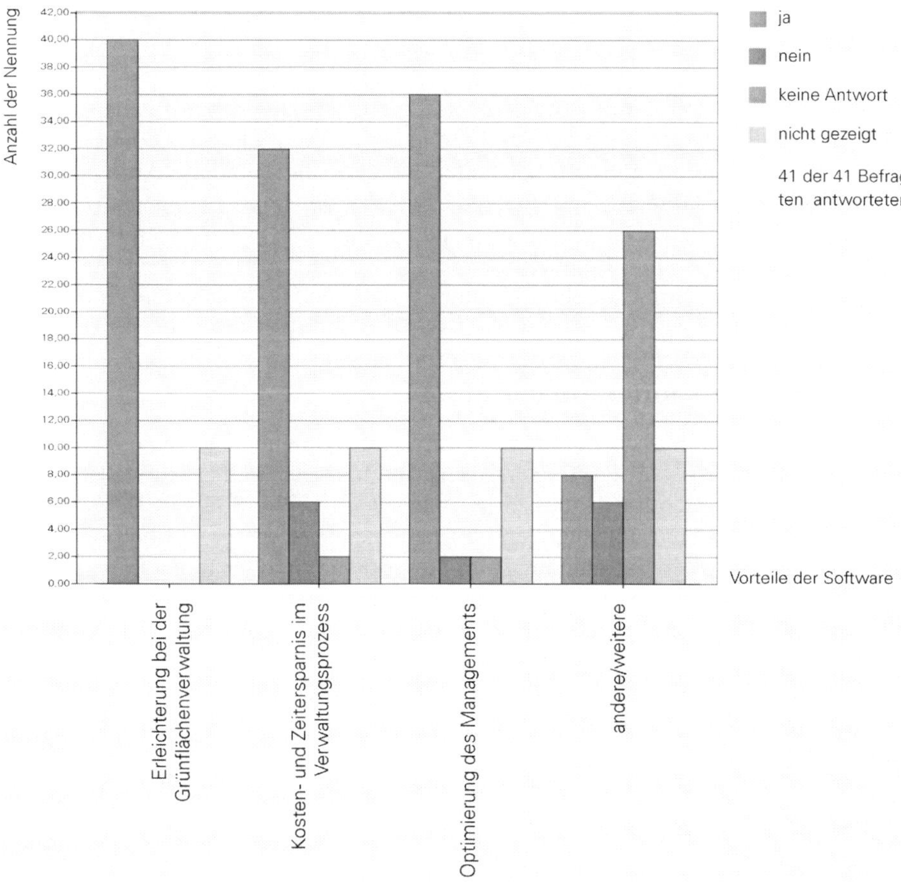

Abb. 1: Vorteile von Fachsoftware

Ungenaue oder lückenhafte Daten liefern falsche Aussagen, oder schließen den Einsatz der Software bereits im Vorfeld aus. Die Befragung zeigt jedoch, dass der Einsatz der Software in der Mehrheit positiv bewertet wird und sich demnach Vorteile durch die Nutzung für die Befragten ergeben.

Besonders positiv wird die Erleichterung bei der Grünflächenverwaltung von den Befragten gesehen – hier gab es die höchsten Zustimmungen. Nachfolgend die Kosten- und Zeitersparnis im Verwaltungsprozess und die Optimierung des Managements. Aspekte wie die Verbesserung der Zusammenarbeit zwischen den einzelnen Ämtern durch Datenaustausch, sowie die gemeinsame Datennutzung, Nachweis und Dokumentation zum Beispiel von geleisteten Arbeiten und statistische Auswertungen wurden ebenfalls genannt. Kritische Anmerkungen und Verneinungen wurden ebenfalls abgegeben. Die Auswertung der Stellungnahmen ergab hier, dass der teils hohe Zeitaufwand für Datenpflege, Betrieb und Unterhaltung der Programme negativ gesehen wird.

3.4 Genutzte Fachsoftware

Fachsoftware für das kommunale Grünflächenmanagement ist Spezialsoftware und wird nur von wenigen Herstellern angeboten. Es wurden acht Programme über eine Recherche ermittelt.

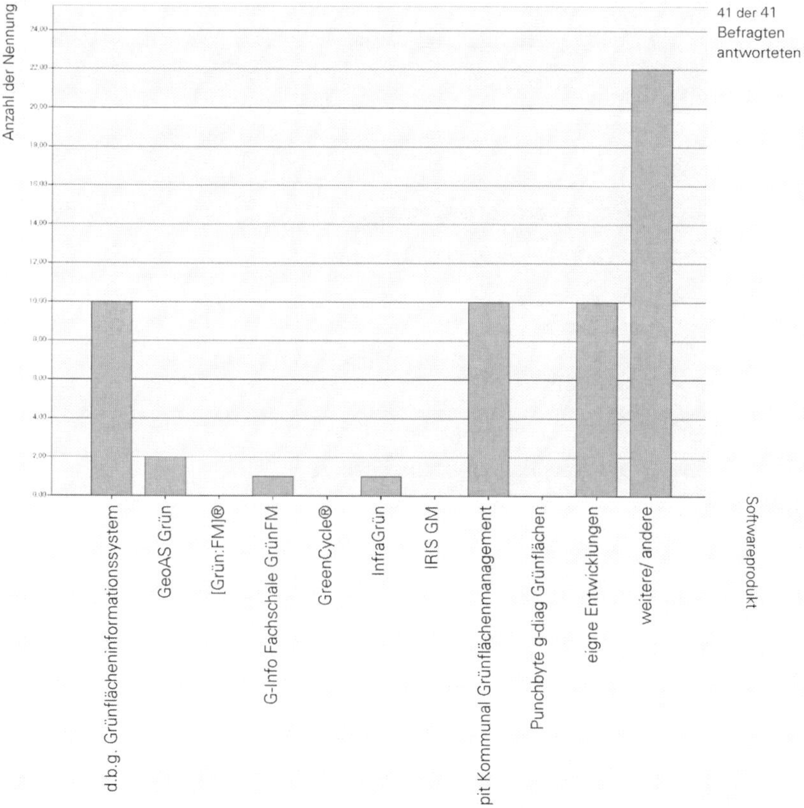

Abb. 2: Von den Befragten genutzte Fachsoftware

Die Mehrheit der Programme besitzt Fachschalen. Zum Grundprogramm werden hier Fachmodule zugeladen, die das Leistungsspektrum des Programms erweitern bzw. weitere Abfragen ermöglichen. Ein Vorteil ist, dass die Software leistungsspezifischer auf den Kunden eingerichtet werden kann. Fachschalen dienen immer Spezialaufgaben – sie stellen eine vertiefte und qualifizierte Abfragemöglichkeit bestimmter Daten dar.

3.5 Einsatzgebiete der Software

Der Einsatz und die Einsatzgebiete der Software können sehr vielfältig sein. Die Vielgestaltigkeit der Aufgaben, von Planung bis Unterhalt, erfordert ein weitreichendes Management. Gestellte Anforderungen, die Breite des Aufgabenspektrums der umzusetzenden Vorgaben und des damit verbundenen Managements und Verwaltung, generieren einen hohen zeitlichen und personellen Aufwand. Es wurde in zwei Blöcken abgefragt, für welche Aufgaben die Software verwandt wird. Der erste Block berührt Punkte des Managements, der zweite ökonomische.

3.5.1 Management

Die Häufigkeit der durchgeführten Abfragen bzw. der damit abgeleisteten Arbeiten zeigt, dass *Datenbereitstellung für Fremdpflege*, *Nachweis der Leistungserbringung*, *Erarbeitung von Ausschreibungen* und *Inventarisierung* besonders häufig über die Programme abgefragt werden.

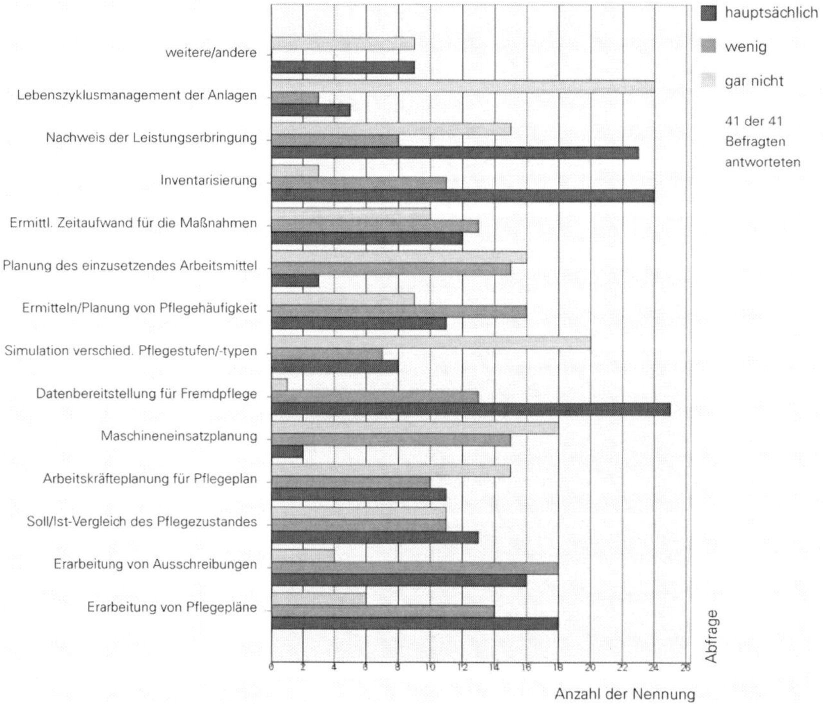

Abb. 3: Einsatzgebiete der Software im Management

Gefolgt von der *Ermittlung des Zeitaufwands für die Maßnahmen, Ermitteln/Planung von Pflegehäufigkeit, Simulation verschiedener Pflegestufen/-typen und der Arbeitskräfteplanung für den Pflegeplan. Lebenszyklusmanagement*, die *Planung des einzusetzenden Arbeitsmittels* und die *Maschineneinsatzplanung* werden am wenigsten mit der Software ermittelt.

3.5.2 Kosten

Bei den Abfragen wird die Option *Pflegekostenkalkulation, Optimierung der Grünflächenpflege und Kostenermittlung* am häufigsten verwendet – die Ermittlung der Kosten steht hier klar im Vordergrund. Weniger abgefragt werden *Umsetzen von Sparaufträgen, Optimierung und Angleichung des Pflegestandards über die ganze Stadt, Errechnung der Folgekosten bei der Ausschreibung von Freiraumwettbewerben, Ergänzung des Grünflächenkatasters und der Pflegepläne mit Kostenelementen, Budgetierung* und *Arbeitsvorbereitung für Werterhaltungsmaßnahmen*. Die *Optimierung eines Bauprojektes aufgrund der Durchrechnung verschiedenen Varianten* und die *Lebenszykluskosten der Anlagen* werden selten abgefragt.

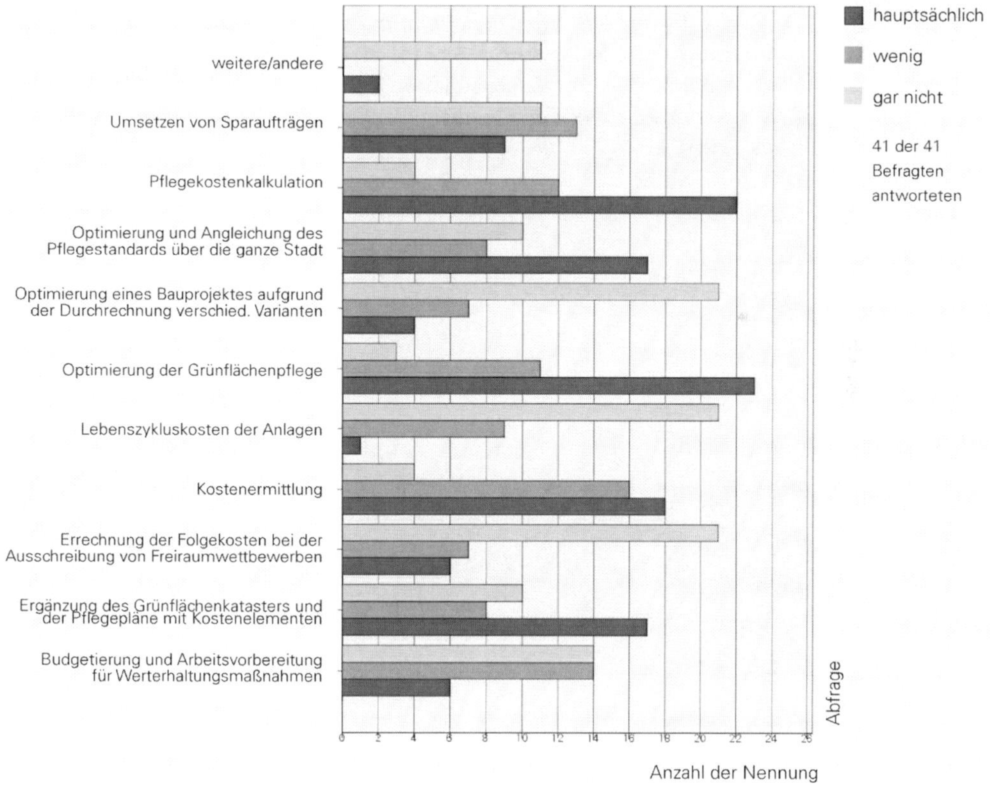

Abb. 4: Einsatzgebiete der Software im Bereich Kostenmanagement

4 Ausblick – Bedarfsentwicklung

34 % der Befragten, die bereits Fachsoftware nutzen, gaben an, sich in nächster Zeit weitere anzuschaffen. Das Aufgabenspektrum ist dem der momentanen Nutzung ähnlich. Aus der Gruppe derer, die bisher keine Fachsoftware nutzen (16 % der Gesamtanzahl) gaben 62 % an, sich in nächster Zeit entsprechende Programme zu kaufen. Die Mehrheit der Befragten sieht einen steigenden Bedarf im Bereich Fachsoftware. Gründe dafür sind in der Einführung der DOPPIK und der Verknappung finanzieller, technischer und personeller Ressourcen zu suchen. Bedarf und Bedarfsentwicklung hängen stark von den Vorteilen ab, den die Nutzer im Einsatz sehen.

Verweise

GALK – Arbeitskreis Organisation und Betriebswirtschaft, Arbeitsgruppe Grünflächeninformationssysteme (2000), *Leitfaden zur Erstellung und Fortschreibung eines Grünflächeninformationssystems* (ohne Ort), 18-20.

Herbarth, S. (2013), *Fachsoftware im Grünflächenmanagement. Ergebnisse einer Umfrage in den kommunalen Ämtern.* In: Stadt+Grün, 62 (1), 41-44.

Hochschule Osnabrück – Fakultät Agrarwissenschaften und Landschaftsarchitektur (2010), *GRÜNE DOPPIK – Werkzeug zur Vermögensbewertung öffentlicher Grün- und Freiflächen.* Osnabrück, 3, 10-11.

Invited Workshop Contributions

Patch-Scape – Pads, Pods, Phones and Spatial Data

Concept for a workshop at the Digital Landscape Architecture DLA Conference 2013

Invited contribution by
Nadia AMOROSO, Andrew HUDSON-SMITH, Mike PHILLIPS,
Chris SPEED and Katharine WILLIS

Abstract

"We should now learn to 'hook up' social channels like we do cable for our televisions. Society does not cover the whole any more than the World Wide Web is really worldwide." (LATOUR, 2005, 242)

The **Patch-Scape** workshop offers a challenging but playful opportunity for participants to generate spatial, social and environmental data derived from the landscape and manage it's transposition into a series of representational modes using digital technology. Using the **Patch-Scape** *Digital Switchboard*, the workshop explores the potential to transpose different data sets into a different 2D and 3D forms.

This paper introduces the strategies, methods and technologies behind the workshop and offers a critical toolkit toward developing trans-media processes for manifesting digital architectures.

1 Introduction

The development of tools for representing, modeling and visualising the natural landscape has tended to focus on either representing information through GIS-based tools, or enabling more realistic visualisations of virtual models. According to HARVEY et al. (2008) representations are both supported by and contribute to models which provide the set of constructs for describing and representing parts of the real world digitally (LONGLEY et al. 2006), whereas 'visualisations' offer a flexible medium for analysing and interacting with real and artificially created environments (MACEACHREN et al., 2001). More recently supporting participatory approaches have explored the potential for user-generated approaches, such as the 'geospatial web' or 'geoweb' which enables users 'to navigate, access, and visualize geo-referenced data as they would in a physical world' (LECLERLC et al., 2001). The possibilities offered by mapping and mashup formats developed from the geospatial web in conjunction with mobile computing are also inspiring new metaphors for collaborative mapping and the description of experiences in geographic spaces. For example, mobility data (RATTI et al., 2006) and sensor data (BURKEY et al., 2006) are increasingly used to obtain different kinds of 'geovisualizations' (GIACCARDI & FOGGLI, 2008). This shift sees users not only choosing how the landscape is represented but increasingly responsible for the data gathering that become part of the ecology (TURNER et al., 2001) of the landscape information.

Instead of being passive viewers of the represented landscape, they become instrumental in how the changing landscape is interpreted. This acknowledges an anthropological approach to landscape, such as that outlined by Ingold, who claims that 'it is a process that is not about 'representing' or capturing what is observed but … coupling the movement of the observer's attention (haptic, aural etc.) with currents of environmental activity' (INGOLD, 2000, 220). The landscape is experienced is at the point in which it is interpreted or translated between the environment and the person (or group of people). In this way the representation or visualization is no longer a way of objectifying or abstracting the landscape's characteristics. Instead the experience of the landscape takes place at the point of transition – recognizing the intricate relationship between humans and the natural landscape as a form of ecology. This acknowledges and gives space for an 'affective' and ecological perspective on landscapes 'that can reveal how we are "affected" by environmental settings, and in turn "affect" the way in which we experience and interpret' the environment (GIACCARDI & FOGGLI, 2008, 174).

In the workshop we take the approach of looking at how the landscape 'affects' and is 'affected' by us. We will explore how multiple affective inputs start to combine together to form a living data ecology, realized in a form of ad-hoc patchwork of inputs and outputs. We will work with a broad range of live user-centred data formats such as GPS traces, social media, location-based images and sensor data. The importance of this patching technique is not in the original data format, but in what happens at the point when it starts to operate as a data ecology and how it becomes 'patched' into a changing landscape.

Part one of the workshop participants will be offered a series of 'input' technologies that will plug data into the switchboard. This may include GPS data gathered from smart phones, photographs and text developed in response to exploring the local area, and the use of environmental sensors that are able to stream data to the switchboard via the internet. The second part of the workshop will allow participants to explore how to 'output' the data by plugging it into different visualization, representation and modeling technologies. This may include use of the Unity 3D game engine, rapid prototyping technologies, storytelling forms and network maps.

We will employ a patching system to switch different inputs and outputs across participants, so that they not only work with data they themselves are creating but also responding to or being 'affected' by that being created by other participants. Through the process of transposition, it is hoped that participants will benefit from the rich potentials that digital media offers into transforming from one meaningful form into another. (and an understanding of how they operate within a dynamic ad-hoc patching of environmental representtations).

2 Input Technologies

2.1 Comob (www.comob.org.uk)

Comob is a method of social and spatial mapping. This free software for the iPhone allows groups of people to see each other's movements represented on screen as circular nodes with lines linking their individual positions (Fig. 1). This data is also sent live to visualisation software that allows observers to see their movement at a distance. Previous projects have mapped and tracked individuals, however Comob proposes that those individual tracks are only part of how we move through space. Use of public space is a social activity, one that we do in relation to other people. Comob allows for observation of how movement through space is a social activity, and proposes that those movements can be used to map relationships to space.

Comob workshops to date (Edinburgh, Manchester, Belfast and Dundee) have used open-ended themes such as pollution, fear and community as a subject to map within a city. These open ended terms are used highly contested when identified within actual landscapes, in other words, what is a beautiful part of the landscape to one person, may be pollution to another. During workshop sessions, groups of participants are introduced to the application and briefed with the task of working as a small team to identify areas that correspond with the theme in the local area and demarcate them by forming a Comob 'shape' around the area.

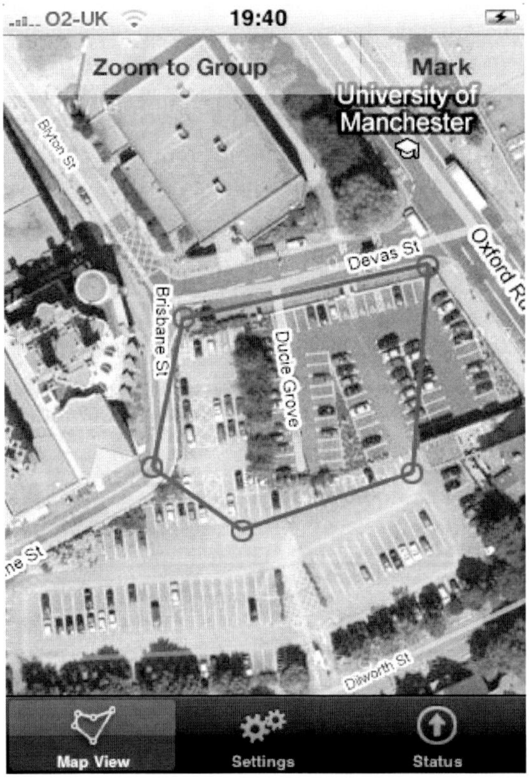

Fig. 1:
Screen Shot from Comob Net iPhone application allowing groups of people to see each others movements and link their individual positions.
Developed by J. Ehnes, H. Ekeus for C. Lowry, W. Mackaness, J. Southern, C. Speed & M. Wright. © 2009

Whilst reflecting upon their experiences of using Comob, participants of workshops have described a numbers of experiences that indicate that the software offers particular insights into interpreting the landscape:

1. Using Comob meant that initial assumptions about an abstract concept such as pollution or fear were immediately brought into question. Litter (for example) might

seem like a straightforward category to pollution but when a group went out to map it they quickly realised that their assumptions were both confirmed and challenged on the ground. Instead of discussing this at a later date, the software encourages discussion and reflection in action and in situ.

2. Having agreed upon an abstract or non visual subject, Comob encourages people to reflect on their individual perceptions of that subject by asking them to decide where it stops and starts by mapping it with their bodies. By seeing how other people were making these decisions, participants questioned their own decisions about abstract ideas.

3. Comob showed a potential for use in the co-ordination of strategic spatial action. Each participant was able to see the rest of the group and co-ordinate their movements. In offering the overview from within an embedded view group movements can be co-ordinated in new ways.

2.2 Tales of Things (www.talesofthings.com)

Tales of Things explores the relationship of personal memories and real world objects (things) and provides, enabled by tagging technologies, a context for sharing of personal and social memories through digital media. Things that are tagged with QR Codes or RFID tags as part of the Tales of Things service become tokens for the access and inscription of memories when they are brought in contact with Tales of Things clients. Tales of Things has been developed to support our research in the ESPRC funded TOTeM (Tales of Things and electronic Memory) project that cordially runs between five different universities across the UK (Edinburgh College of Art, Brunel University, University College London, The University of Dundee, The University of Salford.

The technical architecture of Tales of Things consists of a web application that provides enabling backend services and different clients (browser, mobile phones, RFID readers) that access this service via different API's (Application Programming Interfaces). People that register for a free account on the project website can add new objects to a user-generated object database via a web browser interface or in situ using a client on their mobile phone. During this process people are asked to provide (optional) meta-information (e.g. name, keywords, location) and a story (tale) about the object (thing). Tales can be told using text and any additional media that can be referenced via a URL (Unified Resource Locator). The system is capable of analyzing provided URLs and rendering media from services such as YouTube, Flickr and Audioboo in an integrated media player interface.

When a new object (thing) is created the service creates a unique two-dimensional barcode (QR Code) for the object that can be printed out and attached to the object. People are also able to link the objects using RFID tags. The web interface provides additional functions such as a commenting system, display of the location of things and tales on a map, search, creation of groups, user profiles, email and Twitter notifications. Downloadable mobile clients that can read Tales Of Things QR Codes provide additional functions such as a specific format to present the tales and an interface to add new tales when a barcode has been scanned, these are available for the iPhone and Android platform. Other non-project specific QR Code readers can also be used with our tags and will redirect people to the public URL of the object.

The Tales of Things website also offers members the ability to generate blank QR codes that may be printed out and ready to be associated with material. Technically pre-assigned to an instance in the database, the blank QR codes can be scanned by a mobile client and 'filled' with content directly from the smart phone. The ability to add a photo, story and keywords means that it is very convenient for users to carry blank codes and attach them to objects as they find them in situ and not have to log into a desktop computer.

2.3 Eco-OS (http://www.eco-os.org/)

Like a matryoshka doll i-DAT's Operating Systems recursively collapse in on themselves. somewhere the body sits (Bio-OS.org – a body that is neither ill or super fit, simultaneously an individual and a crowd) located in a physical architecture (Arch-OS.org – an operating system for contemporary architecture/software for buildings) framed by a Social network (S-OS.org – where hapiness lies somewhere at the end of a bell curve and true love can be found in a slice of pie chart) and all sitting snugly within a complex ecology (Eco-OS.org – harvesting data from the environment, bringing the landscape a little closer). Through ECO-OS, an ecological Operating System, the manifestation of human and environmental interactions are literally placed in the broader landscape. Eco-OS further develops the networked sensor model of these Operating Systems through the manufacture and distribution of networked environmental sensor devices. Location aware data harvested from across a landscape is transmitted to the Eco-OS server for processing.

Fig. 2: An ecoid in a tree is worth... developed by and deployed in the Confluence Project developed the North Devon Biosphere, Beaford Arts and i-DAT 2011

Ecoids: are sensor devices (small pods) that can be distributed through an environment (work place, domestic, urban or rural). Based on xbee's the sensors allow environmental data to be collected from the immediate vicinity. The sensors can be connected together through the formation of Wireless Sensor Networks (mesh and Star) that enable the coverage of an extensive territory. Each ecoid has a unique id and its location within a network can be triangulated giving its exact location. Consequently locative content can be tailored to a specific geographical area. Ecoids can also be used to produce content be receiving instructions from Eco-OS. Distributed performance can then be orchestrated across a large territory through light displays or acoustic renditions. Environmental data from light, temperature, humidity, movement, turbidity and flow, etc can be harvested. Calibration and an agreed/aggregated meaning can then be negotiated.

2.4 WiMo: The Technology Probe

WiMo is a smart phone application that enables a way to capture and understand people's emotional response to places. Unlike direct observation such as is common in usability studies, this field study tool allows participants to self-report their response and actions over a period of time in a fairly naturalistic manner.

Fig. 3: Process of defining physical extent of a 'place' (left image: 194sqm, right image: 581 sqm)

The interface is based on a series of three input stages; firstly, the emotional matrix, secondly the definition of the place and finally a text entry screen for a description or note. Once the user has created an emotional tag there are two options for viewing the overview map.

To create an emotional geo-tag at a location the user is first promoted to define the quality of the emotion in the matrix. They tap the screen somewhere within the range of the four axis matrix to select the appropriate emotion. Following this the tag is displayed as a cloud icon overlaid on the real- time position on the map, which the user can customize by defining the approximate area of the cloud so that it matches with the physical extent of the place (Fig 2).

The user then has the option to enter a text description of their response. Once the user submits this description they enter the screen with other user's emotional tags overlaid on the map. They can then choose to view the tags either as pins or as coloured tags on a map background. The colour represents the type of emotion chosen by the user and the extent the physical are of the location they are describing. The user can click on a pin to view the text description associated with the tag as well as the name of the user that created the tag (Fig. 4).

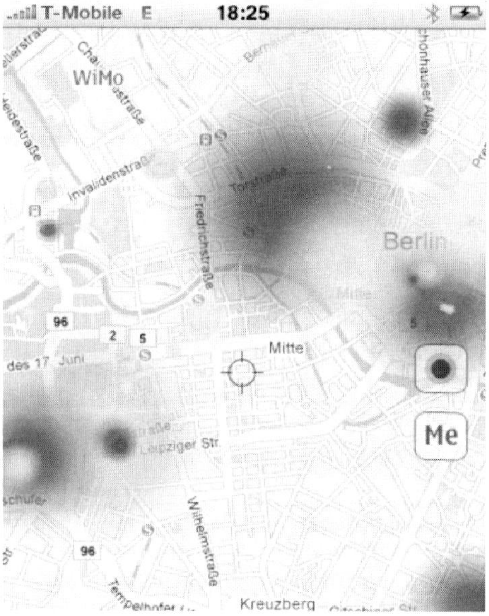

Fig. 4:
Range of emo-tags in Berlin Key:
Yellow = like/comfortable;
Light blue= Like/uncomfortable;
Purple = Dislike/comfortable;
Red = Dislike/uncomfortable

WiMo has been used with participants in a series of case studies. The findings of these studies show that although expressing emotions is a social activity some people also want to record their emotions in a diary format that would remain private. Generally however the opportunity to view a series of feelings overlaid on the space of the city caused people to reflect on the way they viewed the places of their everyday life and to recognize their affective value. Thus we proposed that not only the expression of feelings but the sharing of these emotional response can support a positive sense of community.

3 Output Technologies

3.1 DataAppeal (www.dataappeal.com)

DataAppeal Inc is a data analytics and visualization company interested in helping organizations increase the use and sharing of information to improve high-quality decision making. The DataAppeal™ platform visually represents geo-referenced raw data into spatial-designed maps, providing an instant and clear understanding of the information. The visual representation of the data allows users to quickly analyze the information and start to draw conclusions from previously hidden trends, areas of interest, and through quick comparisons of results. DataAppeal^{TM}, is a web-based application that renders raw location-based data (such as that in spreadsheets) as 3D or animated maps on the Google Earth platform, and the user can actually immerse themselves into their data-map and walk through their data on street view or drill down to the source.

No GIS training is required to use our tool. The user can simply upload a spreadsheet (csv, excel, and shapefiles) and design their data through a palette of visual options. It's quite intuitive to use. Users typically upload all types of data from various industries and sources. For example, environmentally based data is a popular type of data; we have groups that have uploaded and visualized CO_2 levels, smog, and pollution readings. Data regarding primary site-analysis values is another type. We have had landscape architects and students upload assessment value through a matrix data spreadsheet to visualize the site-analysis values along street corridors, river banks, as well as city districts and development sites. They have used our platform to visualize a matrix or scoring value for a given section on the site as part of their SWOT (Strengths, Weaknesses, Opportunities, and Threats) analysis. Some of the data that has been visualized includes: green development infrastructure growth; number of people using a space at a specific time; tree coverage; nightlife activity; cultural venues count; façade beautification, just to name a few.

Fig. 5: Screen Shot from DataAppeal showing CO_2 readings in France. Developed by Dataappeal.com. © 2012

Typically, the data is collected by the users. Sometimes individuals upload data from open-source data providers such as NYC Open Data, Wellbeing Toronto, and other city-based open-data websites.

DataAppeal™ also hosts a community-data-sharing platform, called the "Data Gallery," which provides a data market on our application for registered users. Members who have uploaded and rendered their maps have the option of posting their data publicly on our platform. Other users are then able to browse, view, and access it to enrich their own maps. Landscapes architects and urban planners can use the created data-map as a presentation aid to show their clients areas of concern or interest. From here, these maps can be used as an instrumental device to help make better design decisions.

3.2 Game Engine (http://www.op-sy.com/)

Was once the closest the inside got to the outside was through the creeping of floral patterns onto the living room wallpaper. Penetrating the membrane of the conservatory wall the landscape data feeds now seep into a different kind of fantasy world. The domestic consumption and immersive qualities of the game engine enable a different engagement with the landscape out there in here. This is not a 'Virtual World' constructed on the screen, but something more akin to an environmental dashboard or an Albertian window, but with a different kind of perspective. Something like standing in the rain looking at your mobile phones weather forecast, it is raining, you are getting wet, but somehow the digital representation doesn't feel like that.

Visualising and sonifying the data harvested from the landscape is an essential component of i-DAT's Operating Systems. Normally the preoccupation is with FullDome immersive environments (Dome-OS), as a transdisciplinary instrument for the manifestation of material, immaterial and imaginary worlds. A credible tool for the rendering of these interactive real-time visualisations is the game engine, in particular Blender and Unity 3D.

Fig. 6: Unity 3D environmental data feeds. I-DAT 2012

In this workshop the weapon of choice is Unity 3D mainly because, although not open source, it offers a rapid production pathway. For the FullDome environment the use of a

FullDome/Fisheye library is required but here the flat screen representation will be used. Feeds from the Ecoids will be read into the game engine template allowing simple interactions and visualisations. Alternative feeds (xml formatted) can also be incorporated.

Fig. 7: Blender Atomic Force Microscope Data Landscape, i-DAT 2012

3.3 3D Printing and Augmented Reality

Technology to create a physical representation of a digital input is increasingly moving into general application. Systems such as the Maker Bot Replicator 2 allow rapid prototyping of objects from industrial design through to household implements. The workshop expands this by capturing the moments in time and space via the inputs for 3D landscapes and augmented reality data visualisation.

The Replicator 2 is arguably the first 3D printer aimed at the non-professional / technology inclined user. The system is able to print almost any shape by extruding layers of heated plastic in an additive process. Creating multiple layers allows a 3D model to be created through the sequential addition of print material, in our case polyactic plastic (PLA). Data collected via the workshop will be transformed firstly into a digital landscape with visualisations via various data engines; the challenge is to move the data collected in the real world back into a physical form while maintaining a level of representation.

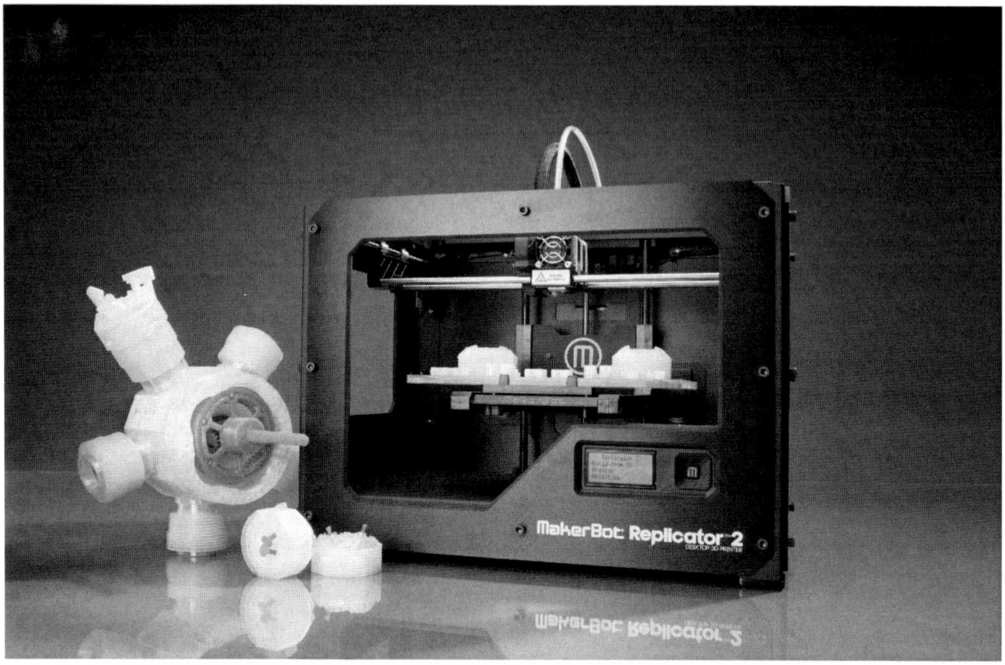

Fig. 8: MakerBot Replicator 2 Desktop 3D Printer

Via use of a 3D mesh of the landscape and links back to Unity via the Augmented Reality sdk of Vuforia (https://developer.vuforia.com/resources/sdk/unity) the workshop aims to transform a 3D printed data mesh into an augmented reality view of the event. As a means of recursive representation, the printed landscape represents output from Comob, Tales of Things and Eco-OS. Integrating an agent based modeling approach along with game engine technology, a reel of polyactic plastic and augmented reality offer the possibility to creating a true digital landscape in physical form – a PLA patch scape.

4 Conclusions and Futures

The Patch-scape toolkit of Input and Output technologies offers a system across which data derived from a landscape is parsed and manifested as new material landscapes. Any distinction between quantitative and qualitative collapses, as hard data is transformed in to soft lines, and fuzzy data create solid forms. In reflecting upon the role of information visualisation Lima's observation hold's true: *"Form doesn't follow data. Data is incongruent by nature. Form follows a purpose, and in the case of Information Visualisation, Form follows Revelation"* (LIMA 2009). The possibilities for reconstructing landscapes according to what we would like to reveal are becoming richer. As data sets are broken down to common parts and new technologies are developed to interpret and recast them into different media, we may better understand how other people interpret environments.

"I believe that the sheer scale and sophistication of what is happening now amount to something quite different: a studied extension of the spatial practices of the human which consists of the production of quite new material surfaces which are akin to life, not objects, and thereby new means of bodying forth: new forms of material intelligence producing a new, more fluid transubstantiation." (THRIFT 2004)

References

Burke, J., Estrin, D., Hansen, M., Parker, A., Ramanathan, N., Reddy, S. & Srivastava, M. B. (2006), *Participatory Sensing.* Paper presented at ACM Sensys 2006, Boulder, Colorado.

Giaccardi, E. & Fogli, D. (2008), *Affective Geographies: Toward Richer Cartographic Semantics for the Geospatial Web.* International Conference on Advanced Visual Interfaces (AVI 2008). New York, ACM Press, 2008, 173-180.

Harvey, G., Mount, N., Aplin, P. & Priestnall, G. (2008), *Introduction to Representing, Modeling, And Visualizing The Natural Environment.* In: Priestnall, P., Harvey, G., Mount, N. & Aplin, P. (Eds.), Representing, Modeling, and Visualizing the Natural Environment. Taylor and Francis.

Ingold, T. (2000), *The perception of the environment.* Routledge.

Latour, B. (2005), *Reassembling the Social: An Introduction to Actor-Network-Theory.* Oxford, Oxford University Press.

Leclerc, Y. G., Reddy, M., Iverson, L., & Eriksen, M. (2001), *The GeoWeb – A New Paradigm for Finding Data on the Web.* Proc. ICC2001, Beijing, China, August 2001.

Lima, A. (2009), *Weblink: Plotting a critical path. Reflections upon Lima lecture by William Owen.* http://madebymany.com/blog/plotting-a-critical-path.

MacEachren, A. M. & Kraak, M. J. (2001), *Research challenges in geovisualization.* Cartography and Geography Information Science.

Longley, P., Goodchild, M., Maguire, D. & Rhind, D. (2006), *Geographic Information Systems and Science.* John Wiley and Sons.

Ratti C., Pulselli, R. M., Williams, S. & Frenchman, D. (2006), *Mobile Landscapes: Using Location Data from Cell-Phones for Urban Analysis* In: Environment and Planning B: Planning and Design, 33 (5), 727-748.

Thrift, N. (2004), *Driving in the City.* Theory, Culture & Society, 21 (4-5), 41-59.

Turner, M. G., Gardner, R. H. & O'Neill, R. V. (2001), *Landscape Ecology in Theory and Practice.* New York, NY, USA, Springer-Verlag.

3D Visualization of Plants

Concept for
Digital Landscape Architecture Workshop, June 2013 during DLA Conference 2013

Invited contribution by
Philip PAAR and Jan Walter SCHLIEP

Objectives of Workshop: The workshop on 3D Plants is focusing on concepts, workflows, tips & tricks and, last but not least, on hands-on demonstrations by the creators of Laubwerk Plants Kits. Jan Walter will demo how he combines procedural plant modeling techniques with handcrafted details. Jan Walter Schliep and Philip Paar will present some of their next generation of 3D plant modeling and editing tools. Furthermore, we want to discuss with the participants about limits and requirements of representing vegetation in 3D visualizations. Are there any specific needs or "mental state" of landscape architects regarding 3D plants, which would demand specific approaches and "looks", e.g. other than in CG architectural visualization?

Target group of workshop: Landscape Architects, students, researchers and practitioners interested in 3D modeling and visualization of gardens, landscapes, and architecture.

Background of workshop: Literally everybody dealing with the computer graphical representation of outdoor scenes – pen and ink style or photo-realistic – visualization or computer games, has to deal with the fact that Mother Nature, especially the flora is very complex. And while users are facing this for decades now, dealing with virtual foliage in 3D applications is still complicated, time-consuming, often painfully slow and usually the final plants still don't look computer generated. Especially in digital landscape architecture practice, photomontage still seems to be the first choice when it comes to the representation of vegetation. Professional CG artists have been forced to chose between static plants that lack versatility, or overly-complicated modeling tools with hundreds of parameters that drastically slow down the creative process.

We present a new approach for landscape architects and CG artists looking for a fluent way to insert authentic, highly-detailed 3D plants into their 3D scenes. With drag-and-drop simplicity, and easy-to-use tools that modify the shape, age, season and level of detail.

Figures: Example of 3D plants in CG landscape architecture visualization.

Variations of different tree species, shapes, and seasons. Landscape design of the future campus of the Technical University of Chemnitz and 3D modeling by landscape architect Thomas Jarosch (haefner-jimenez.de). Project taken from the urban design ideas and realization competition for the conversion and extension of the heritage building "Alte Aktienspinnerei" into the Central Library (rendering by Jan Walter Schliep using VRAYforC4D, April 2013)

Kanzan cherry (*Prunus serrulata 'Kanzan'*) in spring

Globe robinia (*Robinia pseudoacacia 'Umbraculifera'*) in summer

Globe robinia (*Robinia pseudoacacia 'Umbraculifera'*) in fall

Red maple (*Acer rubrum*) in fall

Enduring GeoKnowledge

Concept for a workshop at the Digital Landscape Architecture DLA Conference 2013

Invited contribution by
Manfred MITTLBÖCK and Thomas HEISTRACHER

Constituting a novel approach, "geoKnowledge design" principles, concepts and methods aim at leveraging service-oriented geo-technologies and tools for creating and sharing 'live' (environmental) monitoring data and information pools as formalized knowledge pieces tailored to end users' needs.

"geoKnowledge design" workflow strategy covers the novel trend, opening and sharing of any location-aware content, information or data of public or private origin. Examples will show and explain the suitability of open government data (OGD) and INSPIRE resources. They show how make use of environmental 'live' sensing knowledge (e.g. as Situational-aware content for mobile phone apps) leveraging SOA-based geo-information exchange data and the related processing.

The workflows will be based upon the following iterative design pattern building blocks: **Structure, Integration, Sketch, Analysis, Template, Task & Automation, Presentation and cognitively Re-Assembly.**

Goal Based on the prepared building blocks of "GeoDesign" and, "OpenKnowledge" the participants will learn how to utilize public available resources like INSPIRE and Open-Government data. A particular focus is to fitting time-enabled spatial information to pre-defined end-users' needs. This new concept ultimately allows to fostering time-dependent models and related dynamic validation for the design of data structures, workflows and processes leveraging open information repositories on a technical and organizational level.

3D Geo-Design Mapping Using DataAppeal

Concept for a workshop at the Digital Landscape Architecture DLA Conference 2013

Invited contribution by
Nadia AMOROSO and Haim SECHTER

Abstract

Maps provide layers of geographical data that direct one through the city. They guide the user in space and reveal specific parameters of the city. The "Geo-Design Mapping Workshop" offers an engaging opportunity for participants to transform environmental data derived from site and design it into visually engaging 3D and animated maps using a new simplified GIS web-based application called DataAppeal. Using DataAppeal as the digital platform, participants will take CSV or excel data-files with geo-coordinates and numerical values, and create spatial maps designed with different "presentational looks". This offers a playful opportunity to design the visual display of your data while revealing further insights into your information.

Introduction: Research Behind the Application

Manifesting the Invisibles of the City into New Form

The concept behind the DataAppeal application is grounded on research pertaining to new depictions of the city blended by art, design and data. Imagine a city invisible to the human eye and only manifested by its non-visual urban phenomena. What shape will it take? With the new urban forms represented as images, would they become guides into the city? Revealing forces not visible to the naked eye, the research abstracts textual urban data and transforms them into landscape architectural visions. Examples of these forces in the urban context would include criminal activities, zoning by-laws, population densities, transportation patterns, public surveillance, cell phone usage, air quality readings, and other spatial statistics. The application draws from aspects taken in Nadia Amoroso's book, "The Exposed City: Mapping the Urban Invisibles".

How can the abstract forces shaping urban life be rendered artistically and informatively in the form of alternative 'maps' which represent urban dynamics, not usually accessible to routine professional expertise (such as the urban designer) or the ordinary urban dweller; and in what ways do the new visualizations provide deeper insights into the city than conventional maps are capable of revealing? The research described in "The Exposed City" proposes to test the visual and spatial qualities of new urban forms through the creation of multi-dimensional maps. The research defines and redefines the term map throughout the twentieth century, experimenting with a hybrid of map and image to create alternative visual representations of urban conditions. Using new and alternative modes of graphic display, these maps express otherwise unseen urban data.

Also considered in this research is the capacity of an artistically rendered image (or map) to influence and capture the audience. An image of the city that has artistic merits attracts its viewers with its seductive qualities and is likely to inspire immediate reactions – which can be said to bridge the gap between data, visual art and urban design.

Key Figures of Inspiration: Hugh Ferriss, Edward Tufte, Richard Saul Wurman, MVRDV, James Corner …

In the research, key individuals from the early to late twentieth century whose works are analysed as precedents for the study of maps as both informative and artistic productions are examined. This included the graphical interpretations of the 1916 zoning Ordinance of New York City drawings of Hugh Ferriss, interpreting his sketches a theoretical framework for analyzing the representation of the city through powerful visual documents. Ferriss's drawings are examined to show the ways in which this artist revealed the unknown constraints of the 1916 Ordinance of New York. These drawings also referred to as the 'Evolution of the Set-Back Building', can be considered three-dimensional maps, offering guidance in which the permissible building form as be constructed. The drawings of Ferriss contributed to an all-encompassing awareness of the effects of zoning and its architectural and urban design implications. Presenting Ferriss's maps as catalysts for the innovations which characterize the architectural movements of the early 1920s may seem an exaggerated proposition. However, the fact that his drawings consolidated and provided a clear definition of the major legal and economic shifts within the milieu of early twentieth century American culture cannot be denied. His images spoke to all sets of audiences simply by their familiar architectural form and style.

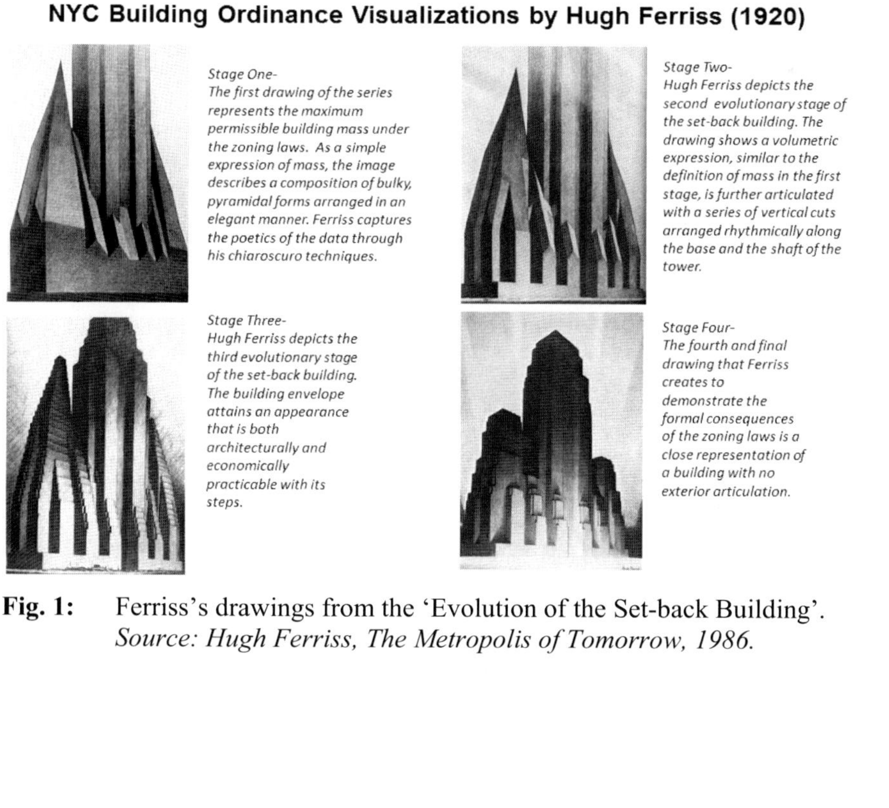

Fig. 1: Ferriss's drawings from the 'Evolution of the Set-back Building'.
Source: Hugh Ferriss, The Metropolis of Tomorrow, 1986.

Other key figures that provided inspiration to the concept behind DataAppeal included the works of the geographer and theorist Kevin Lynch. Active in the 1960s and 70s, Lynch is best known for investigating how people conceived the city. The people made maps. His process is labelled as "cognitive mapping". In a more recent investigation of the potentials of representational media, contemporary landscape architect James Corner portrays mapping as a creative and inventive process and shows the potentials of maps to unfold "hidden" facts of the city. His seminal book with famed photographer Alex MacLean "*Taking Measures Across the American Landscape" (1996), Corner* showcases landscapes, agricultural lands, railway sites, mountainous terrains, wind farms and burning fields (among other subjects) from above while offering further "measures" on the site with his created supplementary *map-drawings* (a combination of map and drawing styles such as collage crafted by Corner as means to interpret the aerial photos, and thus, creating a *map-drawing* of the site).

Other key figures included the works of MVRDV and their celebration of the *datascape.* Similar to Ferriss's visual investigations of the 1916 Ordinance, *datascapes* serve as an architectural language, which turns statistical description into representations of the theoretical envelopes within which built forms can be placed:

"*Datascapes* are visual representations of all quantifiable forces, which can have an influence on the work of the architect or are even able to determine and to steer them."[1]

Another description of the concept of *datascape* include:

"*Datascapes*are visual representations of all the visual and non-visual influences that have an impact on or regulate the work of an architect."[2]

James Corner, former partner with Stan Allen, landscape architectural theorist, and Chair of the Department of Landscape Architecture at the University of Pennsylvania, describes *datascapes* as:

"... imagings constructive and suggestive of new spatial formations,...so 'objectively' constituted (from numbers, quantities, facts, and pure data) that they have great persuasive force in the bureaucratic and management aspects of contemporary city design. They differ from the quantitative maps of conventional planning in that they image data in knowingly selective ways. They are designed not only to reveal the spatial effects of various shaping (e.g., regulatory, zoning, legal, economic, and logistical rules and conditions), but also to construct a particular eidetic argument."[3]

Also, *datascapes* have been described as "...revisions of conventional analytical and quantitative maps and charts that reveal and construct the various shape-forms of forces and processes operating across a given site."[4]

---

1 STEELE (1998, p. 10).

2 Bart Lootmsa expresses his thoughts on the concept of *datascapes* in *Berlage Cahiers 7*, July 1997, p. 39.

3 James Corner, *Operational Eidetics: Forging New Landscapes.* In: Harvard Design Magazine, Fall, 1998, p. 25.

4 Ibid.

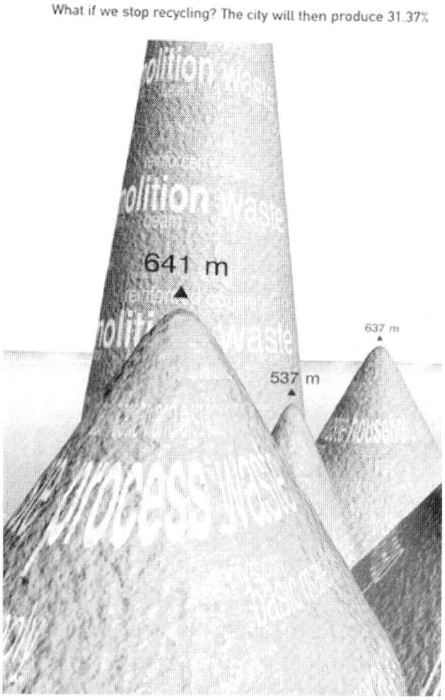

Fig. 2:
Rendering by MVRDV. *Image source: Winy Maas, Metacity/Datatown, 1999.*

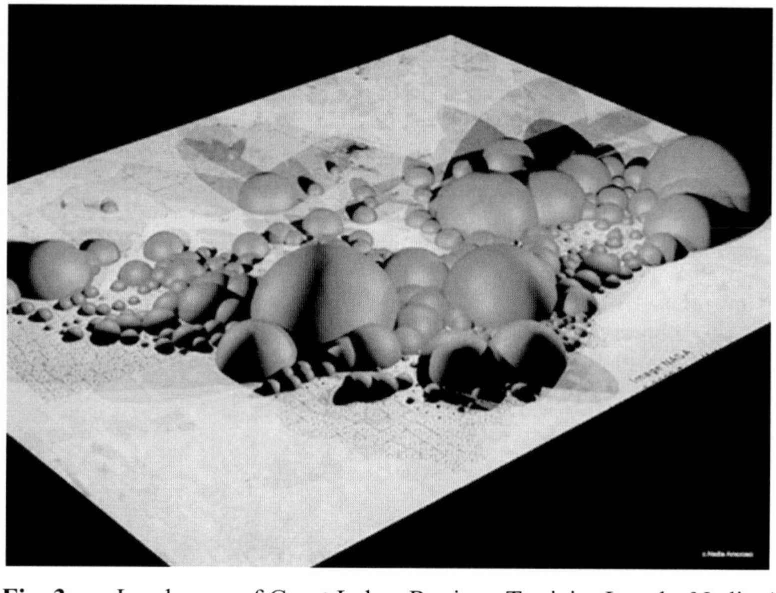

Fig. 3: Landscape of Great Lakes Regions Toxicity Levels, Nadia Amoroso 2010, using 3D Studio MAX, Photoshop

Upon examining these key individuals and their works, the attempt is to create a hybrid "map-landscapes" that visualize "hidden" forces (numerical data) into new 3d mapping styles. Map-landscapes fuses art, spatial form, site-based data to form a new kind of data-scape. In an attempt to inspire better built forms, new modes of representation are proposed.

DataAppeal Application

Grounded in this research-based design principles, we developed a web-based application the transforms data files into beautifully design 3D and animated maps situated on the Google Earth platform. DataAppeal is a simplified mapping and Geographic Information System platform, intended to visualize location-based information, transforming it into business insight, instantly.

Numerical data from simple data file (csv, xls) is proportional rendered into models referenced to a geo-coordinates. The larger the numerical value, the bigger the model is creation. This follows principles of graphical integrity as professed by acclaimed info-graphic guru, Edward Tufte. According to Tufte, graphical integrity is maintained when the visual representation of the data is consistent with the numerical representation.[5] The first principle is that: "The representation of numbers, as physically measured on the surface of the graphic itself, should be directly proportional to the numerical quantities represented."[6] The relative size of the symbol is critical. Size applies to all dimensions involved.

Collectively, all these models create a unified 3d data-map. If the user applies the colour gradient feature, low values are rendered with a light hue, while larger values are rendered with a darker hue. The same dataset category is rendered in the same colour, following design principles suggested by Information Architect, Richard Saul Wurman.

As part of his mapping principles, Wurman suggests, "When and why to use colour" are critical aspects in mapping information visually. He claims that many cartographers are incorrect in their use of colour. "Cartographers have had it wrong. They use different colours to represent the same category such as temperature change. In a weather map, for example, high temperature readings are usually coloured in red (which is fine). Higher temperatures should still use red but in richer intensities. Lower temperature readings should be rendered using the same colour but with a moderate to lighter intensity such as lighter shade of red. Often a blue is used, which is incorrect. Different colours should only express different categories."[7] In the image below, Wurman uses colour – intensification to highlight denser areas in New York City.

[5] TUFTE (2001, p. 55).

[6] Ibid. (p. 56).

[7] Discussion with Wurman on mapping strategies, principles and techniques on July 8, 2006.

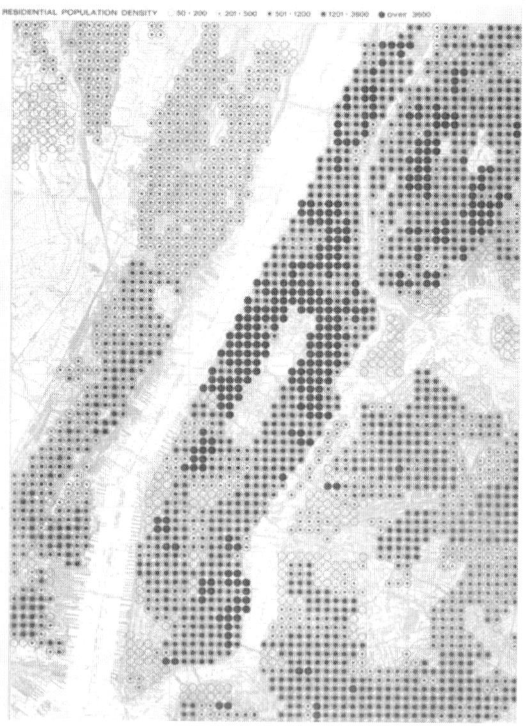

Fig. 4:
Wurman's map on Residential Population Density for New York from Passonneau and Wurman's Urban Atlas (1966) depicting various urban data sets for twenty American cities. *Source: Joseph R. Passonneau and Richard Saul Wurman. Urban Atlas: 20 American Cities; a Communication Study Noting Selected Urban Data at a Scale of 1: 48,000, 1966, section on New York City.*

Some of the main goals that underline the development of the DataAppeal application include:

- FAST-Visualize many rows of data in seconds for instant decision making
- SIMPLE-Provides an intuitive self-service application to analyze data. No previous training required No GIS training is required to use our tool. The user can simply upload a spreadsheet (csv, excel, and shapefiles) and design their data through a palette of visual options. It's quite intuitive to use.
- EMPOWERING-Places information at the fingertips of all departments.
- CREATIVE-Makes the use of information more palatable, ensuring high quality, evidence-based decision making.
- COMPATIBLE-Leverages Google Earth base map to ensure familiarity with tool. Accepts file formats from existing enterprise applications.

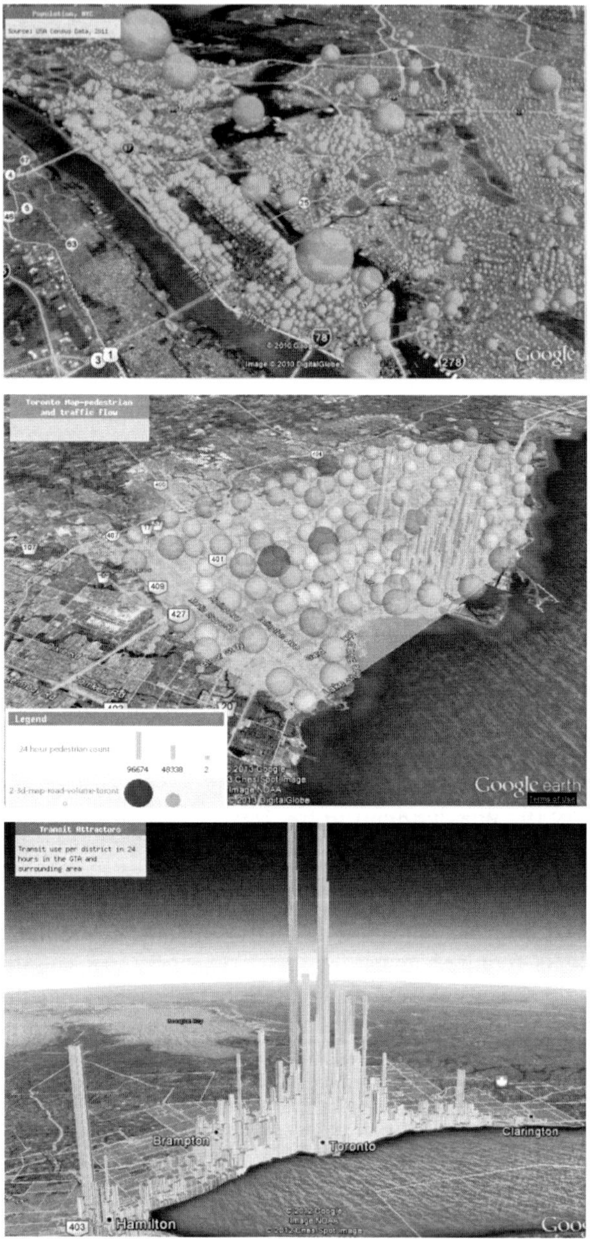

Fig. 5-7: Various data-maps generated and designed using DataAppeal (NYC population, Transit attract-tor in Toronto and surrounding area, 24 pedestrian count and road volume, Toronto)

How can landscape architects and urban planners use DataAppeal?

DataAppeal provides a simplied GIS platform, therefore landscape architects without any GIS training can use the application to get visually engaing site analysis, that can be used to retrieve further insights on the site and can also be used for visual communication purposes. Also, landscape architects can take the raw data about the site typically found on government and natural resource websites, and then upload this data in our tool. Thousands of rows of information will be rendered into a landscape revealing hidden trends in the city. Landscapes architects and urban planners can use the created data-map, as an alternative presentational and research device. These design maps can be used as an instrumental device to help make better design decisions. For example, a group of landscape architecture graduates used DataAppeal to pin point and visualize contaminated site along the Miami River corridor in Miami Florida. They also used tool to highlight abandoned areas along this space. Other elements visualized and analysed including market value and demographics. Using the data map layering features, the landscape architects combined various datasets to "see" and examine brownfield space, parks and urban efforts on natural sites to determine appropriate future development sites. The data-maps helped the designers discover new findings of their location and to draw linkages between various datasets, when maps were overlaid.

Along Washington Ave, in Miami, one group of landscape architects visualized the number of homeless occupying various spots along this corridor. The design maps revealed a high homeless presence near 20th to Dade Blvd and presence of homeless along 13th to Lincoln. Another group, for example, created maps reflecting property and housing values along the strip, while others examined data relating to foot-traffic at various times of the day. Higher volumes were present near popular restaurants, cafes and cultural venues; while at night, key bar and clubs had larger spikes in the model representation. Other 3D mapping representations included façade beautification values and transparency levels in the buildings along Washington Ave. Also, façade beautification efforts seemed to have a direct relationship with housing prices, but not with pedestrian traffic. Others included noise levels with night time activities.

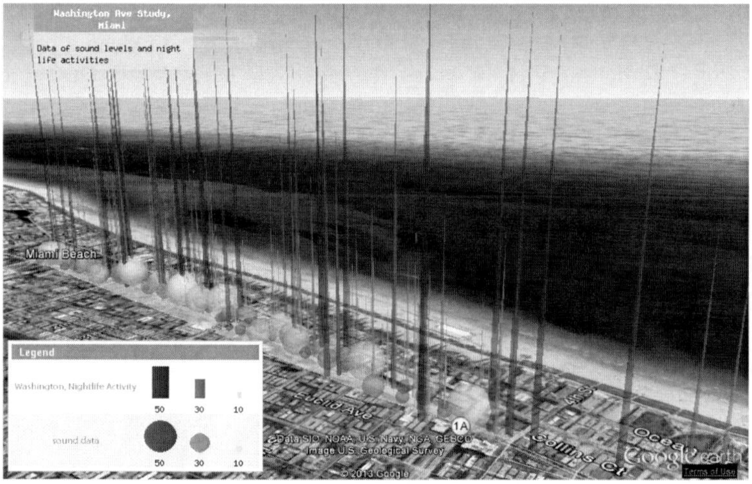

Fig. 8: Washington Ave, Miami Visual Study – Night time activity and sound levels, data-map designed using DataAppeal Application

Landscape architects have also used the tool to visualize municipality's Well-being data. For example, a group of planners and landscape architects obtain typical textual city data on the diabetes rate and green space. Normally, this type of data would be presented in spread sheets, offering no visual impact to the audience.

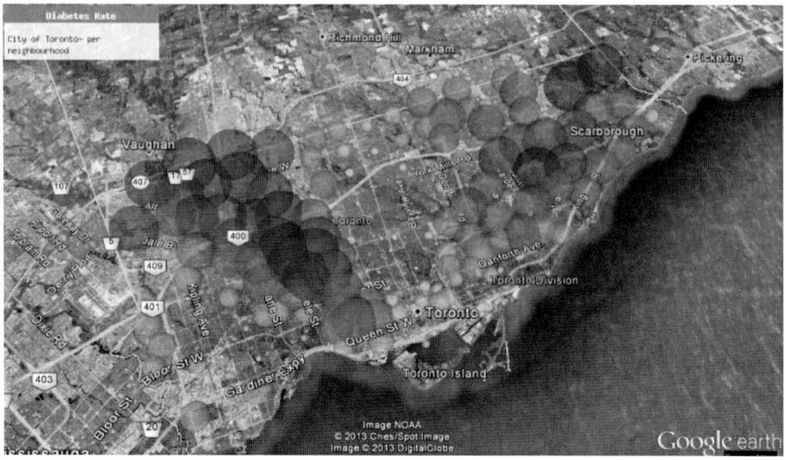

Fig. 9: Diabetes rates data-map, Toronto, designed using DataAppeal application

The landscape architect upload these datasets into the DataAppeal application and quickly design the look of this information. Higher diabetes rates can be quickly revealed throughout the city, the application of colour gradient quickly revealed areas of higher diabetes rates on the periphery of the city. The landscape architect then overlaid this data with cultural venues and green spaces in the city to see if there is any relationship between the datasets and the location. As a visual. It offers a strong case to design or improve various location of the city based on the data representation. Landscape architects can use the images as communication material and as analytical devices.

A recent post regarding data on DataAppeal's Data Gallery pertaining to Hurricane Sandy visualized a representation of the number of trees damaged in New York City and surrounding area. The data-map provided a powerful visual story of the storm's damage. Landscape architects could use this information to determine which areas deserve the most attention or replanting.

Fig. 10: Inside DataAppeal Application showing design control list panel on the right

The DataAppeal Data Gallery is on online store and data sharing platform, in which users can browse and access data and data-map and use them for their own work. As more data becomes available and accessible, people need a way to make sense of it. I think it's important to make the flow of data easier and more widely understood – to create an open-data culture with an ever-growing number of participants.

References

Excerpts of this paper are taken from Nadia Amoroso's PhD thesis and *from The Exposed City: Mapping the Urban Invisibles* (Routledge, 2010).

Amoroso, N. (1998), *Operational Eidetics: Forging New Landscapes*. In: Harvard Design Magazine, Fall 1998, 22-26.

Amoroso, N. (1999), *Eidetic Operations and New Landscapes*. In: Recovering Landscape: Essays in Contemporary Landscape Architecture. Ed. by James Corner. New York, Princeton Architectural Press, 213-252.

Amoroso, N. (2003), *Invisible Urban Forms: Quantitative Data as Inspiration for Design*. In: Archis, 2 (5-6), 98-101.

Amoroso, N. (2005), *Redrawing the Map*. In: The Next American City, 8 (Spring 2005), 44-45.

Amoroso, N. (2010), *The Exposed City: Mapping the Urban Invisibles*. New York, Routledge.

Amoroso, N. & MacLean, a. (1996), *Taking Measures across the American Landscape*. New Haven, CT., Yale University Press.

Corner, J. (1996), *A Landscape at Work*. In: The Chronicle of Higher Education, 43, 15, December 6, B96.

Ferriss, H. (1922), *The New Architecture*. In: The New York Times, 19 Mar. 1922, 52-54.

Ferriss, H. (1922), *Civic Architecture of the Immediate Future*. In: Arts and Decoration, Nov. 1922, 12-13.

Ferriss, H. (1925), *Truth in Architectural Rendering*. In: The Journal of the American Institute of Architects, 13, March 1925, 99-101.

Ferriss, H. (1926), *The Skyscraper Climbs to a New Height*. In: The New York Times, 5 Sept. 1926, SM3.

Ferriss, H. (1929-1973), *Rendering, Architectural*. In: The Encyclopedia Britannica, 14th Edition (revised in 1961 and titled as "Architectural Rendering").

Ferriss, H. (1931), *How Hugh Ferriss Draws: Six Progressive Stages in Rendering a Proposed Building*. In: American Architect, July 1931, 30-33.

Ferriss, H. (1986), *The Metropolis of Tomorrow*. Princeton, NJ, Princeton Architectural Press.

Lynch, K. (1960), *The Image of the City*. Cambridge, MA, The MIT Press.

Lynch, K. (1982), *A Theory of Good City Form*. Cambridge, MA, The MIT Press.

Maas, W. & van Rijs, J. (1998), *Datascape2*. In: Process: Belage Cahier, *7*, 38-45.

Maas, W. (1999), *Metacity/Datatown*. Rotterdam, MVRDV/010 Publishers.

Schumacher, P. (1997), *The Dialectic of the Pragmatic and the Aesthetic: Remarks on the Aesthetic of Datascapes* (lecture at Architectural Association, London). http://www.patrikschumacher.com/aesthetics.htm.

Steele, B. (1998), *Reality Bytes: Datascapes*. In: Daidalos, 69/70, 10.

Tufte, E. R. (1990), *Envisioning Information.* Cheshire, CT, Graphics Press.
Tufte, E. R. (1997), *Visual Explanation: Images and Quantities, Evidence and Narrative.* Cheshire, CT, Graphics Press.
Tufte, E. R. (2001), *The Visual Display of Quantitative Information.* 2nd edition. Cheshire, CT, Graphics Press.

DLA Poster Submission 2013: "My 3D Landscape"

A part of the annual DLA Conference is the poster competition.
The poster theme of 2013 is: **"My 3D Landscape"**. The final deadline for the poster submission is 15 May 2013 (after the submission date of this publication).

We are pleased to show some selected poster contributions, which had been entered earlier by 1st of April 2013:

Title:	Studio Gardens for BUGA 2015 Havelberg
Author(s):	Hushaam Ali Abbasi (home country Pakistan), Wang Fei (home country China), Sun Ting (home country China), Sun Ping (home country China)
Organization:	Master of Landscape Architecture Program, Anhalt University
Country:	Germany

Title:	The Sphere Pavilion
Author:	Sanja Đurković
Organization:	Faculty of Art and Design, Megatrend University, Belgrade
Country:	Serbia

Title:	4Change / Seaside atrium "Palm tree"
Author(s):	Marija Antic (home country Serbia), Alessandro Vino (home country Italy), Dragana Romic (home country Serbia), Natalia Vergara (home country Columbia)
Organization:	Weihenstephan University of Applied Sciences
Country:	Germany

Title:	Studio: Multi Generation Private Park in Dessau-Roßlau (1), Germany
Author(s):	Adeeb Kanaan (home country Jordan), Song Huang Song (home country China), Ji Yi (home country China), Sadiq Ramazan (home country Iraq)
Organization:	Master of Landscape Architecture Program, Anhalt University
Country:	Germany

Title:	Green walls
Author:	Maria Bostenaru Dan
Organization:	"Ion Mincu" University of Architecture and Urbanism
Country:	Romania

Title:	Studio: Multi Generation Private Park in Dessau-Roßlau (2), Germany
Author(s):	Daeyong Kim (home country Korea), Baritoadi buldan rayanda rito (home country Indonesia), Karzan Saadoon (home country Iraq), Michael Ajayi (home country Nigeria)
Organization:	Master of Landscape Architecture Program, Anhalt University
Country:	Germany

Title:	eGovernmentAR – Bildbasierende Kommunikationsmodelle als Schlüssel zum barrierearmen partizipativen eBeteiligungsverfahren
Author(s):	Lisa Rockmann (Anhalt University), Simon Adler (IFF Frauenhofer Institut Magdeburg), Rene Krug (Anhalt University)
Organization:	Faculty of Agriculture, Ecotrophology and Landscape Development & IFF Frauenhofer Institut Magdeburg
Country:	Germany

Accepted poster entry outside the poster theme:

Title:	Concept and construction of a crowd sourcing portal to visualize geocoded messages of a social network based on the example of Twitter
Author(s):	Stefanie Männl, Mareike Wendtland und Martin Becker
Organization:	Master Geoinformatik, Anhalt University
Country:	Germany

Please find the result of the poster competition at www.digital-la.de.

Poster: Studio: Gardens for BUGA 2015 Havelberg

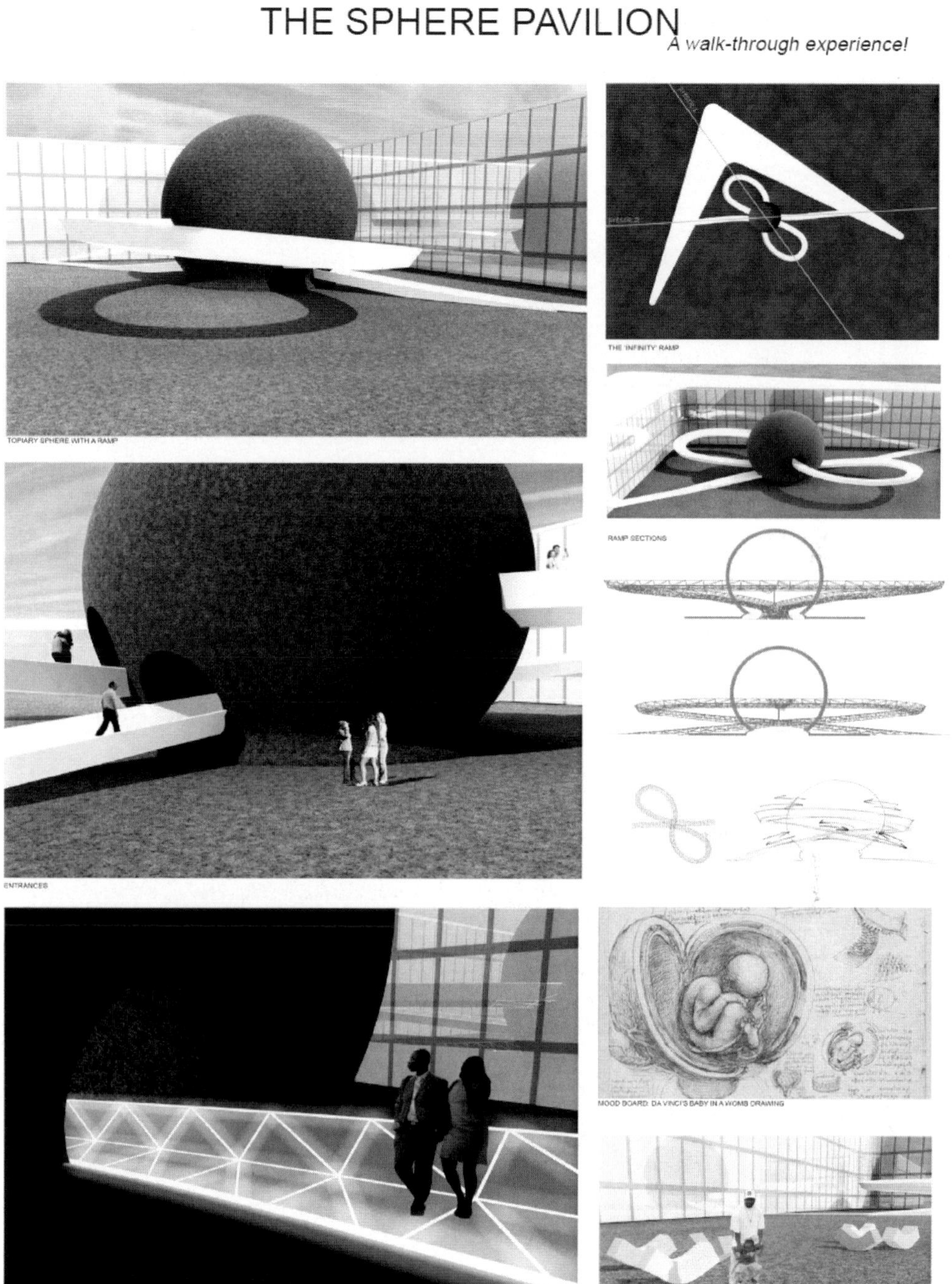

Poster: The Sphere Pavilion

Poster: Seaside atrium "Palm tree"

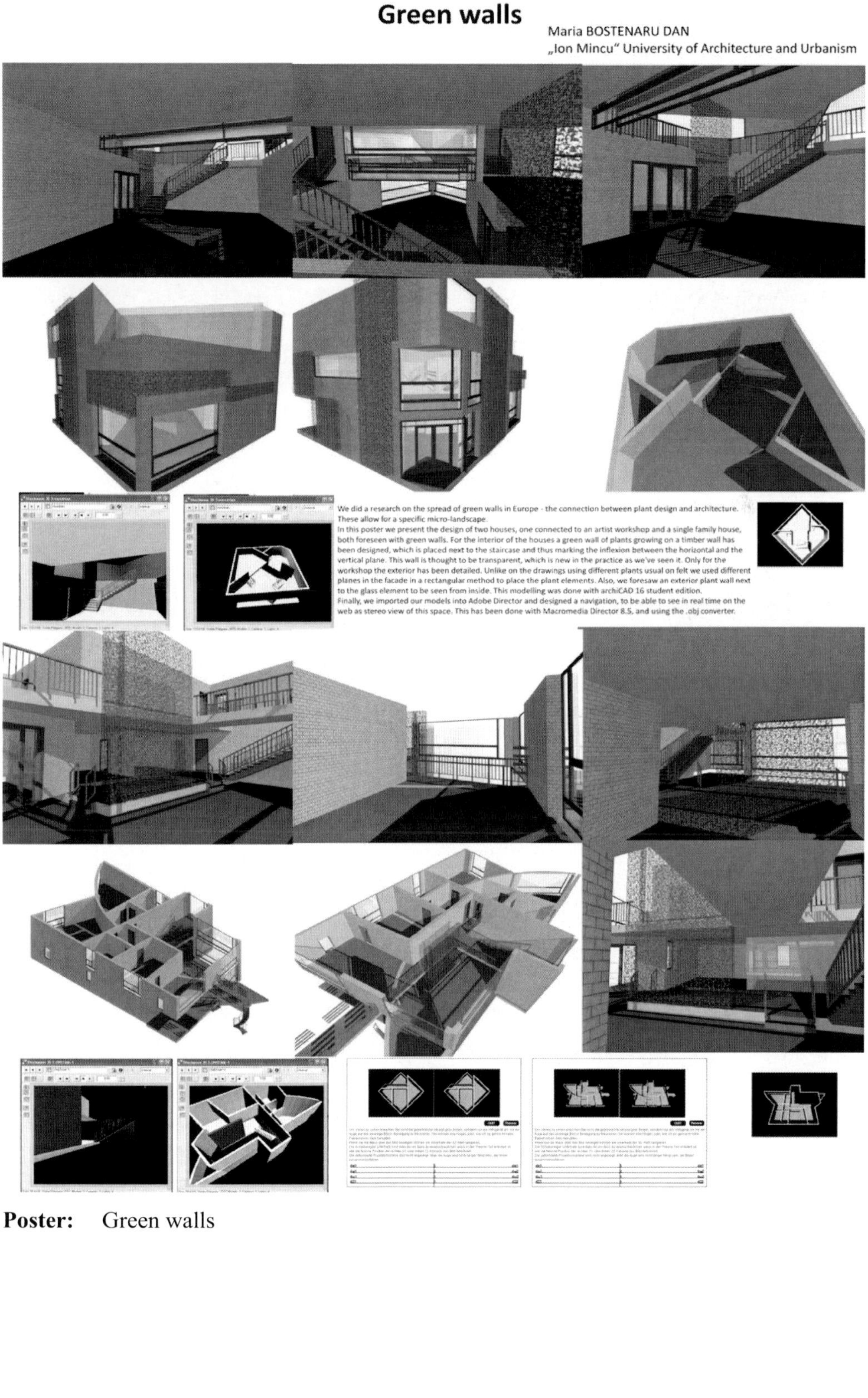

Poster: Green walls

Poster: Studio: Multi Generation Private Park in Dessau-Roßlau (1)

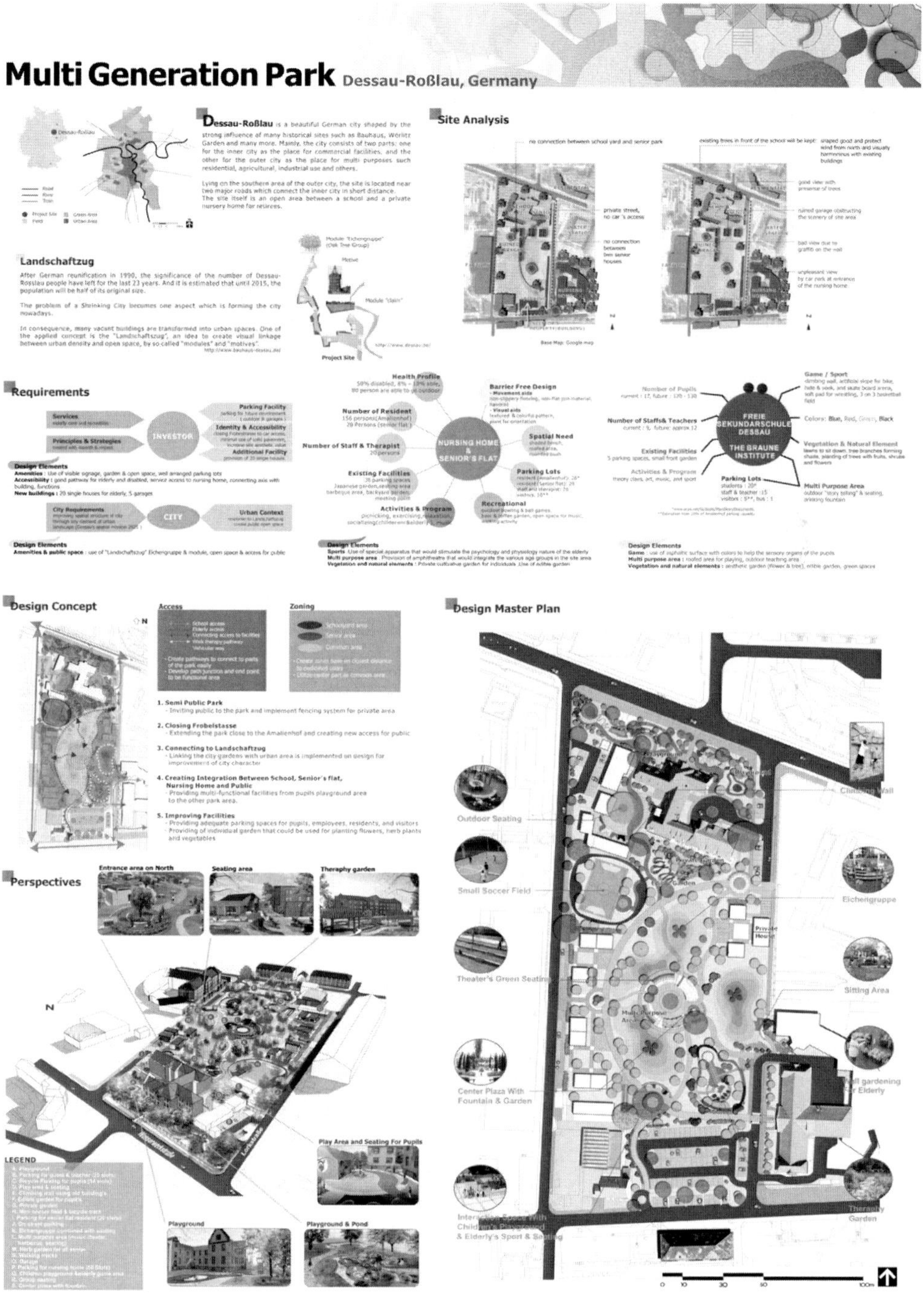

Poster: Studio: Multi Generation Private Park in Dessau-Roßlau (2)

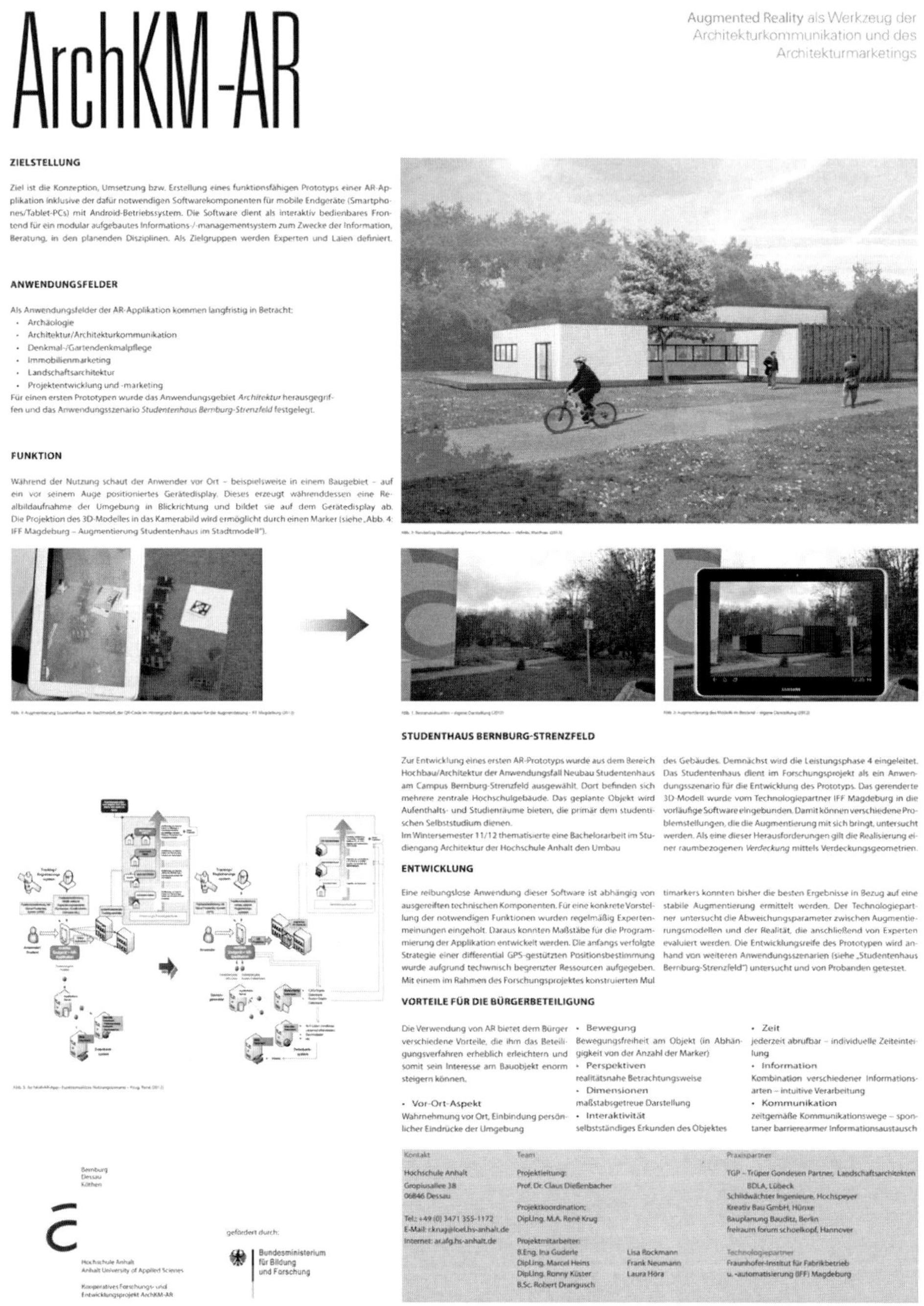

Poster: eGovernmentAR – Bildbasierende Kommunikationsmodelle als Schlüssel zum barrierearmen partizipativen eBeteiligungsverfahren

Crowdsourcing Portal "geo.tweet"

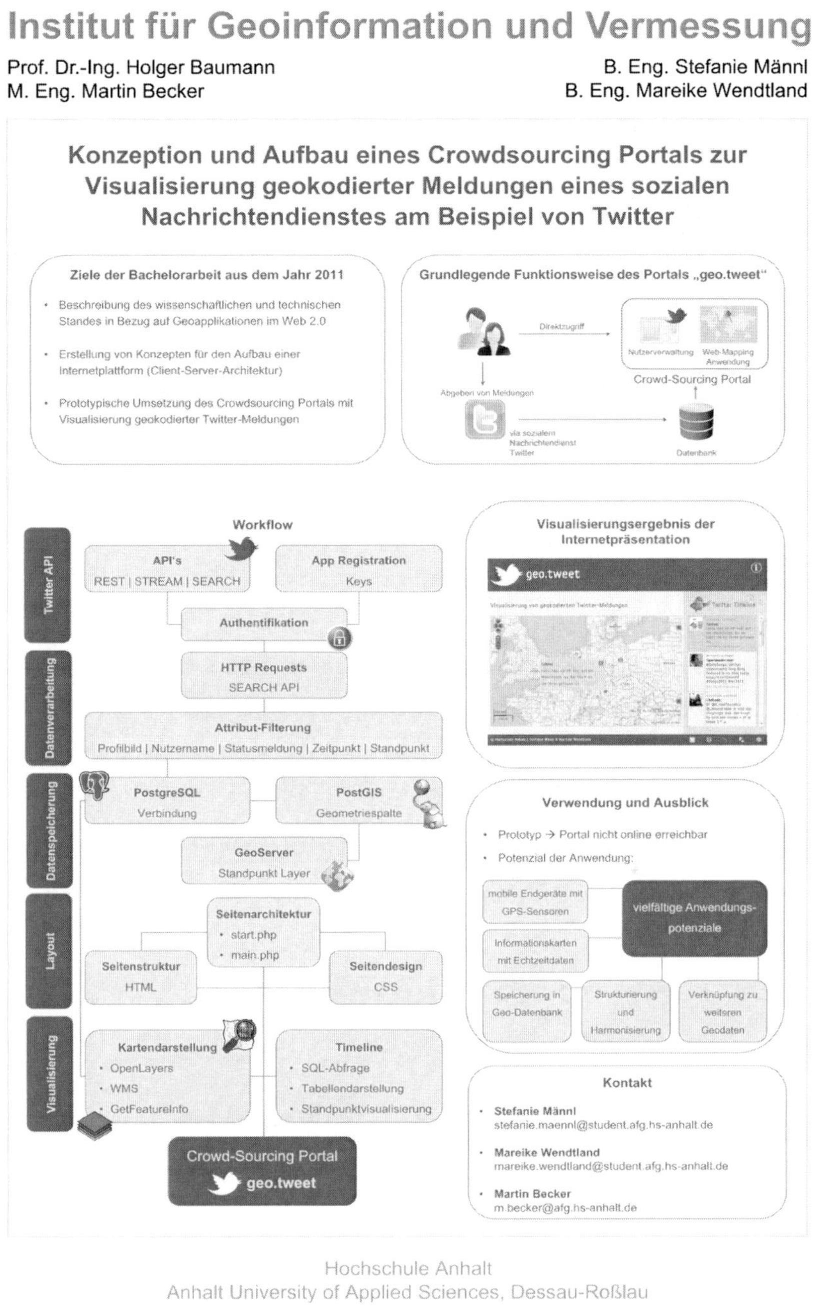

Poster: Concept and construction of a crowd sourcing portal to visualize geocoded messages of a social network based on the example of Twitter

DLA Awards 2013

We are pleased to announce those who have received the awards from the 2013 conference

Digital Landscape Architecture DLA 2013 Award by Review Committee

The following awards are for the full papers, resulting from the blind peer review by the Digital Landscape Architecture Review Committee.

- **SCIENTIFIC EXCELLENCE** was given to

  Prof. Dr. Ian Bishop, University of Melbourne, Australia, for the contribution:
  "Modeling and Analysis of Sequential Landscape Experiences"
  (Highest possible score of 6.0).

The second highest score of 5.67 was given to the following authors:

  Prof. Dr. Jörg Rekittke, *Philip Paar* and *Yazid Ninsalam*, National University of Singapore, Singapore:
  "Brawn and Technology under the Urban Canopy"

The third highest score of 5.5 was given to the following author:

  Prof. Mintai Kim, Virginia Tech, United States for his contribution:
  "Who is Afraid of "Light?"

Congratulation for your excellent papers!

We thank all the reviewers who helped with their recommendations and we compliment the authors for their **SCIENTIFIC EXCELLENCE!**

The DLA 2013 Award for

- **HIGHEST LEVEL OF COMMITMENT IN THE REVIEW COMMITTEE**:
  by chair of the review committee was given to

  Prof. Dr. James Palmer, FASLA, Vermont, United States

  Prof. Dr. Ian Bishop, University of Melbourne, Australia

  and as already in 2012 again to:

  Dr. Ulrike Wissen-Hayek, ETH Zurich, Switzerland

The 2013 DLA awards by vote of the participants and the **results of the DLA 2013 Poster Competition "My 3D Landscape"** will be available online at the beginning of June after the 2013 DLA conference at: www.digital-la.de.

Digital Landscape Architecture DLA 2012 Awards

During the DLA 2012 conference following awards had been given by participants.

The DLA 2012 Award for

- **SCIENTIFIC EXCELLENCE IN LANDSCAPE ARCHITECTURE**:
  by vote of the participants, was given to:
  Dr. Anna Czinkoczky, Corvinus University of Budapest, Hungary

The DLA 2012 Award for

- **BEST PRESENTATION**:
  by vote of the participants, was given to:
  Dr. Mintai Kim, Virginia Tech, United States

The DLA 2012 Award for

- **IT RESEARCH IN LANDSCAPE ARCHITECTURE**:
  by vote of the participants, was given to:
  Dr. Mintai Kim, Virginia Tech, United States

The DLA 2012 Award for

- **IT APPLICATION IN LANDSCAPE ARCHITECTURE:**
  by vote of the participants, was given to:
  Prof. Dr. Jörg Rekittke, National University of Singapore

The DLA 2012 Award for

- **BEST GEODESIGN APPLICATION IN LANDSCAPE ARCHITECTURE**:
  by vote of the participants, was given to:
  Prof. Andrew Lovett, University of East Anglia, United Kingdom

The DLA 2012 Award for

- **BEST PROFESSIONAL CONTRIBUTION TO DIGITAL LANDSCAPE ARCHITECTURE:**
  by vote of the participants, was given to:
  Prof. Dr. Thomas Kolbe, TU Berlin, Germany

The DLA 2012 Award for

- **CONTRIBUTION TO "IT" EDUCATION IN LANDSCAPE ARCHITECTURE**:
  by vote of the participants, was given to:
  Prof. David Goldberg, Penn State University, United States

The DLA 2012 Award for

- **BEST MODERATION:**
  by vote of the participants, was given jointly to:
  Dr. Stephen Ervin, Harvard University, United States
  and
  Prof. Harlind Libbrecht, School of Arts, University College Ghent, Belgium

Committees Contributing to the DLA Conference

Scientific Program Committee of DLA

Anhalt University of Applied Sciences has been organizing and hosting the annual Conference on Information Technology in Landscape Architecture for fourteen years. The growing international interest in this series of conferences, which focuses on the specific needs of applying and developing information technologies for the field of landscape architecture, has lead to the formal establishment of a Scientific Program Committee. Members of this committee have all been giving academic advice on the development of this conference for a number of years now and/or have served as very important reviewers in the blind review process.

The conference committee advises the Conference Director on essential long-term development issues such as:

- the main theme of the annual conference,
- selection of the Local DLA Conference Chair and the location of the next DLA Conference,
- both the content and production of the call for papers and promotional posters,
- selection of relevant keynote speakers, and
- selection of editors of the peer-reviewed proceedings.

The conference committee also advises the Local DLA Conference Chair on:

- the conference program based on the blind evaluations of the abstracts, and
- the resolution of conflicts among the evaluations of different reviewers.

The Local DLA Conference Chair rotates among members of the committee. Among the responsibilities of the Local Conference Chair are:

- management of the review process,
- local logistic arrangements, and
- national fundraising for the conference.

The DLA 2013 was held in Bernburg. The DLA 2014 is scheduled to be held at the Science City Campus, Zurich-Hoenggerberg in Switzerland, and the DLA 2015 is scheduled for the Dessau and Bernburg campuses of Anhalt University in Germany. Universities with a Landscape Architecture program wishing to host the DLA 2016 are being sought and should get in touch with us.

The Scientific Program Committee of the annual Digital Landscape Architecture DLA Conference currently includes the following members:

BUHMANN, Prof. Erich, Landscape Architect BDLA
Chair and Scientific Director of the International Conference on Information Technology in Landscape Architecture: DIGITAL LANDSCAPE ARCHICTECTURE DLA
Chair of the ECLAS Committee on Digital Technology
Anhalt University of Applied Sciences, Hochschule Anhalt, Germany
http://www.digital-la.de | private: atelier.bernburg@t-online.de

BISHOP, Prof. Dr. Ian.
University of Melbourne, Australia,
http://www.geom.unimelb.edu.au/cgism/ian.html | i.bishop@unimelb.edu.au

BILL, Ralf
Rostock University
ralf.bill@uni-rostock.de, http://www.auf-gg.uni-rostock.de/

BLASCHKE, Prof. Dr. Thomas
Center for GeoInformatics, University of Salzburg, Austria
http://www.zgis.at | thomas.blaschke@sbg.ac.at

DÖLLNER, Prof. Dr. Jürgen
Computer Graphics Systems, Hasso-Plattner-Institut at the University of Potsdam, Germany, http://www.hpi3d.de | doellner@hpi.uni-potsdam.de

ERVIN, Dr. Stephen, Landscape Architect FASL
Assistant Dean for Information Technology and Director of the Computer Resources Group at the Graduate School of Design, Harvard University, USA
http://www.gsd.harvard.edu | servin@gsd.harvard.edu

FLAXMAN, Prof. Dr. Michael
Dep. of Urban Studies and Planning, Massachusetts Institute of Technology, USA
http://dusp.mit.edu/ | http://futures.mit.edu | mflaxman@mit.edu

HASBROUCK, Prof. Hope
The University of Texas at Austin, USA
http://soa.utexas.edu/people/profile/hasbrouck/hope | hhasbrouck@austin.utexas.edu

HEHL-LANGE, Dr. Sigrid
The University of Sheffield, United Kingdom,
s.hehl-lange@sheffield.ac.uk

JØRGENSEN, Prof. Dr. Ian, Landscape Architect
Forest & Landscape, University of Copenhagen, Denmark
http://www.land-3d.com | Iajo@life.ku.dk

KIAS, Prof. Dr. Ulrich
Weihenstephan University of Applied Sciences, Freising, Germany
kias@fh-weihenstephan.de

KIEFERLE, Prof. Joachim, Architect
Hochschule RheinMain, Wiesbaden, Germany
http://caad.fab.hs-rm.de | joachim.kieferle@hs-rm.de

KIM, Prof. Dr. Mintai
Virginia Tech, Blacksburg, United States
www.lar.archdesign.vt.edu/mintaikim/ | mintkim@gmail.com

KOLBE, Prof. Dr. Thomas
Institute for Geodesy and Geoinformation Science, TU Berlin
http://www.igg.tu-berlin.de | thomas.kolbe@tu-berlin.de

KRAMER, Henk
Wageningen UR, Alterra
henk.kramer@wur.nl

LANGE, Prof. Dr. Eckart
The University of Sheffield, United Kingdom
http://www.shef.ac.uk/landscape/staff/profiles/eckartlange | e.lange@sheffield.ac.uk

LOVETT, Andrew
University of East Anglia, School of Environmental Sciences, Norwich, United Kingdom, a.lovett@uea.ac.uk

LENZ, Roman
Hochschule für Wirtschaft und Umwelt Nürtingen-Geislingen, HfWU
Nürtingen, Germany, roman.lenz@hfwu.de

MARLOW, Christopher
Ball State University, Department of Landscape Architecture,
College of Architecture and Planning, Muncie, Indiana, 47306, United States
marlow@bsu.edu

PALMER, Prof. Dr. James, Landscape Architect FASL
SUNY Syracuse, New York, USA
palmer.jf@gmail.com

PETSCHEK, Prof. Peter, Landscape Architect
Dep. of Landscape Architecture, HSR Hochschule für Technik Rapperswil, Switzerland
peter@petschek@hsr.ch

PIETSCH, Matthias, M.SC. GIS, Landscape Architect
Anhalt University of Applied Sciences, Hochschule Anhalt, Germany
http://www.kolleg.loel.hs-anhalt.de/gis-seminar/index.htm | pietsch@loel.hs-anhalt.de

REKITTKE, Prof. Dr. Jörg, Landscape Architect
National University of Singapore (NUS), Singapore,
http://www.arch.nus.edu.sg | rekittke@email.de

STEINITZ, Prof. Dr. Carl
Alexander and Victoria Wiley Research Professor, Graduate School of Design,
Harvard University, USA
csteinitz@gsd.harvard.edu

STEMMER, Boris
Kassel University, Germany
stemmer@uni-kassel.de

STREMKE, Assistant Prof. Dr. Sven, Landscape Architect
Wageningen University and Research, Wageningen, The Netherlands
www.NRGlab.net<http://www.NRGlab.net> | sven.stremke@wur.nl

STROBL, Dr. Josef
Department of Geography and Applied Geoinformatics
University of Salzburg, Austria,
http://www.zgis.at

SCHWARZ-V.RAUMER, Dr. Hans-Georg
University of Stuttgart, Germany
svr@ilpoe.uni-stuttgart.de

TAEGER, Prof. Dr. Stefan
Fachhochschule Osnabrück, Fakulät Agrarwissenschaften und Landschaftsarchitektur, Germany
s.taeger@fh-osnabrueck.de

TOMLIN, Prof. Dr. Dana
School of Design, University of Pennsylvania, USA
http://www.cml.upenn.edu | tomlin.dana@verizon.net

ZEILE, Peter, Dr.-Ing.
TU Kaiserslautern | FG CPE, Germany
http://cpe.arubi.uni-kl.de,
zeile@rhrk.uni-kl.de

Review Committee and Thank You from the Editors

Every paper accepted for this publication has been rigorously reviewed by at least two peers from an international panel of scholars. These proceedings are the result of the peer review process of the DLA Conference 2013, held in Bernburg, Germany. A total of forty-eight abstracts were submitted. The anonymous peer review process led to twenty-six full papers being accepted for publication as peer reviewed papers.

The high standards of the reviewers have assured that the papers in this year's proceedings will advance the theory and application of digital methods in landscape architecture.

With fifty excellent reviewers from the different areas of IT in landscape architecture, and an efficient online conference management system hosted by Prof. Joachim Kieferle from Rhein-Main Hochschule in Wiesbaden Germany, we have been able to continuously improve and professionalize the peer review process for the proceedings of the conference series DIGITAL LANDSCAPE ARCHITECTURE DLA. Reviewers were able to retrieve papers from a conference review system and submit commentaries and evaluations directly to a content management system.

For the first step of the blind review process, five blind reviewers were asked to give an initial evaluation for each of the anonymous abstracts assigned to her/him and place it into one of the following six levels of evaluation:

(1) Reject: Content inappropriate for the conference or has little merit.
(2) Probable Reject: Basic flaws in content or presentation or very poorly written.
(3) Marginal Tendency to Reject: Not as badly flawed; major effort necessary to make abstract acceptable but content already well-covered in literature.
(4) Marginal Tendency to Accept: Content has merit, but accuracy, clarity, completeness, and/or writing should and could be improved in time.
(5) Clear Acceptance: Content, presentation, and writing meet professional norms; improvements may be advisable but abstract is acceptable as is.
(6) Must Accept: Candidate for outstanding submission. Suggested improvements still appropriate.

The reviewer was also asked to recommend the particular conference session most appropriate for that paper. Finally, reviewers were asked for written comments addressed to each author and/or the Conference Chair.

Abstracts with an average score of less than 4 were rejected, while those with an average score of 4 were accepted for short presentations.

For the second step, the blind review of the full papers, only two reviewers with in-depth knowledge of the subject matter were assigned per paper. These second reviewers were usually selected from the original group that had evaluated each initial abstract, and they were asked to consider recommendations already made.

As a result of this review, each paper was placed into one of the following categories.

(1) Reject for publication w/o complete reworking.
(2) Probably reject for publication w/o reworking.
(3) Marginal rejection for publication if not reworked.
(4) Marginal acceptance for publication with minor corrections.
(5) Accept for publication with minor corrections.
(6) Fully accept for publication in this form.

In the written suggestions, the reviewers were also asked to check for

- missing references to the literature, and
- irrelevant or mistaken assertions.

English language proof-reading was done continuously throughout this period by Jeanne Colgan, and Michael O. Ajayi assisted in the final English editing of this publication.

Each author received the comments of the reviewers and additional comments by the editors for revising the full paper. When necessary, papers were returned to authors for revision until accepted by both reviewers. In some cases, additional anonymous reviewers were also asked to comment.

Twenty six groups of authors from DLA conferences received a final confirmation letter that their contribution would be published in the peer-reviewed proceedings; Buhmann/ Pietsch/Ervin (Eds.) (2013): Peer Reviewed Proceedings of Digital Landscape Architecture 2013, Anhalt University of Applied Sciences. Wichmann Verlag, Berlin/Offenbach, Germany, June 2013. We are pleased that nearly all positively reviewed authors managed the rigid time schedule to submit the full paper for publication.

We have completed the peer reviewed papers with three chapters of invited contributions in order to show more aspects of the ongoing technical developments. A chapter of invited papers with applications of digital landscape architecture practice and contributions by vendors are followed by selected descriptions of workshops being held at the DLA 2013 and a chapter with selected poster contributions to the DLA 2013.

The more than 250 peer-reviewed pages of this publication and the additional 100 pages of invited contributions on current issues in digital Landscape architecture applications and technological developments will be distributed to all libraries of landscape architecture schools. All professional journals in landscape architecture, urban design and environmental planning will also receive a copy to evaluate.

We would like to give special thanks to all our colleagues who helped serve on the DLA Review Board and by which further develop the level of presentations of this conference and the scientific standards of peer-reviewed proceedings of this conference. We thank all of them for the extensive time spent in reviewing the abstracts and the papers for these proceedings.

The quality of the peer-reviewed papers benefits greatly from the extensive advice given in comments by the blind reviewers to the anonymous authors. We are very flattered by the academic support given by our DLA reviewers.

We hope that the reviewers listed below will also find time to help us in the future.

AGIUS, Carol – MAGI, Malta, http://www.maltagi.org/

ALBERT, Christian, Leibniz University Hannover, Germany, albert@umwelt.uni-hannover.de

BISHOP, Prof. Dr. Ian – University of Melbourne, Australia, http://www.geom.unimelb.edu.au/cgism/ian.html

BRAMLET, Alison – University of Georgia, United States, alisonsb@uga.edu

BECK KOH, Anemone – Landscape Architect Wageningen, Netherlands

BERBEROGLU, Suha – Cukurova University, Landscape Architecture Department, Turkey, suha@cu.edu.tr, http://landgis.cu.edu.tr/index_eng.html

BLASCHKE, Prof. Dr. Thomas – Center for GeoInformatics, University of Salzburg, Austria http://www.zgis.at

BRAMLET, Alison – University of Georgia, United States, alisonsb@uga.edu

BUHMANN, Prof. Erich – Anhalt University of Applied Sciences, Campus Bernburg, Germany, www.landschaftsinformatik.de

BURLEY, Prof. Dr. Jon Bryan – Michigan State University MSU, United States

CANFIELD, Tess – Landscape Architect, United Kingdom, tesscanfield@yahoo.com

CONRAD, Max – Louisiana State University, United States, mconrad@lsu.edu

COTE, Paul, Harvard University Graduate School of Design, United States, http://www.gsd.harvard.edu/gis/manual | pbcote@gsd.harvard.edu

DIAZ, Joaquin – FH Giessen, http://www.fh-giessen.de

DEVIREN, Dr. A. Senem – Istanbul Technical University, Turkey, sd07landscape@yahoo.com

De VRIES, Jeroen – Van Hall Larenstein, The Netherlands

DIESSENBACHER, Prof. Dr. Claus – Department of Architecture, Anhalt University of Applied Sciences, Campus Dessau, Germany

DÖLLNER, Prof. Dr. Jürgen – Hasso Platner Institut, Universität Potsdam, http://www.hpi3d.de

ERVIN, Prof. Dr. Stephen – Director Computer Resources, Assistant Dean for Information Technology Harvard Design School, Boston, United States, http://www.gsd.harvard.edu

ESBAH TUNCAY, Hayriye – Istanbul Technical University, Turkey, hayriyeesbah@yahoo.com

FERGUSON, Bruce – University of Georgia, United States, bfergus@uga.edu http://web.me.com/brucekferguso

FIRNIGI, Anett – Corvinus University of Budapest, Hungary anett.firnigl@gmail.com

FLAXMAN, Prof. Dr. Michael – Massachusetts Institute of Technology, United States, http://futures.mit.edu

FRICKER, Pia Christina, ETH Zurich, Institute of Landscape Architecture, Switzerland, fricker@arch.ethz.ch

GRUNAU, Jens-Peter – University Stuttgart, Institut für Grundlagen der Planung, Fakultät für Architektur und Stadtplanung, Germany, mail@grunau.de

von HAAREN, Prof. Dr. Christina – Institute of Landscape Planning and Nature Conservation, University of Hannover, Germany

HAASE, Prof. Dr.-Ing. Andrea – Master of Architecture Program, Anhalt University of Applied Sciences, Campus Dessau, Germany

HASBROUCK, Prof. Dr. Hope – The University of Texas at Austin, United States http://soa.utexas.edu/people/profile/hasbrouck/hope

HEHL-LANGE, Dr. Sigrid – The University of Sheffield, United Kingdom, s.hehl-lange@sheffield.ac.uk

HEINS, Marcel – Anhalt University of Applied Sciences / Hochschule Anhalt, Germany m.heins@loel.hs-anhalt.de

HERNIK, Józef – University of Agriculture in Kraków, Poland, rmhernik@cyf-kr.edu.pl

HÖFFKEN, Stefan – TU Kaiserslautern, Germany, s.hoeffken@rhrk.uni-kl.de, http://cpe.arubi.uni-kl.de/

JØRGENSEN, Prof. Ian – University of Copenhagen, Forest & Landscape, Denmark, http://www.land-3d.com

KIAS, Prof. Dr. Ulrich – Weihenstephan University of Applied Sciences, Germany, kias@hswt.de

KIEFERLE, Prof. Joachim – University of Applied Sciences Wiesbaden, Germany http://caad.fab.fh-wiesbaden.de

KIRCHER, Prof. Dr. Wolfram – Master of Landscape Architecture Program, Anhalt University of Applied Sciences, Campus Bernburg, Germany

KLEINSCHMIT, Prof. Dr. Birgit – Technische Universität Berlin, Germany, http://www.geoinformation.tu-berlin.de/

KOEPPEL, Hans Werner – Bonn, Germany, http://www.koeppel-home.de

KOPPERS, Prof. Dr. Lothar – Anhalt University of Applied Sciences, Dep. Architecture, Facility Management and Geoinformation, Campus Dessau, Germany

LANGE, Prof. Dr. Eckart – The University of Sheffield, United Kingdom, http://www.shef.ac.uk/landscape/staff/profiles/eckartlange

LAMM, Bettina – Forest & Landscape, University of Copenhagen, Denmark, http://www.sl.life.ku.dk/

LAMMEREN, Ron van -Centrum Landschap, Alterra, Wageningen Netherlands, Ron.vanLammeren@wur.nl.

LINDHULT, Prof. Mark – Master of Landscape Architecture Program, University of Massachusetts, United States, http://www-unix.oit.umass.edu/~lindhult/

MACH, Rüdiger – Mach:Idee, Ing.-Buero R. Mach, Karlsruhe, Germany, http://www.machidee.de

MATTOS, Cristina – ESRI Germany, c.mattos@esri.de

MERTENS, Elke – Hochschule Neubrandenburg, Germany, mertens@hs-nb.de

MOORE, Kathryn – Birmingham City University, United Kingdom, kathryn.moore@bcu.ac.uk

NIEMAN, Thomas – University of Kentucky, United States, tnieman@uky.edu

ORLAND, Brian – Penn State University, United States, boo1@psu.edu, http://larch.psu.edu/faculty/brian-orland

ORTACESME, Veli – Akdeniz University, Turkey, http://peyzaj.ziraat.akdeniz.edu.tr

OZDIL, Taner R. – The University of Texas at Arlington, United States, tozdil@uta.edu, http://www.uta.edu/ra/real/editprofile.php?pid=3021&onlyview=1

OZIMEK, Agnieszka – Cracow University of Technology, Poland, http://www.pk.edu.pl/~aozimek

OZIMEK, Dr. Pawel – Cracow University of Technology, Poland, http://www.pk.edu.pl/~aozimek

PAAR, Philip – TU Berlin, Berlin, Germany, paar@laubwerk.com http://www.laubwerk.com

PALMER, Prof. Dr. James – SUNY Syracuse NY, United States, palmer.jf@gmail.com

PIETSCH, Matthias – Anhalt University of Applied Sciences, Campus Bernburg, Germany, http://www.kolleg.loel.hs-anhalt.de/gis-seminar/index.htm

PINKAU, Prof. Stephan – Master of Architecture Program, Anhalt University of Applied Sciences, Campus Dessau, Germany, s.pinkau@afg.hs-anhalt.de

REKITTKE, Prof. Dr. Jörg – National University of Singapore (NUS), Singapore, http://www.arch.nus.edu.sg

ROLLEY, Stephanie – Kansas State University, United States, srolley@k-state.edu, http://capd.ksu.edu/larcp/

ROSS, Lutz – Technische Universität Berlin, Germany, http://www.geoinformation.tu-berlin.de

ROOS-KLEIN LANKHORST, Janneke – Centrum Landschap, Alterra, Wageningen Netherlands, Janneke.roos@wur.nl

SCHROTH, Olaf – CALP / University of British Columbia UCA, Canada, http://www.calp.forestry.ubc.ca/

SECKIN, Cagatay – Istanbul Technical University, Turkey, cseckin@itu.edu.tr

SHARKY, Bruce – Louisiana State University, United States, bshark2@lsu.edu

SIMON, Alf – University of New Mexico, United States, http://saap.unm.edu

SPEED, Prof. Dr. Chris – Edinburgh College of Art, United Kingdom, http://fields.eca.ac.uk

STREICH, Prof. Dr. Bernd – Department of CAAD and Planning Methods, University of Kaiserslautern, streich@rhrk.uni-kl.de, http://cpe.arubi.uni-kl.de/

TAEGER, Prof. Dr. Stefan – Fachhochschule Osnabrück, Fakultät Agrarwissenschaften und Landschaftsarchitektur. Germany

TOBIAS, Prof. Dr. Kai – Ecological Planning and Environmental Impact Assessment, University of Kaiserslautern, Germany, tobias@rhrk.uin-kl.de

TOMLIN, Dana – University of Pennsylvania, United States, tomlin.dana@verizon.net

TUTUNDZIC, Andreja – University of Belgrade / EFLA, Serbia, andreja.tutundzic@sfb.bg.ac.rs, http://www.sfb.bg.ac.rs,

VUGULE, Kristine – Latvia University of Agriculture, Latvia, kristine.vugule@llu.lv, www.llu.lv

WISSEN, Dr. Ulrike Hayek – ETH Zürich, Switzerland, http://www.irl.ethz.ch/plus/index_DE

ZEILE, Peter, Dr.-Ing. – TU Kaiserslautern | FG CPE, http://cpe.arubi.uni-kl.de, zeile@rhrk.uni-kl.de

ECLAS and LE:NOTRE Committee on Digital Technology

Since its founding in 2002, the European Council of Landscape Architecture Schools, ECLAS, has benefited enormously from the activities of the LE:NOTRE Thematic Network and its spiritual father Prof. Richard Stiles from the Technical University of Vienna. With the help of LE:NOTRE, a strong network, a common internet platform, and numerous meetings have all helped to develop a link between the fast growing numbers of landscape architecture schools in Europe and worldwide. Within LE:NOTRE, working groups have been established. In order to assure continuity after the LE:NOTRE project funding comes to an end, ECLAS has established a subcommittee to follow up these working groups, the ECLAS Committee on Digital Information. All fifty members of the former LE:NOTRE IT working group are being asked to continue in this new committee. At the moment, LE:NOTRE is still offering the "Channel Information Technologies" of the LE:NOTRE platform for communication of this committee.

In 2009, Prof. Erich Buhmann was voted as Chair for the newly established Committee on Digital Technology of the European Council of Landscape Architecture Schools ECLAS. The main objective for ECLAS Committee on Digital Technology is the Improvement and further development of Teaching & Learning with Digital Methods & Tools in Landscape Architecture. This was also the main conference theme of the DLA 2011 in Dessau, Germany.

The annual conference DIGITAL LANDSCAPE ARCHICTECTURE DLA will be an important asset for the work of this committee. Further objectives are being discussed. Besides meetings with the Vitero conference system as needed, a second regular personal meeting is provided by the annual ECLAS conferences.

The contents of past and upcoming DLA conferences are stored on the homepage of the Anhalt University homepage www.digital-la.de or http://www.kolleg.loel.hs-anhalt.de/landschaftsinformatik/dla-conference.html.

The project group "Digital Landscape Architecture Conferences" of the ECLAS / LE:NOTRE Committee on Digital Technology has been established as a communication platform at www.lenotre.com: http://www.le-notre.org/pg/pg_details.php?pg_id=112.

Addresses of Authors

ABU BAKAR, Shamal
University of Putra
Selangor, Malaysia
305, Loudon Rd Apt 611
24060, Blacksburg, U.S.A
shunown@gmail.com

ADLER, Simon
Fraunhofer-Institut für Fabrikbetrieb
und -automatisierung IFF
Sandtorstraße 22
D-39106 Magdeburg, Germany
simon.adler@iff.fraunhofer.de

AMOROSO, Prof. Dr. Nadia
University of Guelph, Ontario
50 Stone road East Guelph
N1G 2W1 Ontario, Canada
namoroso@uoguelph.ca

BAYLAN, Emel
Yuzuncu Yil University,
Faculty of Architecture, Department of
Landscape Architecture
65080, Rektorlugu, Turkey
emel_baylan@hotmail.com

BISHOP, Prof. Dr. Ian
University of Melbourne
Grattan Street
3010 Parkville, Australia
i.bishop@unimelb.edu.au

BRABEC, Prof. Elizabeth
University of Massachussetts
MA 01003, Amherst
Massachussetts, U.S.A
ebrabec@larp.umass.edu

BRANDS, Bart
Karres en Brands
Oude Amersfoortseweg 123
1212 AA Hilversum, Netherlands
info@karresenbrands.nl

BUHMANN, Prof. Erich
Anhalt University of Applied Sciences
Hochschule Anhalt,
Landschaftsinformatik, Solbadstraße 2,
D-06406 Bernburg, Germany
la@loel.hs-anhalt.de
http://www.landschaftsinformatik.de
privat: atelier.bernburg@t-online.de

CETIN, Dr. Mehmet
Temple University
Civil and Environmental Engineering
1947 North.12th Philadelphia
PA 19122 Pennsylvania, U.S.A
mehmet.cetin@temple.edu

CURETON, Paul
Manchester School of Architecture,
Chatham Building, Cavendish Street
M15 6BR Manchester, U.K
paul.cureton@paperlandscapearchitectur
e.co.uk

CZINKÓCZKY, Anna
Corvinus University of Budapest
Villanyi ut. 29-43
1118 Budapest, Hungary
annaczinkoczky@gmail.com

DANGERMOND, Jack
Esri Headquarters
380 New York Street
Redlands, CA 92373-8100, U.S.A.

DÄRR, Mathias
Därr Landschaftsarchitekten
Ernst-Grube-Straße 1
D-06120 Halle (Saale), Germany
freiraum@la-daerr.de

DE WAAL, Renee
Wageningen University
P.O. Box 47
6700AA Wageningen, Netherlands
renee.dewaal@wur.nl

DOBRICK, Christian
West8, Rotherdam
Schiehaven 13M
3024 EC Rotherdam, Netherlands
pr@west8.com

ERVIN, Dr. Stephen M.
Harvard University
Massachusetts Hall
Cambridge, MA 02138, U.S.A
servin@gsd.harvard.edu
http://www.gsd.harvard.edu

FORMOSA, Dr. Saviour
University of Malta, Department of Criminology, Faculty for Social Wellbeing, Room 113, Humanities A (Laws,Theology,Criminology), MSD 2080 Msida, Malta
saviour.formosa@um.edu.mt

GORMUS, Sevgi
Bartin University, Faculty of Forestry
Department of Landscape Architecture
Orman Fakültesi, Peyzaj Mimar
74100 Bartin, Turkey
sevgigormus@gmail.com

GUTIERREZ, Maria
Anton & Ghiggi Landschaft Architektur
Limmatstrasse 204
CH-8005 Zurich, Switzerland
Kolings16@gmail.com

HAHN, Howard
Kansas State University
201B Seaton Court
66506 Manhattan, U.S.A
hhahn@ksu.edu

HEINS, Marcel
Anhalt University of Applied Sciences
Hochschule Anhalt
Strenzfelder Allee 28
D-06406 Bernburg, Germany
heins@loel.hs-anhalt.de
http://www.landshaftinformatik.de

HEISS, Dr. Michael
IP SYSCON
Mary Somerville Straße 1
D-28359 Bremen, Germany
mh@ipsyscon.de

HERBARTH, Stephan
Beuth Hochschule für Technik Berlin
Luxemburger Straße 10
D-13353 Berlin, Germany
stepher@gmx.de

HOOFTMAN, Eelco
GROSS.MAX
EH 1 3EG Edinburgh, Scotland
mail@grossmax.com

HUDSON-SMITH, Dr. Andy
Centre for Advanced Spatial Analysis
UCL Faculty of the Built Environment
90 Tottenham Court Road London
W1T 4TJ, London, England
a.hudson-smith@ucl.ac.uk

JAROSH, Thomas
Hafner/Jimenez Berlin
Schwedter Straße 263
D-10119 Berlin, Germany
haefner@haefner-jimenez.de

JOYE, Ruben
Hogeschool Gent
School of Arts,
University College Ghent
Jozef Kluyskensstraat 2
BE-9000 Gent, Belgium
ruben.joye@hogent-be

KALBERG , Heiki
University of Tartu,
Institute of Technology, Nooruse St 1
50411 Tartu, Estonia
heiki@artes.ee

KEMPENAAR, Annet
Wageningen University
Postbus 47
6700AA, Wageningen, Netherlands
annet.kempenaar@wur.nl

KIM, Dr. Mintai
Virginia Tech
120D Burruss Hall (MC 0190)
VA 24061 Blacksburg, U.S.A
mintkim@vt.edu

KUMAKURA, Dr. Eiko
Tokyo Institute of Technology
4259-G5-2, Midori-ku, Nagatsutacho
2268502 Yokohama-City, Japan
kumakura.e.aa@m.titech.ac.jp

MÄNNL, Stefanie
Anhalt University of Applied Sciences
Hochschule Anhalt
Strenzfelder Allee 28
D-06406 Bernburg, Germany
Stefanie.maennl@student.afg.hsanhalt.de

MITTLBOECK, Dr. Manfred
Key Researcher – Studio iSPACE
Research Studios Austria
Forschungsgesellschaft mbH
Schillerstrasse 25
A-5020 Salzburg, Austria
manfred.mittlboeck@researchstudio.at

NESSEL, Andrew
SUNY ESF
847 Westmoreland Ave, Apt 4
13210 Syracuse, U.S.A
andrew.nessel@gmail.com

OZIMEK, Dr. Agnieszka
Cracow University of Technology
Warszawska 24
31-155, Krakow, Poland
aozimek@pk.edu.pl

PAAR, Phillip
Laubwerk GmbH
Geschäftsführer/Managing Director
Torstr. 75
D-10119 Berlin, Germany
paar@laubwerk.com
http://laubwerk.com

PALMER, Prof. Dr. James
FASLA, USA
177 Ferguson Avenue
5401, Burlington, Vermont, U.S.A
palmer.jf@gmail.com

PHILLIPS, Prof. Mike
i-DAT Director of Research
Plymouth University Drake Circus,
Plymouth PL4 8AA, Devon, U.K
mike.phillips@plymouth

PIETSCH, Matthias
Anhalt University / Hochschule Anhalt
Strenzfelder Allee 28
D-06406 Bernburg, Germany
http://www.kolleg.loel.hs-anhalt.de/gis-seminar/index.htm
m.pietsch@loel.hs-anhalt.de

RADIES, Christian
Technical University of Munich
Arcisstraße 21
D-80333 Munich, Germany
christiane.radies@gmail.com

REIN-CANO, Martin
TOPOTEK1, Berlin
Sophienstraße 18
D-10178 Berlin, Germany
topotek1@topotek.de

REKITTKE, Prof. Dr. Jorg
National University of Singapore
School of Design and Environment
Department of Agriculture
4 Architecture Drive
117566 Singapore
rekittke@nus.edu.sg

ROCKMANN, Lisa
Anhalt University of Applied Sciences
Strenzfelder Allee 28
D-06406 Bernburg, Germany
lisarockmann@yahoo.de

SCHLIEP, Jan Walter
Laubwerk GmbH
Tucherstraße 17d
90562 Heroldsberg, Germany
jan-walter.schliep@laubwerk.com

SCHOELKOPF, Petra
freiraumforum
Freie Landschaftsarchitektin
Isernhagener Str. 45
D-30163, Hannover, Germany
schoelkopf@freiraumforum.de

SCHROTH, Olaf
University of British Columbia (UBC)
2045-2424 Main Mall
V6T IZ4 Vancouver, Canada
schrotho@interchange.ubc.ca
olaf_schroth@gmx.de

SCHWARZ-V.RAUMER, Hans-Georg
Institut für Landschaftsplanung und
Ökologie,·Universität Stuttgart
Keplerstraße 11
D-70175 Stuttgart, Germany
svr@ilpoe.uni-stuttgart.de

SPEED, Prof. Dr. Chris
University of Edinburgh
Lauriston Place
EH3 9DF, Edinburgh Scotland
c.speed@ed.ac.uk

STEINITZ, Prof. Dr. Carl
Harvard University
Massachusetts Hall
Cambridge, MA 02138, U.S.A
http://www.gsd.harvard.edu
Steinitz@gsd.harvard.edu

STOKMAN, Antje
Institut für Landschaftsplanung und
Ökologie,·Universität Stuttgart
Keplerstraße 11
D-70175 Stuttgart, Germany
antje.stokman@ilpoe.uni-stuttgart.de

STRAUB, Prof. Dietmar
University of Manitoba
Department of Landscape Architecture,
Faculty of Architecture
R3T 2N2 Winnipeg, Canada
Dietmar.Straub@ad.umanitoba.ca

SZÜCS, Linda
Georg-August University Göttingen
Büsgenweg 3.
D-37077 Göttingen, Germany
lszuecs@gwdg.de

TULLOCH, David
Cook College, Rutgers University
93 Lipman Drive
NJ 089018524 New Brunswick, U.S.A
dtulloch@crssa.rutgers.edu

WILLIS, Dr. Katherine
Plymouth University,
Room 402, Roland Levinsky Building,
Drake Circus
PL4 8AA Plymouth, Devon, England

ZEILE, Dr. Peter
Technical University of Kaiserslautern
Gottlieb-Daimler-Straße 47
D-67663 Kaiserslautern, Germany
line@-rhrk.uni kl.de

21 May – 23 May
Zurich, Switzerland

DIGITAL LANDSCAPE ARCHITECTURE DLA 2014

ETH Zurich
Swiss Federal Institute of Technology

CALL FOR PAPERS and POSTERS

15th International Conference on Information Technologies in Landscape Architecture

The 15th annual International Conference on Information Technology in Landscape Architecture: Digital Landscape Architecture DLA 2014 will be held in Zurich, Switzerland, 21 May – 23 May, 2014. We would like to invite you to contribute to this conference on teaching of new technologies in landscape architecture. Please submit "Extended Abstracts" (minimum one page, maximum two pages text including key references and optional additional selected graphics) as pdf file online via a web-based submission system at

→ www.digital-la.de → DLA Conference 2014 by October 15, 2013

for review and possible inclusion in the program as presentation or as poster.

The conference proceedings are published as fully reviewed papers by Wichmann Verlag, Heidelberg as following: "Girot/Grêt-Regamey/Wissen Hayek/Fricker/Buhmann (Eds.): Peer Reviewed Proceedings Digital Landscape Architecture 2014, ETH Zurich, Swiss Federal Institute of Technology, Wichmann Verlag, Berlin/Offenbach, May 2014".

Main Theme and Suggested Topics:
The DLA Program Committee cordially invites you to submit proposals for original, unpublished presentations focusing on the conference's main theme
"Landscape Architecture and Planning – Developing Digital Methods in GeoDesign"
in one of the other following areas:

1) **GeoDesign Concepts**
2) **From Point Clouds to Geo-Based Design**
3) **Interactive Virtual Landscapes and Visual Landscape Assessment**
4) **Teaching GeoDesign in Landscape Design and Landscape Planning**
5) **GeoDesign Case Studies in Regional Planning**
6) **GeoDesign Case Studies in Urban Design**
7) **Mobile Devices for GeoDesign**
8) **Landscape Information Model (LIM) and Standardization**

Important dates for the review process are:

Abstracts due:	**October 15, 2012**
Notification of acceptance:	November 15, 2013
Full manuscript draft due:	December 20, 2013
Reviewed manuscript due:	February 15, 2014
Conference:	May 21 – May 23, 2014

Invited Keynote Speakers:

Prof. Dr. Jürgen Döllner
Hasso-Plattner-Institut, University of Potsdam, Germany – (req.)

Prof. Dr. Marc Pollefeys
Computer Vision and Geometry Lab, ETH Zurich, Switzerland – (req.)

Prof. Dr. Carl Steinitz
Graduate School of Design, Harvard University, Cambridge, MA, USA – (conf.)

Prof. Antje Stokman
Institute for Landscape Planning and Ecology, University of Stuttgart, Germany – (req.)

Clément Willemin
BASE Bien Aménager Son Environnement / Build a Super Environment, Paris, France – (req.)

Invited Session Chairs:

Prof. Dr. Jürgen Döllner
Hasso-Plattner-Institut, University of Potsdam, Germany – (req.)

Dr. Mike Flaxman,
Geodesign Technologies Inc., San Francisco, USA – (req.)

Prof. Dr. Eckart Lange
Department of Landscape, University of Sheffield, UK – (req.)

Prof. Dr. Andrew Lovett
School of Environmental Sciences, University of East Anglia, UK – (conf.)

Prof. Peter Petschek
Department of Landscape Architecture, Hochschule f. Technik Rapperswil, Switzerland – (conf.)

Prof. Dr. Jörg Rekittke
Department of Architecture, National University of Singapore, Singapore – (conf.)

The conference language is English. The conference will be held at the Science City Campus ETH Hönggerberg (Zurich) in Switzerland.

For further information please contact:

Yvonne Largiardèr
ETH Zurich
Institut für Raum- und Landschaftsentwicklung (IRL), HIL H 51.1
Wolfgang-Pauli-Str. 15, CH-8093 Zürich
largiader@nsl.ethz.ch

Conference proceedings: dla_proceedings2014@ethz.ch

Conference organization: dla2014@ethz.ch

Local Chair for 2014: Prof. Dr. Adrienne Grêt-Regamey and Prof. Christophe Girot

DLA Conference Scientific Director: Prof. Erich Buhmann

The concept of the DLA event is developed by the Scientific Program Committee of the Digital Landscape Architecture DLA conference. The DLA is organized in cooperation with long term partners such as AGIT Symposium und Fachmesse Angewandte Geoinformatik of University of Salzburg, the LE:NOTRE Thematic Network of the Technical University of Vienna and the German Professional Association of Landscape Architects, BDLA. The conference serves as platform for the ECLAS Committee Digital Information.

More information on our series of conferences on Digital Landscape Architecture at **http://www.digital-la.de**

Committee
Digital Information of

ECLAS

EUROPEAN COUNCIL OF
LANDSCAPE ARCHITECTURE
SCHOOLS

CALL FOR PAPERS and POSTERS

15th International Conference on Information Technologies in Landscape Architecture